Studies in Classification, Data Analysis, and Knowledge Organization

More information about this series at http://www.springer.com/series/1564

Francesco Palumbo • Angela Montanari •
Maurizio Vichi
Editors

Data Science

Innovative Developments in Data Analysis and Clustering

Springer

Editors
Francesco Palumbo
Department of Political Sciences
University of Naples Federico II
Napoli, Italy

Angela Montanari
Department of Statistical Sciences Paolo Fortunati
Alma Mater Studiorum, University of Bologna
Bologna, Italy

Maurizio Vichi
Department of Statistical Sciences
Sapienza University of Rome
Rome, Italy

ISSN 1431-8814 ISSN 2198-3321 (electronic)
Studies in Classification, Data Analysis, and Knowledge Organization
ISBN 978-3-319-55722-9 ISBN 978-3-319-55723-6 (eBook)
DOI 10.1007/978-3-319-55723-6

Library of Congress Control Number: 2017942955

Printed on acid-free paper

This Springer imprint is published by Springer Nature
The registered company is Springer International Publishing AG
The registered company address is: Gewerbestrasse 11, 6330 Cham, Switzerland

Preface

On 4 July 1985 in Cambridge (UK), six classification societies gave birth to the International Federation of Classification Societies; the *Società Italiana di Statistica* (SIS) played an active propulsive role in the IFCS constitution. During the 30 years of the IFCS life, many other classification societies from all around the world have joined the IFCS. Through the active participation of its members, the SIS has actively and enthusiastically contributed to the IFCS growth. In 1997 the first conference of the Classification and Data Analysis Group of the SIS was hosted by the University of Chieti-Pescara. Following this long story of presence, in the occasion of IFCS's 30th birthday, under the IFCS presidency of Maurizio Vichi, the Classification and Data Analysis Group of the SIS and the Department of Statistical Sciences P. Fortunati of the *Alma Mater Studiorum* University of Bologna were proudly willing to organize the IFCS conference in Bologna. The conference organizer was Angela Montanari (University of Bologna) and Francesco Palumbo (University of Naples Federico II) served as chair of the scientific program committee. The conference was held between 5 and 8 July 2015.

Scholars from many different countries attended the conference. The commitment of the local organizing committee and the earnestness of the Scientific Program Committee ensured a successful and worthwhile conference. We are grateful to the members of the Scientific Program Committee: A. Cerioli (ClaDAG, Italy), D. Choi (KCS, Korea), C. Cuevas Covarrubias (SOLCAD, Mexico), N. Dean (BCS, UK), A. Ferligoj (SSS, Slovenia), P. Giudici (ClaDAG, Italy), C. Hennig (BSC, UK), T. Imaizumi (JCS, Japan), B. Lausen (GfKl, UK), P. McNicholas (CS, Canada), M. Nadif (SFC, France), A. Okada (JCS, Japan), I. Papadimitriou (GSDA, Greece), J. Pociecha (SKAD, Poland), A. Sbihi (MCA, Marocco), B. Scotney (IPRCS, Ireland), F. Sousa (CLAD, Portugal), D. Steinley (CS, USA), I. Van Mechelen (VOC, Belgium), and J. Vermunt (VOC, the Netherlands). IFCS has long-standing tradition of cooperation and exchange with other scientific statistical societies; in occasion of the IFCS conference in Bologna, V. Esposito Vinzi (France) and P. Groenen (the Netherlands) were invited to join the Scientific Program Committee as delegates of the ISBIS and IASC societies, respectively.

More than 200 contributions were organized into specialized sessions, contributed paper sessions, and one poster session. Moreover, five keynote lectures were given by eminent colleagues on different topics of data analysis and classification. The opening plenary session was devoted to the IFCS birthday celebration and a special session celebrated the 25th anniversary of the publication of the book on Generalized Additive Models by Hastie and Tibshirani.

Thanks to the collaboration with the publisher Springer and to its interest and attention to the IFCS activities during these 30 years, according to the longly consolidated tradition, the present post proceeding volume has been edited after the conference.

The scientific community unanimously considers *data science* as one of the most promising fields where to direct scientific research in the next years. However, already in occasion of the fifth IFCS conference, which was held in the year 1996 in Kobe (Japan), the related proceedings volume was entitled *Data Science, Classification, and Related Methods* (Hayashi et al. eds; Springer Japan, publisher). To emblematize the line of continuity along the IFCS, in occasion of the 30th birthday conference, we have decided to entitle the volume *Data Science: Innovative Developments in Data Analysis and Clustering*. The volume is a collection of full papers submitted after the conference. Papers were selected after a peer-review process, according to the high-quality standards of the series.

The volume is made of 27 contributions organized in three parts including contributions on:

- Classification methods for high-dimensional data
- Clustering methods and applications
- Multivariate methods and applications

Bologna, Italy — Angela Montanari
Napoli, Italy — Francesco Palumbo
Roma, Italy — Maurizio Vichi
November 2016

Acknowledgments

We are indebted to many people who allowed the success of the IFCS 2015 conference with their commitment. This book represents the final outcome of all the work done for the organization of the conference, during the days of the conference and after the end of it. First, we are grateful to the Department of Statistical Sciences of the University of Bologna who hosted the conference. In particular our thanks are addressed to the members of the organizing committee: L. Anderlucci, S. Bianconcini, S. Cagnone, L. De Angelis, G. Galimberti, A. Lubisco, M. Lupparelli, P. Monari, L. Stracqualursi, and C. Viroli. Two more additional thanks are for Laura Anderlucci, who has taken care of the conference web site, and for Paola Monari who, discreetly but effectively, has made all her experience and shrewdness in the organization available for the success of the conference.

We are also indebted to our colleagues that have collaborated in the review process of this volume:

Andrea Cerioli,
Claudio Conversano,
Pasquale Dolce,
Leonardo Grilli,
Patrick Groenen,
Christian Hennig,
Tadashi Imaizumi,
Berthold Lausen,
Antonello Maruotti,
Paul McNicholas,
Fionn Murtagh,
Mohamed Nadif,
Akinori Okada,
Domenico Piccolo,
Giancarlo Ragozini,
Iven Van Mechelen,
José Fernando Vera,
Rosanna Verde,
Vincenzo Esposito Vinzi,
Domenico Vistocco.

Last but not least, we are also indebted with SAS Institute, Springer, APT Servizi Regione Emilia Romagna and Ascom Bologna for their financial support to the conference.

Contents

Contributors

Patrizia Agati Department of Statistical Sciences, University of Bologna, Bologna, Italy

Olivier Alata Hubert Courien Laboratory, UMR 5516, Jean Monnet University, Saint-Étienne, France

Slavka Bodjanova Texas A&M University-Kingsville, Kingsville, TX, USA

Eva Boj Facultat d'Economia i Empresa, Universitat de Barcelona, Barcelona, Spain

Anne-Laure Boulesteix Department of Medical Informatics, Biometry and Epidemiology, University of Munich, München, Germany

Stefania Capecchi Department of Political Sciences, University of Naples Federico II, Naples, Italy

Gunnar Carlsson Department of Mathematics, Stanford University, Stanford, CA, USA

Theodore Chadjipadelis Aristotle University of Thessaloniki, Aristotle University Campus, Thessaloniki, Greece

Marcello Chiodi Department of Economics, Management and Statistics, University of Palermo, Palermo, Italy

Teresa Costa Facultat d'Economia i Empresa, Universitat de Barcelona, Barcelona, Spain

Renato Cordeiro de Amorim School of Computer Science, University of Hertfordshire, Hatfield, UK

Patrizia de Luca Dipartimento di Scienze Economiche Aziendali Matematiche e Statistiche, Università di Trieste, Trieste, Italy

Francesca Di Salvo Department of Economics, Management and Statistics, University of Palermo, Palermo, Italy

Vincent Egan Department of Psychiatry and Applied Psychology, University of Nottingham, Nottingham, UK

Abdellatif El Idrissi LabTIC Laboratory, ENSA-Tangier, Tangier, Morocco

Elaine Fehrman Men's Personality Disorder and National Women's Directorate, Rampton Hospital, Retford, Nottinghamshire, UK

Mario Fordellone Sapienza University of Rome, Rome, Italy

Josep Fortiana Facultat de Matemàtiques, Universitat de Barcelona, Barcelona, Spain

Krzysztof Gałuszka Department of Finance, University of Economics in Katowice, Katowice, Poland

Álvaro Gómez-Losada Department of Statistics and Operational Research, University of Seville, Seville, Spain

Alexander N. Gorban Department of Mathematics, University of Leicester, Leicester, UK

Tomasz Górecki Faculty of Mathematics and Computer Science, Adam Mickiewicz University, Poznań, Poland

Myriam Hatz Department of Medical Informatics, Biometry and Epidemiology, University of Munich, München, Germany

Khadidja Henni Department of Computer Science, University of Sciences and Technologies Oran "Mohamed Boudia" USTO-MB, Bir El Djir, Algeria

Jaroslav Hlinka Institute of Computer Science of the Czech Academy of Sciences, Prague, Czech Republic

National Institute of Mental Health, Klecany, Czech Republic

Lynette A. Hunt University of Waikato, Hamilton, New Zealand

Tadashi Imaizumi School of Management and Information Sciences, Tama University, Tokyo, Japan

Jan Kalina Institute of Computer Science of the Czech Academy of Sciences, Prague, Czech Republic

National Institute of Mental Health, Klecany, Czech Republic

Jadwiga Kostrzewska Department of Statistics, Cracow University of Economics, Cracow, Poland

Maciej Kostrzewski Department of Econometrics and Operational Research, Cracow University of Economics, Cracow, Poland

Agnieszka Kozera Poznań University of Life Sciences, Poznań, Poland

Mirosław Krzyśko Faculty of Mathematics and Computer Science, Adam Mickiewicz University, Poznań, Poland

Daniela Lucini BIOMETRA Department, University of Milan, Milano, Italy

Aleksandra Łuczak Poznań University of Life Sciences, Poznań, Poland

Vladimir Makarenkov Département d'informatique, Université du Québec à Montréal, Montreal, QC, Canada

Geoffrey J. McLachlan Department of Mathematics, University of Queensland, St Lucia, QLD, Australia

Facundo Mémoli Department of Mathematics, The Ohio State University, Columbus, OH, USA

Department of Computer Science, The Ohio State University, Columbus, OH, USA

Evgeny M. Mirkes Department of Mathematics, University of Leicester, Leicester, UK

Boris Mirkin Department of Data Analysis and Machine Intelligence, National Research University Higher School of Economics, Moscow, Russia

Department of Computer Science and Information Systems, Birkbeck University of London, London, UK

Odysseas Moschidis University of Macedonia, Thessaloniki, Greece

Ahmed Moussa LabTIC Laboratory, ENSA-Tangier, Tangier, Morocco

Awaz K. Muhammad Department of Mathematics, University of Leicester, Leicester, UK

Fionn Murtagh University of Derby, Derby, UK

Goldsmiths University of London, London, UK

Miki Nakai Department of Social Sciences, College of Social Sciences, Ritsumeikan University, Kyoto, Japan

Atsuho Nakayama Tokyo Metropolitan University, Hachioji-shi, Japan

Shu Kay Ng School of Medicine and Menzies Health Institute Queensland, Griffith University, Nathan, QLD, Australia

Massimo Pagani BIOMETRA Department, University of Milan, Milano, Italy

Francesco Palumbo Federico II University of Naples, Napoli, Italy

Pietro Patricola Department of Economics, Management and Statistics, University of Palermo, Palermo, Italy

Barbara Pawełek Department of Statistics, Cracow University of Economics, Cracow, Poland

Domenico Piccolo Department of Political Sciences, University of Naples Federico II, Naples, Italy

Alejandro Ribeiro Department of Electrical and Systems Engineering, University of Pennsylvania, Philadelphia, PA, USA

Rosaria Romano University of Calabria, Cosenza, Italy

Gabriella Schoier Dipartimento di Scienze Economiche Aziendali Matematiche e Statistiche, Università di Trieste, Trieste, Italy

Santiago Segarra Department of Electrical and Systems Engineering, University of Pennsylvania, Philadelphia, PA, USA

Rosaria Simone Department of Political Sciences, University of Naples Federico II, Naples, Italy

Nadia Solaro Department of Statistics and Quantitative Methods, University of Milan-Bicocca, Milano, Italy

Luisa Stracqualursi Department of Statistical Sciences, University of Bologna, Bologna, Italy

Nadia Tahiri Département d'informatique, Université du Québec à Montréal, Montreal, QC, Canada

Brigitte Vannier Receptors, Regulation and Tumor Cells (2RCT), University of Poitiers, Poitiers, France

Waldemar Wołyński Faculty of Mathematics and Computer Science, Adam Mickiewicz University, Poznań, Poland

Feliks Wysocki Poznań University of Life Sciences, Poznań, Poland

Lynda Zaoui LSSD Laboratory, Department of Computer Science, University of Science and Technology, Oran, Algeria

Part I
Classification Methods for High Dimensional Data

Missing Data Imputation and Its Effect on the Accuracy of Classification

Lynette A. Hunt

Abstract Multivariate data sets frequently have missing observations scattered throughout the data set. Many machine learning algorithms assume that there is no particular significance in the fact that a particular observation has an attribute value missing. A common approach in coping with these missing values is to replace the missing value using some plausible value, and the resulting completed data set is analysed using standard methods. We evaluate the effect that some commonly used imputation methods have on the accuracy of classifiers in supervised leaning. The effect is assessed in simulations performed on several classical datasets where observations have been made missing at random in different proportions. Our analysis finds that missing data imputation using hot deck, iterative robust model-based imputation (IRMI), factorial analysis for mixed data (FAMD) and Random Forest Imputation (MissForest) perform in a similar manner regardless of the amount of missing data and have the highest mean percentage of observations correctly classified. Other methods investigated did not perform as well.

1 Introduction

Many of the multivariate data sets collected today would have unobserved or missing observations scattered throughout the data set. These missing values can have no particular pattern of occurrence. Despite the frequent occurrence of missing data, many machine learning algorithms assume that there is no particular significance in the fact that a particular observation has an attribute value missing:— the value is simply unknown, and the missing value is handled in a simple way.

With many classification algorithms, a common approach that is used is to replace the missing values in the data set with some plausible value and the resulting completed data set is analysed using standard algorithms. The procedure that replaces the missing values using some value is known as imputation.

L.A. Hunt
University of Waikato, Hamilton, New Zealand
e-mail: lah@waikato.ac.nz

F. Palumbo et al. (eds.), *Data Science*, Studies in Classification, Data Analysis, and Knowledge Organization, DOI 10.1007/978-3-319-55723-6_1

Some algorithms treat the missing attribute as a value in its own right, whilst other classifiers such as Naïve Bayes ignore the missing data:—If an attribute is missing, the likelihood is calculated on the observed attributes and there is no need to impute a value. Decision trees, such as C4.5 and J48 [12, 13, 23], cope with missing values for an observation by notionally splitting the observation into pieces with a weighted split, and sending part down each branch of the tree.

However, the effect the treatment of missing values has on the performance of classifiers is not well understood. The estimation of the missing values can introduce additional biases into the data depending on the imputation method used and affect the classification of the observations.

This paper analyses the effect that several commonly used methods of imputation have on the accuracy of classification when classifying data that has a known classification. In Sect. 3, we review the mechanisms that can lead to data being missing, and in Sect. 2, we review the basic strategies for handing missing data. In Sect. 5, the imputation methods considered in this paper are examined and in Sect. 6, the effect the imputation methods have on the classification accuracy of several data sets is assessed.

2 Missing Data Mechanisms

The treatment of missing data indicators as random variables which were subsequently assigned a distribution was proposed by Rubin [16]. Depending on the distribution of the indicator, three basic mechanisms were defined by Rubin [16]:

1. Missing completely at random (MCAR).
   If the missingness does not depend on the values of the data, either missing or observed, then the data are MCAR.
2. Missing at random (MAR). If the missingness depends only on the data that are observed but not on the components that are missing, the data are MAR.
3. Not missing at random (NMAR). If the distribution of the missing data depends on missing values in the data matrix, then the mechanism is NMAR.

Knowledge of the mechanism that led to the values being missing is important in choosing an appropriate analysis to use for the data [10]. Hence it is important to consider how the classifier handles the missing data to avoid bias being introduced into the knowledge induced from that classifier.

3 Strategies for Handling Missing Data

There are several basic strategies that can be used to deal with missing data in classification studies. Some of these methods were developed in the context of sample surveys and can have some disadvantages in classification.

3.1 Complete Case Analysis

Complete case analysis (also known as elimination) is an approach in which observations that have any missing attributes are deleted from the data set. This strategy may be satisfactory with small amounts of missing data. However with large amounts of data, it is possible to lose considerable sample size. The critical concern with this strategy is that it can lead to biased estimates as it requires the assumption that the complete cases are a random subsample of the original observations. The completely recorded cases frequently differ from the original sample.

3.2 Available Case Analysis

Available case analysis is another approach that can be used. As this procedure uses all observations that have values for a particular attribute, there is no loss of information as all cases are used. However, the sample base changes from attribute to attribute depending on the pattern of missing data, and hence any statistics calculated can be based on different numbers of observations. The main disadvantage to this approach is that the procedure can lead to covariance and correlation matrices that are not positive definite, see, for example, [7]. This approach is used, for example, by Bayesian classifiers.

3.3 Weighting Procedures

Weighting Procedures are another approach to dealing with missing data. This approach is frequently used in the analysis of survey data. In survey data, the sampled units are weighted by their design weight which is inversely proportional to the probability of selection. Weighting procedures for non-response modify the weights in an attempt to adjust for non-response as if it were part of the sample design.

3.4 Imputation Procedures

Imputation procedures in which the missing data values are replaced with some value is another commonly used strategy for dealing with missing value. These procedures result in a hypothetical 'complete' data set that will cause no problems with the analysis. Many machine learning algorithms are designed to use either a complete case analysis or an imputation procedure.

Imputation methods often involve replacing the missing values with estimated values based on information that is in the data set. Many of the imputation methods are restricted to coping with one type of variable (i.e. either categorical or continuous) and make assumptions about the distribution of the data or subsets of variables. The performance of classifiers with imputed data in unreliable, and it is hard to distinguish situations in which the methods work from those in which they fail. When imputation is used, it is easy to forget that the data is incomplete [6]. However, imputation methods are commonly used in classification algorithms. There are many options that are available for imputation.

Imputation using a model-based approach is another popular strategy for handling missing data. A predictive model is created to estimate the values to be imputed for the missing values. With regression imputation, the attribute with missing data is used as the response attribute, and the remaining attributes are used as input for the predictive model. Maximum likelihood estimation using the EM algorithm [5] is one of the recommended missing data techniques in the methodological literature. This method assumes that the underlying model for the observed data is Gaussian.

Rather than imputing a single value for each missing data value, multiple imputation procedures are also commonly used. With this method, the missing values are imputed with values drawn randomly (with replacement) from a fitted distribution for that attribute. This is repeated a number, N, of times. The classifier is applied to each of the N "complete" data sets and the misclassification error is calculated. The misclassification error rates are averaged to provide a single misclassification error estimate and also estimate variances of the error rate.

Iterative regression imputation is not restricted to data having a multivariate normal distribution and can cope with mixed data. For the estimation, regression methods are usually applied in an iterative manner where each iteration uses one variable as an outcome and the remaining variables as predictors. If the outcome has any missing values, the predicted values from the regression are imputed. Iterations end when all variables in the data frame have served as an outcome.

4 Methods Used to Deal with Missing Values

The methods used in this paper for imputing the missing values are now described.

4.1 Mean and Median Imputation

Imputation of the missing value by either the mean, median or mode for the attribute are commonly used imputations. These types of imputation ignore any relationships between the variables. For mean imputation, it is well known that this method of imputation will underestimate the variance covariance matrices for that data [17].

The authors [10] also point out that with mean imputation the distribution of the "new values" is an incorrect representation of the population values as the shape of the distribution is distorted by adding values at the mean. Both mean and median imputation can only be used on continuous attributes. For categorical data, the mode is often imputed whilst using either mean or median imputation.

4.2 Hot Deck Imputation

Hot deck imputation is another imputation method that is commonly used, especially in survey samples, and it can cope with both continuous and categorical attributes. Hot deck imputation involves replacing the missing values using values from one or more similar instances that are in the same classification group. There are various forms of hot deck imputation commonly used. Random hot deck imputation involves replacing the missing value with a randomly selected value from the pool of potential donor values. Other methods known as deterministic hot deck imputation involve replacing the missing values with those from a single donor, often the nearest neighbour that is determined using some distance measure. Hot deck imputation has an advantage in that it does not rely on model fitting for the missing value that is to be imputed and thus is potentially less sensitive to model misspecification than an imputation method based on a parametric model. Further details on hot deck imputation can be found, for example, in [2].

4.3 kth Nearest Neighbour Imputation

The *k*th nearest neighbour algorithm is another method that can be used for imputation of missing values. This approach can predict both categorical and continuous attributes and can easily handle observations that have multiple missing values. This approach takes into account the correlation structure of the data. The algorithm requires the specification of number of neighbours, k, and the distance function that is to be used. The algorithm searches through the entire data set looking for most similar instances.

4.4 Iterative Model-Based Imputation

EM based stepwise regression imputation was proposed by Templ et al. [20] as a method for handling missing data. This technique for coping with missing values is an iterative model-based imputation (IRMI) that uses standard and robust methods. This algorithm has the advantage that it can cope with mixed data. In the first step of the algorithm, the missing values are initialised either using mean or KNN imputation. The attributes are sorted according to the original amount of missing values. After the attributes are sorted, we have

$$M(x_1) \geq M(x_2) \geq M(x_3) \geq \cdots \geq M(x_p) \quad (1)$$

where $M(x_j)$ represents the amount of missing values for attribute j and where x_j is now the jth column of the data matrix. The algorithm proceeds iteratively with one variable acting as the response variable and the remaining variables as the predictors in each step of the algorithm.

The authors [20, 22] compared their algorithm with that of IVEMARE [14], an algorithm that also performs iterative regression imputation. IRMI has advantages over IVEWARE with regard to the stability of the initial values, the robustness of the imputed values and the lack of requirement at least one fully observed variable [22]. With IRMI imputation, the user can also use least trimmed squares (LTS) regression (see, for example, [15]), MM estimation ([24] and M estimation [8]).

4.5 *Factorial Analysis for Mixed Data Imputation*

Imputation of missing values for mixed categorical and continuous data using the principal component method "factorial analysis for mixed data" (FAMD)' was proposed by Josse and Husson [9]. See also [3]. This algorithm imputes the missing values using an iterative FAMD algorithm that uses the EM algorithm or a regularised FAMD algorithm where the method is regularised.

4.6 *Random Forest Imputation*

Random forest imputation was proposed by Stekhoven and Bühlmann [19] as a method to deal with missing values with mixed type data. This approach uses an iterative imputation scheme in which a random forest is trained on the observed values in the first stage, the missing vales are predicted and then the algorithm proceeds iteratively. The algorithm begins by making an initial guess for the missing values in the data matrix. This data matrix is the imputed data matrix. The guesses for the missing values could be obtained using mean imputation or some other imputation method. In the first stage, a random forest is trained on the observed values. The missing values are then predicted using the random forest that was trained on the observed values, and the imputed matrix is updated. This procedure is repeated until the difference between the updated imputed data matrix and the previous imputed data matrix increases for the first time for both the categorical and the continuous types of variables. For continuous variables, the performance of the imputation is assessed using a normalised root mean squared error [11], and for categorical values, the proportion of falsely classified entries over the categorical missing values is used. Good performance of the algorithm gives a value that is close to 0 and bad performance gives a value close to 1. The algorithm proposed by Stekhoven and Bühlmann [19] is implemented in the R package MissForest [18]. This package also gives an estimate of the imputation error that is based in the out-of-bag error estimate from random forest.

5 The Analysis

Four classical machine learning data sets listed in Table 1 were taken. Note that the prostate cancer data set of [4] listed in [1] contained the information collected from 506 individuals; however, there were some missing values for some observations. This paper reports a complete case classification of the 12 pretrial attributes where individuals who had missing values in any of the pretrial attributes were omitted from further analysis, leaving 475 of the original 506 individuals.

For each data set, missing values were created such that the probability p of an attribute being missing was independent of all other data values where $p = 0.10$, 0.20, 0.30 and 0.50. This was repeated 20 times. The missing values generated in this fashion are missing completely at random, and the missing data mechanism is ignorable [10]. As some of the amounts of missing data are fairly extreme, it should be a good test of the types of imputation and the effect on the accuracy of a classifier.

The missing values in each data set were imputed using mean imputation, median imputation, k nearest neighbour (*kNN*) imputation with $k = 5$, hot deck imputation (HotD), iterative regression imputation (IRMI), the principal component method "factorial analysis for mixed data" (FAMD) and random forest imputation (MissForest). The missing values were imputed using the R packages VIM [21], HotDeckImputation, missMDA and MissForest. For data sets containing mixed data, the mode was imputed for the categorical attributes when using mean imputation for the continuous attributes.

The resulting "complete" data sets were analysed using the WEKA [23] experimenter with ten repetitions of tenfold cross-validation using several commonly used machine learning classifiers listed in Table 2. The mean percent of observations

Table 1 Datasets analysed

Datasets	Number of observations	Number of attributes	Type of attributes	Number of classes
Fisher's Iris data	150	4	Continuous	3
Pima Indian data	768	8	Mixed	2
Prostate cancer data	506[a]	12	Mixed	2
Wine data	178	13	Continuous	3

[a] This paper analyses the 475 complete cases

Table 2 Classifiers used

Classifier	Function
JRip	Uses a Ripper algorithm for fast efficient rule induction
J48	Implements C4.5 [12] decision tree learning
Naïve Bayes	Implements the standard probabilistic Naïve Bayes classification
Logistic	Builds linear logistics regression models
IBk	*k* nearest neighbours classifier
Logistic model trees (LMT)	Builds logistic model trees

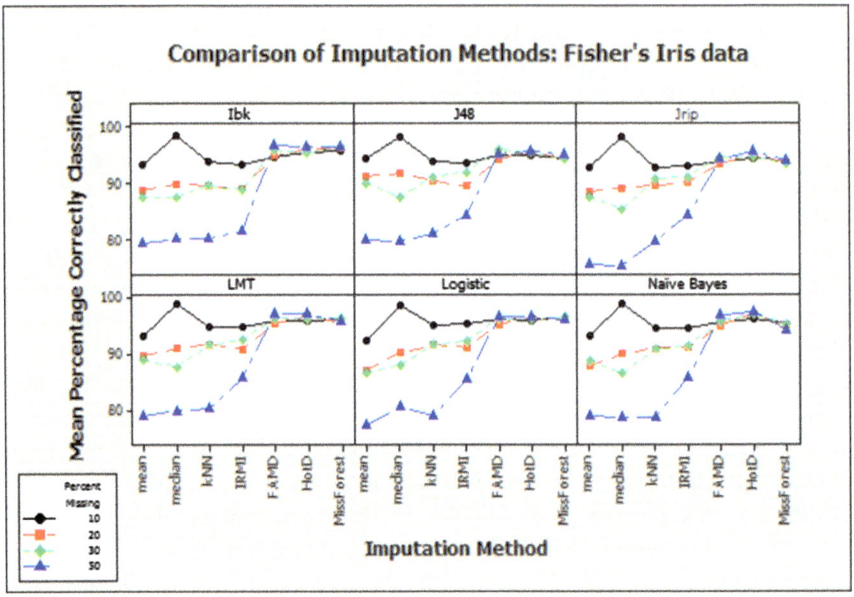

Fig. 1 Comparison of the imputation methods for Fisher's Iris data

correctly classified was recorded for each of the four amounts of missingness and each of the datasets analysed (see Figs. 1, 2, 3, 4).

It can be seen in Fig. 1 that, as the percentage of missing values increased, imputing the missing values in Fisher's Iris data using mean, median, *KNN* and IRMI imputation resulted in a decrease in the percentage of observations that were correctly classified. However with this data set, using FAMD, MissForest and Hot Deck imputation gave similar percentages of the observations correctly classified regardless of the amount of missingness in the data. The percentage correctly classified using FAMD, MissForest and Hot Deck imputation was similar to that for the complete data.

Figure 2 shows that mean, median and *KNN* imputation had similar percentages of observations correctly classified for each of the four missing data percentages, with lower percentages correctly classified as the amount of missingness in the data

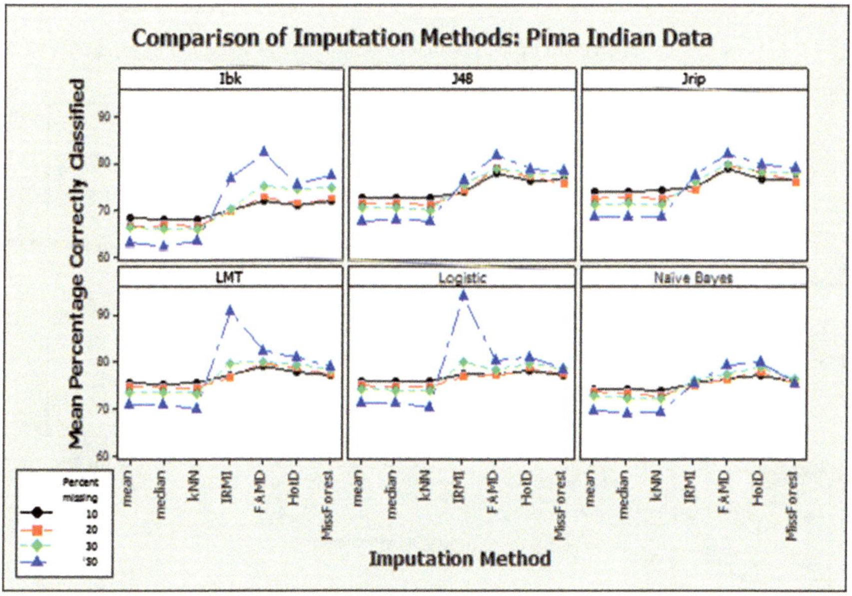

Fig. 2 Comparison of the imputation methods for Pima Indian data

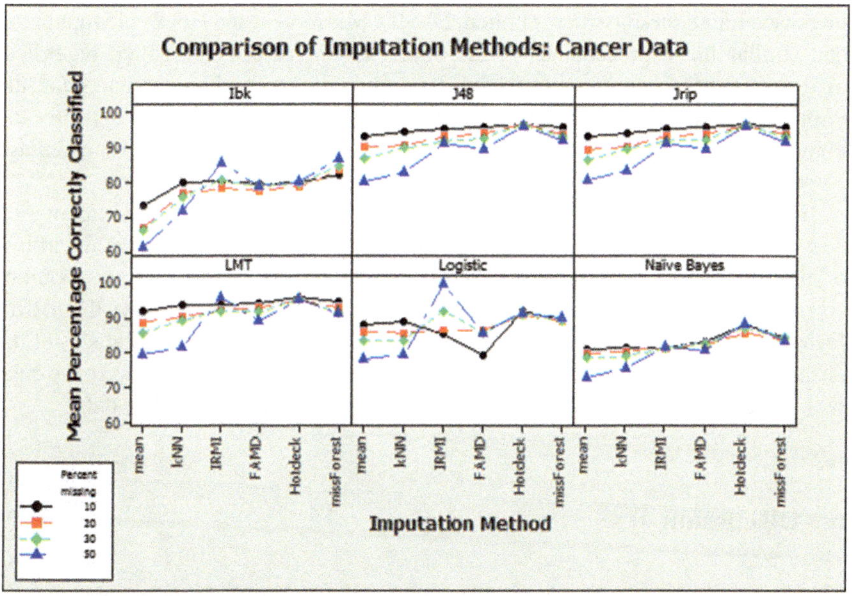

Fig. 3 Comparison of the imputation methods for prostate cancer data

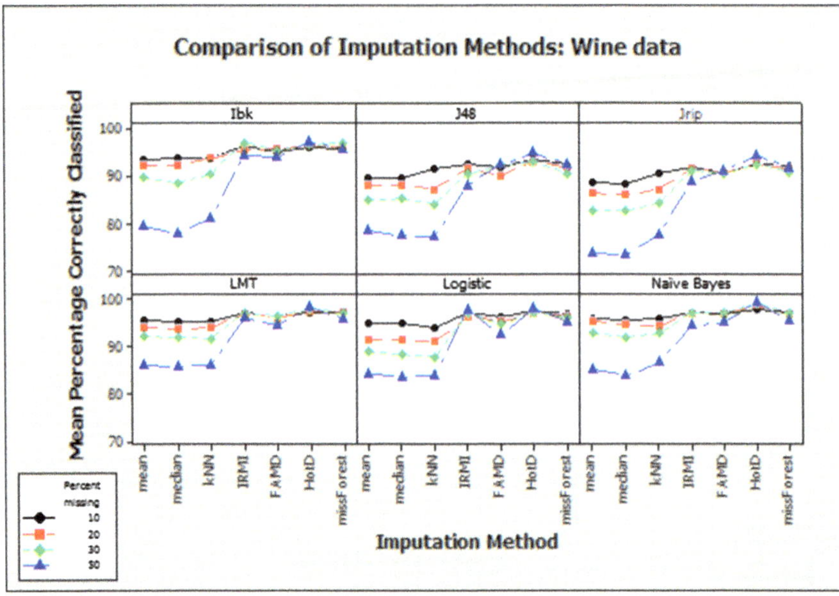

Fig. 4 Comparison of the imputation methods for wine data

increased for all the classifiers applied. FAMD, MissForest and Hot Deck imputation had similar mean percentages of the observations correctly classified regardless of the amount of missingness in the data. Overall for the Pima Indian data, the imputations using IRMI, FAMD, MissForest and Hot Deck gave consistently the highest mean percentage of observations correctly classified, with the percentage correctly classified similar to that for the complete data.

It can be seen in Fig. 3 that FAMD, MissForest and Hot Deck imputation had similar mean percentages of the observations correctly classified regardless of the amount of missingness in the data with the percentage of observations correctly classified similar to that for the complete data. Figure 4 shows that IRMI, FAMD, MissForest and Hot Deck imputations had similar mean percentages of the observations correctly classified regardless of the amount of missingness in the data, with the percentage correctly classified similar to that for the complete data.

6 Discussion

For all datasets analysed, we see that mean, median and kNN imputation have a similar mean percentage of observations correctly classified. We also see that the percentage of observations correctly classified when using the mean, median or kNN to impute the missing observations decreased as the percentage of missing data increased. In general, hot deck, IRMI, FAMD and missing Forest imputation had the highest mean percentage of observations correctly classified, and performed in a similar manner regardless of the amount of missing data imputed.

The investigations have shown that the type of method used for imputing missing values in a data set can have an effect on the accuracy of classification. Future research needs to be undertaken on the effect of imputation on the accuracy of classification on data that has more than three classes.

References

1. Andrews, D.F., Herzberg, A.M.: Data. A Collection of Problems from Many Fields for the Student and Research Worker. Springer, New York (1985)
2. Andridge, R.R., Little, R.J.A.: A review of hot deck imputation for survey nonresponse. Int. Stat. Rev. **78**, 40–64 (2010)
3. Audigier, V., Husson, F., Josse, J.: A principal components method to impute mixed data. Adv. Data Anal. Classif. **10**, 5–26 (2016). doi:10.1007/s11634-014-0195-1
4. Byar, D.P., Green, S.B.: The choice of treatment for cancer patients based on covariate information: application to prostate cancer data. Bull. Cancer Paris **67**, 477–488 (1980)
5. Dempster, A.P., Laird, N.M., Rubin, D.B.: Maximum likelihood from incomplete data via the EM algorithm (with discussion). J. R. Stat. Soc. B **39**, 1–38 (1977)
6. Dempster A.P., Rubin D.B.: Introduction, Incomplete Data in Sample Surveys (Volume 2): Theory and Bibliography. Madow, W.G., Olkin, I., Rubin, D.B. (eds.), pp. 3–10. Academic, New York (1983)
7. Everitt, B.S., Dunn G.: Applied Multivariate Data Analysis. Edward Arnold, London (2001)
8. Huber, P.J.: Robust Statistics. Wiley, New York (1981)
9. Josse, J., Husson, F.: Handling missing values in exploratory multivariate data analysis methods. J. de la Soc. Fr. de Stat. **153**(2), 1–21 (2012)
10. Little R.J.A., Rubin, D.B.: Statistical Analysis of Missing Data. Wiley New York (1987, 2002)
11. Oba, S., Sato, M., Takemasa, I., Monden, M., Matsubara, K., Ishii, S.: A Bayesian missing value estimation method for gene expression profile data. Bioinformatics **19**(16), 2088–2096 (2003). doi:10.1093/bioinformatics/btg287
12. Quinlan, J.R.: C4.5: Programs for Machine Learning. Morgan Kaufmann, San Mateo (1993)
13. Quinlan, J.R.: Improved use of continuous attributes in C4.5. J. Artif. Intell. Res. **4**, 77–90 (1996)
14. Raghunathan, T.E., Lepkowski, J.M., Van Hoewyk, J., Solenberger, P.: A multivariate technique for multiply imputing missing values using a sequence of regression models. Surv. Methodol. **27**(1), 85–9 (2001)
15. Rousseeuw, P., Van Driessen, K.: Computing LTS regression for large data sets. Data Min. Knowl. Disc. **12**, 29–45 (2006). doi:10.1007/s10618-005-0024-4
16. Rubin, D.B.: Inference and missing data. Biometrika **63**, 581–593 (1976)
17. Santos, R.: Effects of imputation on regression coefficients. Proc. Sect. Surv. Res. Methods Am. Stat. Assoc. 140–145 (1981)
18. Stekhoven, D.J.: Using the missForest package. https://stat.ethz.ch/education/semesters/ss2013/ams/.../missForest-1.2.pdf (2012)
19. Stekhoven, D.J., Bühlmann, P.: MissForest – non-parametric missing value imputation for mixed-type data. Bioinformatics **28**(1), 112–118 (2011)
20. Templ, M., Kowarik, A., Filzmoser, P.: EM-based stepwise regression imputation using standard and robust methods. Research Report cs-2010-3, Department of Statistics and Probability Theory, Vienna University of Technology (2010)
21. Templ, M., Alfons, A., Kowarik, A., Prantner, B.: VIM: Visualization and imputation of missing values (2011). http://CRAN.R-project.org/package=VIM. R package version 3.0.0

22. Templ, M., Kowarik, A., Filzmoser, P.: Iterative stepwise regression imputation using standard and robust methods. Comput. Stat. Data Anal. **55**, 2793–2806 (2011)
23. Witten, I.H., Frank, E., Hall, M.: Data Mining: Practical Machine Learning Tools and Techniques, 3rd edn. Morgan Kaufmann, San Francisco (2011)
24. Yohai, V.J.: High breakdown-point and high efficiency estimates for regression. Ann. Stat. **15**, 642–656 (1987)

On Coupling Robust Estimation with Regularization for High-Dimensional Data

Jan Kalina and Jaroslav Hlinka

Abstract Standard data mining procedures are sensitive to the presence of outlying measurements in the data. Therefore, robust data mining procedures are highly desirable, which are resistant to outliers. This work has the aim to propose new robust classification procedures for high-dimensional data and algorithms for their efficient computation. Particularly, we use the idea of implicit weights assigned to individual observation to propose several robust regularized versions of linear discriminant analysis (LDA), suitable for data with the number of variables exceeding the number of observations. The approach is based on a regularized version of the minimum weighted covariance determinant (MWCD) estimator and represents a unique attempt to combine regularization and high robustness, allowing to down-weight outlying observations. Classification performance of new methods is illustrated on real fMRI data acquired in neuroscience research.

1 Robustness and Regularization of Classification Methods

Classification methods (classifiers) have the aim to automatically assign new data to one of K groups ($K \geq 2$) based on decision rules constructed over a training data set. Sensitivity (non-robustness) of standard classifiers to the presence of outlying measurements (outliers) in the data has been repeatedly reported as a serious problem [3] and robust classification methods have been proposed as alternatives, which are resistant to the presence of outliers [8].

Linear discriminant analysis (LDA) as a standard (supervised) classification method assumes the data in each group to come from a Gaussian distribution, while the covariance matrix Σ is the same across groups. Its pooled estimator denoted by S is singular for high-dimensional data with $n < p$ or even $n \ll p$. For such data, which commonly appear in a variety of applications (e.g., in medicine, molecular

J. Kalina (✉) • J. Hlinka
Institute of Computer Science of the Czech Academy of Sciences, Pod Vodárenskou věží 2, 182 07 Prague, Czech Republic

National Institute of Mental Health, Topolová 748, 250 67 Klecany, Czech Republic
e-mail: kalina@cs.cas.cz; hlinka@cs.cas.cz

F. Palumbo et al. (eds.), *Data Science*, Studies in Classification, Data Analysis, and Knowledge Organization, DOI 10.1007/978-3-319-55723-6_2

genetics, chemometrics, or econometrics), regularized versions of LDA have been proposed to avoid the curse of dimensionality. They have become popular tools with a clear comprehensibility.

One common approach to regularized LDA is known as shrunken centroid regularized discriminant analysis (SCRDA) [4]. In this context, regularization brings benefits from both the computational and statistical point of view [13], which is true for $n < p$, as well as for $n > p$ with a relatively small n [5]. Its results may be superior to approaches based on a prior dimensionality reduction performed by selection of the most relevant variables.

However, regularized versions of LDA are sensitive to the presence of outlying values in the data. Unfortunately, most robust versions of LDA, which have been proposed within the framework of robust statistics, are computationally infeasible for $n < p$ [3, 8]. Xanthopoulos et al. [19] estimated high-dimensional covariance matrices allowing for measurement errors in the observed data. The resulting estimates are robust (insensitive) only to noise, but not robust to the presence of outliers. Robust procedures for high-dimensional data have been considered for regression models, including the proposal of a canonical correlation analysis [18] or partial least squares [7]. In the context of estimating a covariance matrix Σ of multivariate data, nonparametric correlation coefficients have been investigated by Croux and Öllerer [2] under the assumption that Σ^{-1} is sparse, which allows interesting applications in the area of graphical modeling. None of these approaches however exploits the idea of coupling the robustness with regularizing the estimated covariance matrix.

This paper exploits principles of robust statistics with the aim to propose new robust classification methods for high-dimensional data. We work with methods which are robust in terms of the breakdown point, which can be characterized as a global measure of robustness of an estimator against severe outliers in the data [9]. Methods with a high breakdown point are commonly denoted as highly robust. We presented a detailed overview of regularized versions of LDA in [11], however without considering robustness aspects. On the other hand, our previous work [10] on robust classification methods cannot be applied to high-dimensional data. Only the current paper exploits a unique coupling of regularization for $n \ll p$ and statistical robustness, which is based on implicit weighting, and thus ensures a high breakdown point.

In Sect. 2 of this paper, several new robust regularized methods for high-dimensional data are proposed based on down-weighting less reliable observations. The following Sects. 3 and 4 illustrate various methods on two real data sets and bring a detailed discussion of the results. Finally, Sect. 5 concludes the paper.

2 Classification Analysis Based on the Regularized Minimum Weighted Covariance Determinant Estimator

In this section, we propose several different robust versions of regularized LDA together with a discussion of their efficient computation. First, the robust regularized estimates of the covariance matrix and the means will be defined.

2.1 Estimation of the Covariance Matrix

In the whole paper, we assume n observations with p variables observed in K different groups

$$X_{11}, \dots, X_{1n_1}, \dots, X_{K1}, \dots, X_{Kn_K}, \tag{1}$$

where $p > K \geq 2$ and $n = \sum_{k=1}^{K} n_k$.

Chen et al. [1] proposed regularized M-estimation of the population mean and covariance matrix of multivariate data based on a popular M-estimator of Tyler [16] and applied it to the task of mining wireless sensor data. While M-estimation represents a popular approach to robust estimation of parameters, it does not possess a high breakdown point in the multivariate model [9].

The minimum weighted covariance determinant (MWCD) estimator is one of highly robust estimators of the mean and at the same time of the covariance matrix Σ of multivariate data [14]. The estimate of the mean has the form of a weighted mean and the estimate of Σ has the form of a weighted covariance matrix. Prior to the computation, the user must specify magnitudes of weights, while the weights themselves are assigned to individual observations after an optimal permutation. Linearly decreasing weights in the form

$$w_i^* = 1 - \frac{i-1}{n}, \quad i = 1, \dots, n, \tag{2}$$

if standardized to have the sum equal to 1, represents a simple and reasonable choice and will be considered also in the example of Sects. 3 and 4. The estimator remains to be reliable for data containing a large percentage of outliers [14].

While robust LDA based on the MWCD estimator was proposed in [10], the next sections propose classification methods based on the regularized MWCD estimator of the covariance matrix Σ in the form

$$\tilde{S}_{MWCD} = \lambda S_{MWCD} + (1 - \lambda)T, \quad \lambda \in (0, 1), \tag{3}$$

where a given target matrix T is symmetric positive definite of size $p \times p$. Such regularization ensures $\tilde{S}_{MWCD}$ to be regular and positive definite even for $n \ll p$. The simplest choices for T are the identity matrix $T = \mathscr{I}_p$ or a diagonal (nonidentity) matrix

$$T = \bar{s}\,\mathscr{I}_p, \tag{4}$$

where $\bar{s} = \sum_{i=1}^{p} S_{ii}/p$. Within the classification procedures defined below, a suitable value of λ may be found by a cross validation in the form of a grid search over all possible values of $\lambda \in (0, 1)$.

2.2 Estimation of the Means

Based on general principles of regularization [4, 5], we propose to consider also the regularization of the means to improve the classification performance of the robust regularized LDA. While all of the classification methods, which will be newly proposed in this paper, consider the pooled covariance matrix to be estimated by the same regularized MWCD estimator $\tilde{S}_{MWCD}$, we will consider different ways for estimating the means of each of the K groups. The MWCD estimator of each of the mean will be denoted by $\bar{X}_{k,MWCD}$ for $k = 1, \ldots, K$.

The regularized MWCD-means will be now defined for the k-th group for $k = 1, \ldots, K$. They will be defined as shrinkage estimators in various norms including the L_2, L_1, and L_0 norm using a fixed value of the regularization parameter. We use the notation $\bar{X}^{MWCD}$ for the overall MWCD-mean across groups, $(x)_+$ for the positive part of $x \in \mathbb{R}^p$, and $\mathbb{1}(B)$ for the indicator function of a random event B.

Definition 1 (Robust Regularized Means)

1.

$$\bar{X}^{(2)}_{k,MWCD} = \delta^{(2)}\bar{X}_{k,MWCD} + (1 - \delta^{(2)})\bar{X}_{MWCD}, \quad \delta^{(2)} \in \mathbb{R} \tag{5}$$

2.

$$\begin{aligned} \bar{X}^{(1)}_{k,MWCD} &= \mathsf{sgn}(\bar{X}_{k,MWCD})\left(|\bar{X}_{k,MWCD}| - \delta^{(1)}\right)_+ \\ &= \mathsf{sgn}(\bar{X}_{k,MWCD}) \max\left\{|\bar{X}_{k,MWCD}| - \delta^{(1)}, 0\right\}, \quad \delta^{(1)} \in \mathbb{R} \end{aligned} \tag{6}$$

3.

$$\bar{X}^{(0)}_{k,MWCD} = \bar{X}_{k,MWCD} \cdot \mathbb{1}\left[|\bar{X}_{k,MWCD} > \delta^{(0)}|\right], \quad \delta^{(0)} \in \mathbb{R} \tag{7}$$

All the estimators of Definition 1 can be interpreted as biased (Stein's shrinkage) versions of the MWCD-mean, while the biasedness allows to improve the mean square error [5]. The shrinkage within the estimator (6) is known as soft thresholding, while (7) is known as hard thresholding, where the latter corresponds to the solution of L_0 regularization [6].

2.3 MWCD-RLDA

The first of the novel methods proposed in this paper, which is denoted as MWCD-RLDA, assigns an observation $Z = (Z_1, \ldots, Z_p)^T$ to group j if

$$(\bar{X}_{k,MWCD} - Z)^T (\tilde{S}_{MWCD})^{-1} (\bar{X}_{k,MWCD} - Z) - 2 \log \pi_k \tag{8}$$

over $k =, 1, \ldots, K$ is minimal exactly for j, where π_k denotes the prior probability of observing an observation from the k-th group for $k = 1, \ldots, K$. Equivalently, the classification rule can be also expressed by means of the robust and regularized linear discriminant score

$$\ell_k^* = (\bar{X}_{k,MWCD})^T (\tilde{S}_{MWCD})^{-1} Z - \frac{1}{2} (\bar{X}_{k,MWCD})^T (\tilde{S}_{MWCD})^{-1} \bar{X}_{k,MWCD} + \log \pi_k \tag{9}$$

and an observation Z is assigned to group j if $\ell_j^* > \ell_k^*$ for every $k \neq j$.

The situation with equal regularized linear discriminant scores $\ell_k^* = \ell_{k'}^*$ for $k' \neq k$ does not deserve a separate treatment, because it occurs with a zero probability for data coming from a continuous distribution. We can say that the method is based on a deformed (regularized) Mahalanobis distance between a new observation Z and the mean of each group. Because $\tilde{S}_{MWCD}$ depends on $\lambda \in (0, 1)$, its suitable value should be found by cross validation.

Because both (9) and (8) are rather obscure from the computational point of view, we propose to avoid computing the inverse matrix by solving a set of linear equations within the following algorithm based on eigendecomposition of the robust regularized covariance matrix.

Algorithm 1 avoids computing the inverse of $\tilde{S}_{MWCD}$. Instead, the group assignment in (8) is done in a more efficient way, which easily follows from

$$\begin{aligned} &(\bar{X}_{k,MWCD} - Z)^T (\tilde{S}_{MWCD})^{-1} (\bar{X}_{k,MWCD} - Z) - 2 \log \pi_k \\ &= (\bar{X}_{k,MWCD} - Z)^T \tilde{Q} \tilde{D}^{-1} \tilde{Q}^T (\bar{X}_{k,MWCD} - Z) - 2 \log \pi_k \\ &= \| \tilde{D}^{-1/2} \tilde{Q}^T (\bar{X}_k^{k,MWCD} - Z) \|^2 - 2 \log \pi_k . \end{aligned} \tag{13}$$

Possible improvements of Algorithm 1 in terms of computational stability include:

1. A possible tailor-made approach for the specific choice $T = \mathscr{I}_p$.
2. Replacing the eigendecomposition by the Cholesky decomposition of $\tilde{S}_{MWCD}$ in the form $\tilde{S}_{MWCD} = LL^T$, where L is a nonsingular lower triangular matrix. Then,

Algorithm 1 MWCD-RLDA for a general T based on eigendecomposition.

1. For a given $\delta \in (0,1)$, compute the matrix

$$A = [\bar{X}_{1,MWCD} - Z, \ldots, \bar{X}_{K,MWCD} - Z] \tag{10}$$

   of size $p \times K$.
2. Compute $\tilde{S}_{MWCD}$ with a fixed $\lambda \in (0,1)$.
3. Compute and store the eigenvalues of $\tilde{S}_{MWCD}$ in the diagonal matrix $\tilde{D}$, and compute and store the corresponding eigenvectors of $\tilde{S}_{MWCD}$ in the orthogonal matrix $\tilde{Q}$.
4. Compute the matrix

$$B = \tilde{D}^{-1/2}\tilde{Q}^T A \tag{11}$$

   and assign Z to group k, if

$$k = \operatorname*{argmax}_{j=1,\ldots,K} \left\{||B_j||^2 - 2\log \pi_j\right\}, \tag{12}$$

   where $||B_j||^2$ is the Euclidean norm of the j-th column of B.
5. Repeat steps 1 to 4 with different values of λ and find the classification rule with the best classification performance.

an efficient computation may exploit that

$$\begin{aligned} &(\bar{X}_{k,MWCD} - Z)^T(\tilde{S}_{MWCD})^{-1}(\bar{X}_{k,MWCD} - Z) - 2\log \pi_k \\ &= (\bar{X}_{k,MWCD} - Z)^T L^{-T} L^{-1}(\bar{X}_{k,MWCD} - Z) - 2\log \pi_k \\ &= \|L^{-1}(\bar{X}_k - Z)\|^2 - 2\log \pi_k. \end{aligned} \tag{14}$$

3. Using the truncated eigendecomposition instead of the (standard) eigendecomposition. Let us recall the latter in the form

$$\tilde{S}_{MWCD} = \sum_{i=1}^{r} d_i q_i q_i^T, \tag{15}$$

   where r is rank of $\tilde{S}_{MWCD}$, $d_1, \ldots, d_r$ are nonzero eigenvalues, and $q_1, \ldots, q_r$ corresponding eigenvectors. The truncated eigendecomposition replaces (i.e., approximates) the expression (15) by

$$\tilde{S}_{MWCD} \approx \sum_{i=1}^{s} d_i q_i q_i^T = \tilde{Q}_* \tilde{D}_* \tilde{Q}_*^T, \tag{16}$$

   where $\tilde{Q}_*$ has size only $p \times s$ and $\tilde{D}_*$ only $s \times s$ for a specified $s < r$.

2.4 Other Classification Methods

We propose several other classification methods which, in contrary to MWCD-RLDA, consider also regularizing the means of each of the groups of the data. They are denoted as MWCD-RLDA2, MWCD-RLDA1, or MWCD-RLDA0, which correspond to regularizing the means in the L_2, L_1, or L_0 norm, respectively. Within each classification method, suitable values of the regularization parameters λ and of (as the case may be) $\delta^{(l)}$ for $l \in \{0, 1, 2\}$ can be found by leave-one-out cross validation. We use the notation $\mathsf{diag}(A)$ to denote the diagonal matrix containing diagonal elements of A. The linear discriminant rules of the novel methods are defined as modifications of (9).

Definition 2 (MWCD-RLDA2)

$$\tilde{\ell}_k^{(2)} = (\bar{X}_{k,MWCD}^{(2)})^T(\tilde{S}_{MWCD})^{-1}Z - \frac{1}{2}(\bar{X}_{k,MWCD}^{(2)})^T(\tilde{S}_{MWCD})^{-1}\bar{X}_{k,MWCD}^{(2)} + \log \pi_k. \tag{17}$$

Definition 3 (MWCD-RLDA1)

$$\tilde{\ell}_k^{(1)} = (\bar{X}_{k,MWCD}^{(1)})^T(\tilde{S}_{MWCD})^{-1}Z - \frac{1}{2}(\bar{X}_{k,MWCD}^{(1)})^T(\tilde{S}_{MWCD})^{-1}\bar{X}_{k,MWCD}^{(1)} + \log \pi_k. \tag{18}$$

Definition 4 (MWCD-RLDA0)

$$\tilde{\ell}_k^{(0)} = (\bar{X}_{k,MWCD}^{(0)})^T(\tilde{S}_{MWCD})^{-1}Z - \frac{1}{2}(\bar{X}_{k,MWCD}^{(0)})^T(\tilde{S}_{MWCD})^{-1}\bar{X}_{k,MWCD}^{(0)} + \log \pi_k. \tag{19}$$

Definition 5 (MWCD-PAM)

$$\begin{aligned} \ell_k^{PAM} = {} & (\bar{X}_{k,MWCD}^{(1)})^T(\mathsf{diag}\{\tilde{S}_{MWCD}\})^{-1}Z - \\ & -\frac{1}{2}(\bar{X}_{k,MWCD}^{(1)})^T(\mathsf{diag}\{\tilde{S}_{MWCD}\})^{-1}\bar{X}_{k,MWCD}^{(1)} + \log \pi_k. \end{aligned} \tag{20}$$

An efficient computation of the new methods can be performed by an analogy of Algorithm 1. If the classification rule based on (20) is formulated by means of the Mahalanobis distances, the formula (13) reduces to a simple form

$$\sum_{i=1}^{p} \frac{(\bar{X}_{ki} - Z_i)^2}{\tilde{S}_{i,MWCD}^2}, \tag{21}$$

where $\bar{X}_k = (\bar{X}_{k1}, \ldots, \bar{X}_{kp})^T$ and $\tilde{S}_{i,MWCD}^2$ denotes the i-th diagonal element of $\tilde{S}_{MWCD}$. MWCD-PAM represents a robust counterpart of the Prediction Analysis of Microarrays [15], where the latter is nothing else than a diagonalized LDA with means regularized in the L_1 norm.

Here, MWCD-RLDA1 can be interpreted as a robust counterpart of SCRDA [4]. Because MWCD-RLDA1 contains an intrinsic variable selection in (6), it is especially suitable if the data set contains a small set of dominant (very relevant) variables. On the other hand, MWCD-RLDA2 can be recommended if the data contain a large number of variables with a small effect on the classification, but without any clearly dominant small subset of variables.

3 Example: Brain Activity Data

A data set on the spontaneous activity of various parts of the brain will be now analyzed, which has been captured by means of fMRI neuroimaging. We have participated on a neuroscience research of the spontaneous brain activity in the resting state (i.e., resting-state brain networks). Our aim now is to illustrate the behavior of the newly proposed classification methods.

The brain activity of $n = 24$ probands is measured by means of fMRI under seven different situations. One of them can be characterized as a resting state, i.e., rest without any stimulus. Besides, the probands were watching each of six different movies while the brain activity was measured. The fMRI divides the brain into 90 regions and we are interested only in values of correlation coefficients between a pair of brain regions. In this context, the correlation coefficient evaluates a (functional) connectivity between the two regions. Thus, we consider $p = 90 * 89/2 = 4005$ variables containing values of correlation coefficients for each of the 24 probands.

The task is to learn a classification rule allowing to discriminate between two groups (resting state and movie) over 24 individuals, i.e., all movies together are considered to be one class. This is a classification to two groups with $p = 4005$ variables. The resting-state group contains 24 observations, but the group corresponding to any movie contains $6 * 24 = 144$ observations. In common applications, fMRI measurements are known to be usually contaminated by noise as well as outliers. It is also true with our data and therefore robust methods are highly desirable for their analysis.

We performed the computations in *R* software. Standard machine learning methods are used with default settings of their parameters. For various regularized versions of LDA, we choose the target matrix T as either $T = \mathscr{I}_p$ or as (4). The results of leave-one-out cross classification are overviewed in Table 1. Performance of classifiers is measured by means of their accuracy, i.e., number of correctly classified cases divided by the total number of cases.

SCRDA as one of available regularized LDA versions turns out to perform reliably, while its classification rule is based only on 81 variables. Also the newly proposed robust LDA versions yield a very good performance. We do not find major differences in the classification performance of robust and non-robust various

Table 1 Results of the examples of Sects. 3 and 4

Classification method	Classification accuracy	
	Brain data	AIM data
SCRDA	1.00	0.86
MWCD-RLDA	1.00	0.86
MWCD-RLDA1	1.00	0.86
MWCD-PAM	0.98	0.77
SVM (Gaussian kernel)	1.00	0.85
Multilayer perceptron	Infeasible	Infeasible
Number of principal components	10	20
PCA $\Longrightarrow$ LDA	1.00	0.83
PCA $\Longrightarrow$ SCRDA	1.00	0.83
PCA $\Longrightarrow$ MWCD-RLDA with $T = \mathscr{I}_p$	1.00	0.84
PCA $\Longrightarrow$ MWCD-RLDA with (4)	1.00	0.84
PCA $\Longrightarrow$ MWCD-RLDA2	1.00	0.84
PCA $\Longrightarrow$ MWCD-RLDA1	1.00	0.84
PCA $\Longrightarrow$ MWCD-RLDA0	1.00	0.84
PCA $\Longrightarrow$ MWCD-PAM	0.96	0.75

Various classification methods are compared, while their classification accuracy is evaluated in a leave-one-out cross validation study

regularized versions of LDA. This can be explained by the fact that the data do not contain a remarkable percentage of outliers. Also the SVM method gives a perfect classification rule, while a multilayer perceptron with one hidden layer is computationally infeasible due to $n \ll p$ in the implementation in R software.

Additionally, we investigated the effect of dimensionality reduction by means of principal component analysis (PCA) on the classification performance. There seems no remarkable small group of genes responsible for a large portion of variability of the data and the first few principal components seem rather arbitrary. All the novel robust methods have a good classification ability if applied on principal components. Thus, the classification results after reducing the dimensionality bring other arguments in favor of the regularization approaches used in this paper.

In order to investigate the performance of various classification methods on data contaminated by noise, we generated proband-independent noise generated from normal distribution $\mathsf{N}(0, \sigma^2)$ for various values of σ. The noise was added to all measurements for each proband and classification rules are learned over this contaminated data set. Such contamination was repeated 100 times and the classification performance of various methods was evaluated for each case. We give the averaged values of the classification accuracy computed over the 100 cases in Table 2 only for selected classifiers, because their computation is rather demanding.

The results of the classification performance of various methods on data artificially contaminated by noise show an evidence of robustness of SCRDA. The larger the value of σ, the more influential outliers are present in the contaminated data set. Indeed, the reduction of the classification performance of the standard data mining

Table 2 Results of the brain activity analysis on data artificially contaminated by normally distributed outliers $\mathsf{N}(0, \sigma^2)$ for different values of σ

Classification method	Classification accuracy		
	$\sigma = 0.1$	$\sigma = 0.2$	$\sigma = 0.3$
SVM (Gaussian kernel)	1.00	0.99	0.98
Classification tree	0.99	0.98	0.98
SCRDA	1.00	1.00	1.00
MWCD-RLDA1 with $T = \mathscr{I}_p$	1.00	1.00	1.00
PCA $\Longrightarrow$ LDA	0.99	0.99	0.99

The classification accuracy is evaluated by a leave-one-out cross validation. PCA is used with a fixed number of ten principal components

methods is not caused by the noise itself, but rather by severe outliers. SCRDA and the novel robust versions of LDA turn out to yield reliable results. The robustness of SCRDA to noise has not however been systematically investigated although it has been recommended as a promising alternative to the SVM [4].

Further, the classification rule distinguishing between the resting state and a particular movie is constructed, which is again a classification to 2 groups with $p = 4005$ variables. This time, each of the groups contains 24 observations. We computed SVM, SCRDA, and PCA $\Longrightarrow$ LDA for 6 different tasks, namely classification between the resting state and movie 1; between the resting state and movie 2, etc. In a leave-one-out cross validation study, every method yields a 100% classification accuracy in all the seven classification tasks. For the sake of comprehensibility, it is important that MWCD-RLDA1 turns out to be based only on a small number of variables, namely 1, 1, 2, 3, 3, and 7 variables. These are the most relevant sets of variables for the particular classification task, while SVM and PCA exploit observed values from the whole set of p variables. If PCA is performed keeping ten principal components, each of the considered classifiers keeps the 100% classification accuracy for each of the six classification tasks.

Additionally, classification between pairs of movies (e.g., classification between movie 1 and movie 2) yields results with 100% classification accuracy, while the number of variables contributing to the classification rule of MWCD-RLDA1 is between 2 and 30 for each of the tasks.

4 Example: Cardiovascular Genetic Study

Further, we illustrate the performance of the novel robust classifiers on the data from a cardiovascular genetic study performed in the Center of Biomedical Informatics in Prague in the years 2006–2011. From the point of view of analyzing the data, the very aim originally was to reduce the dimensionality [12], i.e., to find a small set of genes responsible for the development of the acute myocardial infarction (AMI). Gene expressions of $p = 38{,}590$ were measured across the whole genome. These correspond to the activity of individual genes leading to synthesis of proteins and consequent biological processes. The BeadChip microarray technology was used to acquire the data over $n = 95$ individuals, including 46 AMI patients and 49 controls.

Robust versions of regularized LDA perform well. We find it a success that the method is computationally feasible for such p at all due to the high dimensionality, which severely complicates a potential identification of outliers. While the SVM classifier formally gives a perfect classification result, it suffers from a heavy overfitting, not only because its optimization of parameters tends to a very local optimum for $n < p$, but mainly because the SVM contains too many support vectors and does not capture the multivariate structure of the data. Although it is designed as a black box, we can say that it classifies each new observation with a too strong emphasis on its nearest neighbors.

If the classification rule is learned only over the set of the 20 principal components, robust versions of regularized LDA are able to slightly outperform available (non-robust) classifiers. There seems however no difference among the individual robust methods of Sect. 2, because only negligible values of the regularization parameter for the means are selected for each of the methods and the effect of this regularization is negligible itself.

5 Conclusions

Some of the standard methods of data mining or multivariate statistics are computationally infeasible for high-dimensional data, while others suffer from a numerical instability and lack of robustness to noise or outlying measurements. Therefore, this paper proposes new robust classification methods for high-dimensional observations, i.e., assuming the number of variables to exceed the number of observations. We combine robustness to the presence of outliers with regularized estimation of the population means and covariance matrix of the multivariate data in a unique way.

We propose several robust classifiers, which are based on implicit weighting of individual observations in Sect. 2. The methods are based on a regularized version of a robust covariance matrix, while also the mean of each group is computed by means of a robust regularized estimator. At the same time, implicit weights ensure a high breakdown point with respect to a larger percentage of outliers in a variety of other situations [17]. All the methods require intensive computations. Efficient algorithms allow the methods to be computed even for $n \ll p$.

The robust regularized versions of LDA can be interpreted as modifications of robust LDA corrected for small sample sizes. At the same time, we point out the connection to the shrinkage statistical approach, following Stein's result of estimating the mean of multivariate normal data [5].

We consider all of the newly proposed robust methods to be comprehensible. Particularly, let us discuss the classification rule of MWCD-RLDA1. It assigns an observation Z based on a deformed Mahalanobis distance between Z itself and a (robust) centroid of each of the K groups. Such variables contribute the most to the classification rule, which are most relevant for the separation among groups. Also the implicit weights assigned to individual observations allow a clear interpretation. They deform the Mahalanobis distance, while less reliable observations (potential outliers) obtain small or negligible weights.

In addition, we analyzed two high-dimensional data sets. The fMRI data come from a brain research study, which has the aim to investigate connections among brain parts during a resting state. Results of various classification methods show distinct differences between the resting and non-resting state. At the same time, different movies shown to the set of 24 probands turn out to activate different connections between pairs of brain parts. Future neuroscience research is intended to search for a small set of variables allowing to distinguish schizophrenic patients from control individuals based only on the fMRI measurements of the brain in the resting state. The cardiovascular genetic data set with a dimensionality even larger ($p = 38{,}590$) compared to the fMRI data shows a slight advantage of the newly proposed methods compared to available classifiers. The analysis of this data set allows to detect a predisposition for infarction based only on gene expressions.

Concerning the limitations of our analysis, both SCRDA and its robust counterparts are reliable under an implicit assumption that the variability is not substantially different across variables. Still, the methods seem to yield reliable results although this assumption is violated in the data.

To summarize practical recommendations based on the example, the new robust methods seem to perform reliably for high-dimensional data with a small number of observations. The level of noise in the original data seems to be moderate and the advantage of robust methods compared to non-robust ones is not revealed primarily after adding an artificial contamination to the data. SCRDA itself turns out to be reasonably robust, which can be explained as an effect of the regularization reducing the influence of noise in the data. The main result of the examples is however the reliability of the newly proposed methods for both original and contaminated data sets.

Acknowledgements The work is supported by the project "National Institute of Mental Health (NIMH-CZ)", grant number CZ.1.05/2.1.00/03.0078 of the European Regional Development Fund, Neuron Fund for Support of Science, and the Czech Science Foundation project No. 13-23940S.

References

1. Chen, Y., Wiesel, A., Hero, A.O.: Robust shrinkage estimation of high dimensional covariance matrices. IEEE Trans. Signal Process. **59**, 4097–4107 (2011)
2. Croux, C., Öllerer, V.: Robust and sparse estimation of the inverse covariance matrix using rank correlation measures. Technical Report, KU Leuven (2015)
3. Filzmoser, P., Todorov, V.: Review of robust multivariate statistical methods in high dimension. Anal. Chim. Acta **705**, 2–14 (2011)

4. Guo, Y., Hastie, T., Tibshirani, R.: Regularized discriminant analysis and its application in microarrays. Biostatistics **8**, 86–100 (2007)
5. Hastie, T., Tibshirani, R., Friedman, J.: The Elements of Statistical Learning, 2nd edn. Springer, New York (2009)
6. Herlands, W., De-Arteaga, M., Neill, D., Dubrawski, A.: Lass-0: sparse non-convex regression by local search (2016, submitted)
7. Hoffmann, I., Serneels, S., Filzmoser, P., Croux, C.: Sparse partial robust M regression. Chemom. Intel. Lab. Syst. **149**, 50–59 (2015)
8. Hubert, M., Rousseeuw, P.J., Van Aelst, S.: High-breakdown robust multivariate methods. Stat. Sci. **23**, 92–119 (2008)
9. Jurečková, J., Sen, P.K., Picek, J.: Methodology in Robust and Nonparametric Statistics. CRC Press, Boca Raton (2012)
10. Kalina, J.: Highly robust statistical methods in medical image analysis. Biocybern. Biomed. Eng. **32**(2), 3–16 (2012)
11. Kalina, J.: Classification analysis methods for high-dimensional genetic data. Biocybern. Biomed. Eng. **34**, 10–18 (2014)
12. Kalina, J., Zvárová J.: Decision support systems in the process of improving patient safety. In: Bioinformatics: Concepts, Methodologies, Tools, and Applications, pp. 1113–1125. IGI Global, Hershey (2013)
13. Pourahmadi, M.: High-Dimensional Covariance Estimation. Wiley, Hoboken (2013)
14. Roelant, E., Van Aelst, S., Willems, G.: The minimum weighted covariance determinant estimator. Metrika **70**, 177–204 (2009)
15. Tibshirani, R., Hastie, T., Narasimhan, B.: Class prediction by nearest shrunken centroids, with applications to DNA microarrays. Stat. Sci. **18**, 104–117 (2003)
16. Tyler, D.E.: A distribution-free M-estimator of multivariate scatter. Ann. Stat. **15**, 234–251 (1987)
17. Víšek, J.Á.: Consistency of the least weighted squares under heteroscedasticity. Kybernetika **47**, 179–206 (2011)
18. Wilms, I., Croux, C.: Robust sparse canonical correlation analysis. BMC Systems Biology **10**, 72 (2016)
19. Xanthopoulos, P., Pardalos, P.M., Trafalis, T.B.: Robust Data Mining. Springer, New York (2013)

Classification Methods in the Research on the Financial Standing of Construction Enterprises After Bankruptcy in Poland

Barbara Pawełek, Krzysztof Gałuszka, Jadwiga Kostrzewska, and Maciej Kostrzewski

Abstract In the literature devoted to applications of multivariate statistical analysis to finance, the issue of bankruptcy forecasting is dealt with at length, but few papers concern the statistical evaluation of financial standing of companies after they have been declared bankrupt. The examination of their way out from the insolvency problem may be a source of valuable information, useful for the assessment of the probability that other bankrupt enterprises achieve success as a result of the execution of restructuring proposals. The purpose of this article is to present a proposal to use selected classification methods when studying the financial standing of companies after the declaration of bankruptcy in comparison with the situation of financially sound companies. The logit model and the classification tree were used to classify companies. The evaluation of the classification efficiency was based on the following measures: sensitivity, specificity and AUC. In the study, both univariate (Tukey's criterion) and multivariate (projection depth function) methods for detecting outliers were considered. The study covered construction companies in Poland in the years 2005–2009.

B. Pawełek (✉) • J. Kostrzewska
Department of Statistics, Cracow University of Economics, 27 Rakowicka Street, 31-510 Cracow, Poland
e-mail: barbara.pawelek@uek.krakow.pl; jadwiga.kostrzewska@uek.krakow.pl

K. Gałuszka
Department of Public Finance, University of Economics in Katowice, 50 1 Maja Street, 40-287 Katowice, Poland
e-mail: krzysztof.galuszka@ue.katowice.pl

M. Kostrzewski
Department of Econometrics and Operational Research, Cracow University of Economics, 27 Rakowicka Street, 31-510 Cracow, Poland
e-mail: maciej.kostrzewski@uek.krakow.pl

F. Palumbo et al. (eds.), *Data Science*, Studies in Classification, Data Analysis, and Knowledge Organization, DOI 10.1007/978-3-319-55723-6_3

1 Introduction

The problem of corporate bankruptcy is an important issue in economic sciences. The establishment of new companies and the closure of business by some of the existing companies are natural phenomena occurring in a free-market economy. The cessation of business as a consequence of declared corporate bankruptcy attracts the particular interest of academics, business practitioners and financial institutions. Such interest can be justified by, among other matters, the serious socio-economic consequences of corporate bankruptcy, and it promotes the development of methods for predicting the risk of corporate bankruptcy.

In the literature concerning applications of multivariate statistical analysis to finance, the issue of bankruptcy forecasting is dealt with at length, but few papers concern the forecasting of repeated corporate bankruptcy (Fig. 1).

The purpose of this article is to present a proposal to use selected classification methods when studying the financial standing of companies after the declaration of bankruptcy in comparison with the situation of financially sound companies. The research hypothesis is that classification methods used in forecasting corporate bankruptcy are effective tools for assessing the financial standing of companies after the declaration of bankruptcy. The novelty in the article is the use of methods applied in bankruptcy forecasting to assess the financial standing of companies trying to get out of the problem of insolvency after they have been declared bankrupt by the courts.

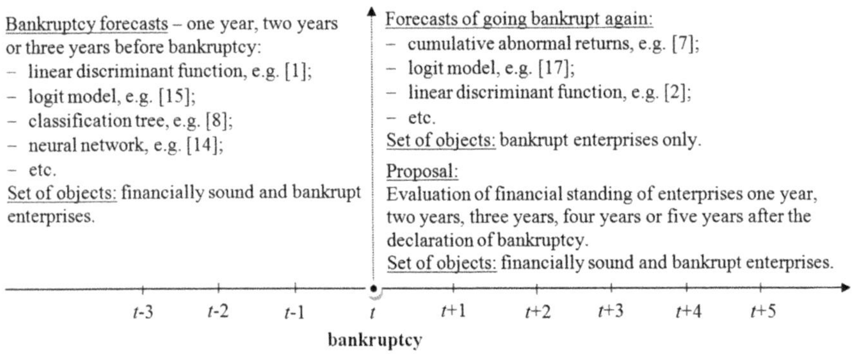

Fig. 1 Brief review of enterprise bankruptcy examinations

2 State of Research

Literature on the subject contains review papers concerning the issue of forecasting bankruptcy (e.g. [9, 20]). In the paper [9], the following criteria of the comparative analysis of papers related to the forecasting of enterprise bankruptcy based on an unbalanced sample have been adopted: type of database, approach applied to evaluate the classification effectiveness, evaluation measures of the classification effectiveness and application of statistical tests. The typical features of this type of research have been indicated after an analysis of more than 140 papers from the years 2000–2013. The basis for considerations is primarily the actual data concerning the economies of certain states; databases contain mainly up to 1000 objects, and then from 1000 to 10,000 objects; a division into training and testing datasets or a cross-validation test is often applied; such measures as the overall effectiveness, type I and type II errors, as well as, increasingly, the AUC measure are primarily used to evaluate the classification effectiveness; no statistical tests are applied.

On the other hand, in the paper [20] about 140 publications from the years 1966–2014 were reviewed. They concerned the subject under discussion in terms of definitions of unfavourable financial standing and corporate bankruptcy, considered methods for forecasting corporate bankruptcy, approaches adopted to construct a research sample, as well as procedures applied to select exogenous variables for the model. The authors highlighted, among other matters, the variety of definitions of poor financial standing of a company. They found that in theoretical considerations, different levels of adverse financial situation are distinguished, whereas empirical studies are generally limited to an analysis of two conditions: a financially sound company and a company declared bankrupt. The applied methods of bankruptcy forecasting can be divided into, for example, univariate and multivariate methods, traditional statistical models and methods based on artificial intelligence, static and dynamic models, etc. In empirical studies, the sector to which companies belong is taken into account when selecting a sample. Sets underlying empirical analyses differ in structure (balanced and unbalanced samples). The selection of an initial set of exogenous variables of the model (financial ratios and non-financial variables) is carried out based on various criteria, including expert opinions, incidence in other studies, availability of financial data, etc. A reduction in the set of potential variables is based on both qualitative and quantitative criteria.

A separate trend in forecasting corporate bankruptcy is the predicting of repeated corporate bankruptcy (e.g. [2, 7, 17]). To justify the usefulness of conducting this type of research, arguments presented by the authors of the paper [2] can be cited. According to them, about 18.25% of companies in the USA land in court once again (or more than once) after a court has declared their bankruptcy open for arrangement. They claim that the Altman Z"-score model may be useful for a court to predict the success or failure of further activity of bankrupt companies. They believe that both courts and persons responsible for preparing and conducting the restructuring of companies should use statistical methods applied to predicting corporate bankruptcy as a supplement to traditional analysis. Those methods can be used to evaluate the restructuring plan and monitor the situation after the bankruptcy in order to introduce adjustments to the plan. In their study, the authors analysed only companies with a court verdict divided into those which had landed in court only once and those which are 'recidivists'. A return to court means that the restructuring has failed in terms of both the concept and socio-economic costs. They emphasised the importance of the early warning system in order to reduce the likelihood of repeated bankruptcy, often preceded by long and costly restructuring.

An alternative approach to the one proposed in the paper [2] is a statistical evaluation of the financial standing of companies in the years following the declaration of bankruptcy compared with the situation of financially sound companies. The examination of companies' paths to get out of the insolvency problem may be a source of valuable information, useful for the assessment of the likelihood that other bankrupt enterprises achieve success as a result of the execution of restructuring proposals.

The results of such an assessment may be helpful in selecting appropriate remedial programmes for companies with a solvency issue. However, caution is advised when interpreting the received results. Tax law regulations, the policies of financial institutions, etc. may affect the assessment of the economic situation of a company and its future on the market. It is also worth remembering that differences in accounting, particularly the degree of implementation of international accounting standards, make it difficult or even impossible to compare data between countries. This, in turn, makes it necessary to conduct research on the financial standing of companies after the declaration of bankruptcy separately for each country.

3 Data and Research Procedure

Bankruptcy can be equated with bankruptcy declared by the courts. In Poland to 31 December 2015, there were two types of bankruptcy: arrangement bankruptcy and liquidation bankruptcy. The courts are guided by the principle according to which bankruptcy proceedings must be conducted in such a way that the creditors' claims can be satisfied as far as possible, and the debtor's enterprise can be preserved. One should be aware that within a particular country there may be no comparability or very limited comparability of data among companies that are subject to the obligation to have their financial statements audited and companies exempt from this obligation. In Poland, some companies with unfavourable financial ratios do not go bankrupt and even 'bloom' due to tax law regulations allowing them to utilise tax losses in the case of mergers of companies based on a company with losses instead of the one with profits. On the other hand, financial institutions are sceptical about companies with unfavourable values of financial ratios and they strive to quickly recover their funds by bringing such companies to liquidation bankruptcy. Thus, a company in a worse financial situation can survive, and the one in a better financial situation may go bankrupt, if the latter has been more

Table 1 Financial ratios

Symbol	Description	Symbol	Description
R_{01}	Current liquidity ratio	R_{08}	Net profitability
R_{02}	Quick liquidity ratio	R_{09}	ROE
R_{03}	Cash ratio	R_{10}	ROA
R_{04}	Total debts to assets	R_{11}	Accounts receivable turnover
R_{05}	Debt to equity	R_{12}	Fixed asset turnover
R_{06}	Long-term debt to equity	R_{13}	Total asset turnover
R_{07}	Gross profitability	R_{14}	Operation cost to sales revenues

indebted to financial institutions. The above observations indicate the difficulties occurring when trying to evaluate companies' financial standing after the declaration of bankruptcy.

The data used in this analysis have been downloaded from the website of the Emerging Markets Information Service (http://www.emis.com). The research objects were 369 construction companies in Poland, five of them were bankrupts (B_1–B_5). Court verdicts were passed between November 17, 2003 and August 30, 2004. The study used 14 financial ratios broken down into the groups of the following indicators: liquidity (R_{01}–R_{03}), liability (R_{04}–R_{06}), profitability (R_{07}–R_{10}) and productivity (R_{11}–R_{14}) (Table 1). The financial data were taken from the period 2005–2009. The following designations were adopted: *NB*—financially sound company (i.e. the one which had not been declared bankrupt before 2009) and *B*—bankrupt company. Due to a small number of bankrupt companies, it was not possible to create a test sample.

An unbalanced set was the basis of empirical studies. In the case of this type of set, the problem of low classification efficiency of bankrupt enterprises occurs more often than in the case of analysing balanced sets within the framework of the considered methods of predicting corporate bankruptcy. Apart from a small share of bankrupt enterprises in the examined sets, this may be caused, among other factors, by the occurrence of atypical objects among financially sound companies [16].

Papers devoted to the forecasting of enterprise bankruptcy present considerations related to the occurrence of outliers among the data. Proposed solutions to this problem oscillate from ignoring [19], through substitution or removal of outliers (e.g. [6, 16, 18, 22]), to the use of robust methods.

An atypical financially sound company is understood here as an object with outlying financial ratios. It was assumed that an outlier is a value that seems to differ significantly from other elements of the group in which it occurs (e.g. [3, 10]). Companies so defined may be characterised by both a very good financial situation and a poor financial situation, similar in terms of many indicators to the situation of companies declared bankrupt. The detecting and removing of atypical financially sound companies from a set of objects also has a substantive justification. Economic practice shows that companies that have a poor financial standing (i.e. those characterised by unfavourable financial ratios) may not fulfil the obligation to file a petition for bankruptcy. In the absence of such a petition also on the part of creditors, such companies may exist on the market and influence the situation of the entire industry. Therefore, the set of financially sound companies was cleaned of atypical financially sound objects. The following were used to detect outliers:

- A univariate method based on Tukey's criterion [21]
- A multivariate method based on a projection depth function [23]

Financial ratios of typical financially sound companies reflect the sector's financial standing, which depends, for example, on the economic situation in Poland.

For each year, the procedure based on Tukey's criterion [21] had the following stages:

- For each financial ratio, the first (Q_1) and third (Q_3) quartiles and the quartile deviation (Q) were calculated. The analysis used financial ratios of financially sound companies.
- Values outside the range: $< Q_1 - 1.5Q, Q_3 + 1.5Q >$ were regarded as outliers.
- A financially sound company was considered atypical if at least one of the values of the financial ratios had been regarded as an outlier.

The concept of data depth is an issue of non-parametric resistant multivariate statistical analysis, developed within the framework of an exploratory data analysis [11]. It enables one to determine a linear order of multivariate observations with the use of a multivariate median, defined as a multivariate centre of the observation set [24]. There are many proposals of functions called depth functions (e.g. Euclidean depth function, Mahalanobis' depth, Tukey's depth, projection depth and Student's depth) assigning a positive number to each observation originating from a certain distribution, which number is a measure of the observation divergence from the centre, due to this distribution.

In the case of applying a projection depth function [11], 10% of all financially sound companies which were furthest from the multivariate centre designated for financially sound companies were considered atypical financially sound companies in a given year.

Selected methods of classification applied to predicting corporate bankruptcy (e.g. [5, 12]) were used in the statistical assessment of the financial standing of companies:

- A logit model
- A classification tree based on CART algorithm

It is worth stressing that the first mentioned method belongs to the group of statistical techniques, whereas the second is the data mining method. Selecting variables is one of the crucial issues while constructing models. The backward stepwise method was used for the logit model taking into account the analysis of the correlation between the explanatory variables. The following technique was applied: in the case the algorithm is interrupted, the correlation matrix is analysed, the most correlated variable is omitted and the backward stepwise method is applied again. In order to create classification trees, the CART algorithm, which simultaneously triggers variables reduction, was used. In classification trees, the Gini Index was employed to assess the quality of the obtained splits of objects in the nodes. The tree pruning was based on the cost-complexity criterion.

To evaluate the classification effectiveness of the considered methods, the following was applied (e.g. [4]):

- Sensitivity (percentage of bankrupt enterprises which had been correctly classified)
- Specificity (percentage of financially sound companies which had been correctly classified)
- AUC measure (area of the figure under the ROC curve, where the ROC curve demonstrates sensitivity as the 1-specificity function)

The calculations were made in the R, Statistica and Excel programs.

4 Results of Empirical Research

The results of the classification of bankrupt companies with the use of the logit model are shown in Table 2. The identification of bankrupt companies characterised by a financial standing significantly different from the standing of the majority of financially sound companies was based on the result of both point estimation and interval estimation of the probability of belonging to the class of financially sound companies. Point probabilities for each enterprise have been calculated by substituting observed values of financial ratios for individual enterprise into the logit model. Confidence intervals for theoretical (predicted) probabilities have been obtained according to [13, p. 244] taking transformation of the dependent variable in the logit model on board.

On the basis of the obtained results, it can be concluded that after removing atypical objects from the set of financially sound companies, the logit model was characterised by generally higher values of the AUC measure. This proves the higher classification effectiveness of this model in comparison with the model estimated based on data not cleaned of atypical financially sound companies. Also, the specificity measure of this model was at a very high level, which means that few financially sound companies were classified as companies with a financial standing similar to the financial situation of companies that had a problem with solvency in the past. Therefore, in the following part of the paper, the interpretation is limited to the results obtained from the logit model estimated on the basis of sets cleaned of atypical financially sound companies.

Figures 2 and 3 present point estimates and 95% confidence intervals estimated for the probability of belonging to the class of financially sound companies, calculated for the considered bankrupt enterprises. The charts show a low precision of the obtained probability estimates in the studied years, mainly in the case of the bankrupt enterprises B_1, B_2, B_4 and B_5 in the years 2007–2009 (with Tukey's

Table 2 Results for the logit model

Outliers	Year	Sample size	Ratios	B^{a}_{-}	B^{b}_{-}	Sensitivity[a]	Specificity[a]	AUC[a]
Not removed	2005	369	$R_{04}R_{13}$	B_3	$B_1B_3B_4$	0.2	1.000	0.959
	2006	369	$R_{06}R_{07}R_{08}$	–	B_2B_3	0.0	1.000	0.833
	2007	369	$R_{05}R_{06}$	B_1	B_1	0.2	0.997	0.767
	2008	369	R_{09}	B_4	B_4	0.2	1.000	0.599
	2009	369	R_{14}	–	–	0.0	1.000	0.826
Removed —Tukey's criterion	2005	188	R_{04}	$B_1B_2B_3B_4$	$B_1B_2B_3B_4$	0.8	1.000	0.976
	2006	205	R_{05}	B_1B_2	$B_1B_2B_4$	0.4	1.000	0.733
	2007	197	$R_{05}R_{06}R_{07}R_{08}$	B_1B_2	$B_1B_2B_4B_5$	0.4	0.995	0.981
	2008	176	$R_{05}R_{06}R_{10}R_{14}$	$B_1B_2B_4B_5$	$B_1B_2B_4B_5$	0.8	1.000	0.961
	2009	188	$R_{06}R_{14}$	B_1B_5	$B_1B_2B_4B_5$	0.4	0.995	0.940
Removed —depth function	2005	333	R_{04}	$B_1B_2B_3B_4$	$B_1B_2B_3B_4$	0.8	1.000	0.955
	2006	333	$R_{04}R_{05}R_{08}$	$B_1B_2B_4$	$B_1B_2B_3B_4$	0.6	1.000	0.912
	2007	333	$R_{05}R_{06}R_{08}R_{14}$	B_1B_4	$B_1B_2B_4$	0.4	1.000	0.920
	2008	333	$R_{05}R_{06}R_{11}R_{14}$	B_4	$B_1B_2B_4$	0.2	1.000	0.976
	2009	333	$R_{08}R_{09}$	B_1	B_1B_4	0.2	1.000	0.794

B_{-}—Bankrupt enterprises, which have not been classified as financially sound ones

[a]According to the point probability of belonging to the class of financially sound companies

[b]According to the lower limit of the 95% confidence interval for the probability of belonging to the class of financially sound companies

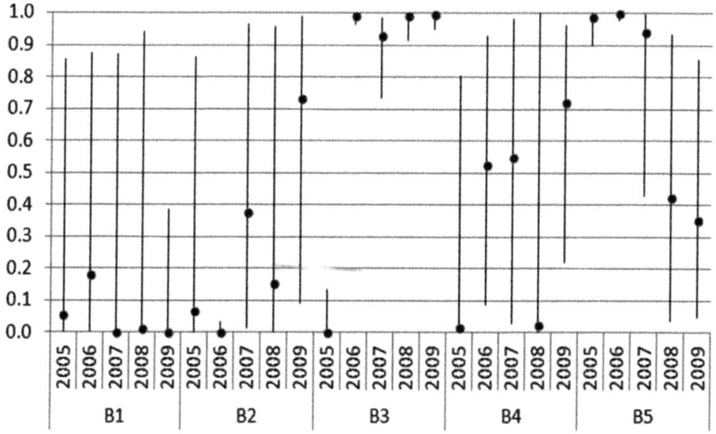

Fig. 2 Point and interval estimates ($1 - \alpha = 0.95$) of the probability of belonging to the class of financially sound companies on the basis of the logit model based on a set cleared of atypical financially sound objects according to Tukey's criterion

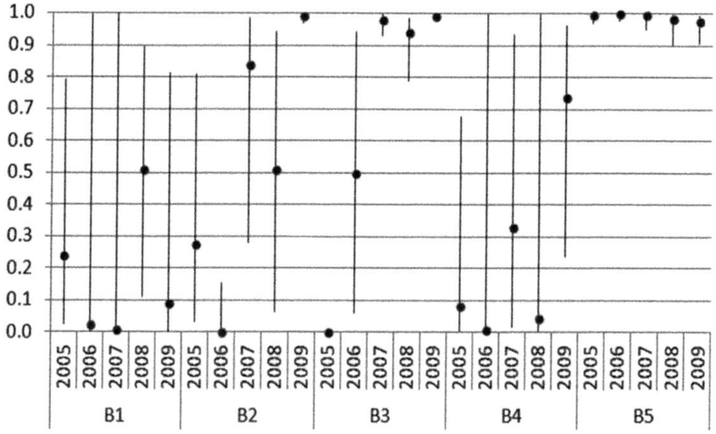

Fig. 3 Point and interval estimates ($1 - \alpha = 0.95$) of the probability of belonging to the class of financially sound companies on the basis of the logit model based on a set cleared of atypical financially sound objects according to the projection depth function

criterion). Only results obtained for bankrupt B_3 can be generally regarded as accurate.

Due to the considerable uncertainty associated with the obtained results, it was decided to apply a very strong criterion for the assessment of the financial standing of bankrupt companies. It was assumed that the lower limit of the 95% confidence interval for the probability of belonging to the class of financially sound companies would be decisive as regards the recognition of a given bankrupt company as a company with a financial standing similar to that typical of financially sound companies.

The adoption of Tukey's criterion enabled the authors to indicate bankrupt enterprise B_3 as a company with a financial standing similar to that observed in the group of financially sound companies in the years 2006–2009. In the case of applying the projection depth function, the financial standing of this bankrupt company was assessed as good in the years 2007–2009. Therefore, it can be concluded that it was a company that had coped best with the problem of insolvency among the considered bankrupt enterprises after having been declared bankrupt by a court. Another bankrupt company whose financial standing was assessed as generally good was company B_5. With the use of the projection depth function, this concerned the entire period of study, but in the case of Tukey's criterion, it applied only to the period from 2006 to 2007. B_1, B_2 and B_4 were categorised as bankrupt enterprises whose financial standing diverged from the standing characteristic of financially sound companies.

The classification results of bankrupt companies with the use of the classification tree based on CART algorithm are shown in Table 3.

The constructed classification trees are characterised by excellent classification efficiency of bankrupt enterprises (Table 3). However, financially sound companies are classified as bankrupt enterprises (i.e. companies which had a solvency issue in the past) to a greater extent than in the case of the logit model. This problem is relevant to both sets not cleaned of atypical financially sound companies and sets that have been cleaned of outliers. Also, the AUC measure shows high classification efficiency in the case of this method. The achieved results also confirm the robustness of classification trees to the presence of atypical observations, since

Table 3 Results for the classification tree

Outliers	Year	Sample size	Ratios	Sensitivity	Specificity	AUC
Not removed	2005	369	$R_{04}R_{14}$	1.0	0.995	0.998
	2006	369	$R_{04}R_{05}R_{12}$	1.0	0.967	0.984
	2007	369	$R_{03}R_{11}R_{12}$	1.0	0.926	0.963
	2008	369	$R_{02}R_{07}R_{11}R_{12}$	1.0	0.956	0.978
	2009	369	$R_{01}R_{03}R_{07}R_{12}$	1.0	0.967	0.984
Removed —Tukey's criterion	2005	188	$R_{04}R_{12}$	1.0	1.000	1.000
	2006	205	$R_{02}R_{04}R_{12}$	1.0	0.985	0.993
	2007	197	$R_{03}R_{11}R_{12}R_{14}$	1.0	0.906	0.985
	2008	176	$R_{02}R_{07}R_{11}R_{12}$	1.0	0.912	0.956
	2009	188	$R_{01}R_{03}R_{12}$	1.0	0.934	0.967
Removed —depth function	2005	333	$R_{01}R_{13}$	1.0	0.976	0.988
	2006	333	$R_{04}R_{06}R_{12}$	1.0	0.976	0.988
	2007	333	$R_{03}R_{11}R_{12}$	1.0	0.902	0.951
	2008	333	$R_{02}R_{07}R_{11}R_{12}$	1.0	0.936	0.968
	2009	333	$R_{01}R_{03}R_{12}$	1.0	0.945	0.973

Table 4 Decision rules in classification trees constructed on sets not cleaned of atypical financially sound objects (sample size = 369)

Year	Decision rules	Probability
2005	$R_{04} > 1.045 \rightarrow B$	0.667
	$R_{04} \leq 1.045 \wedge R_{14} \leq 51.345 \rightarrow B$	1.000
	$R_{04} \leq 1.045 \wedge R_{14} > 51.345 \rightarrow NB$	1.000
2006	$R_{12} > 5.005 \rightarrow NB$	1.000
	$R_{12} \leq 5.005 \wedge R_{04} \leq 0.545 \rightarrow NB$	1.000
	$R_{12} \leq 5.005 \wedge R_{04} > 0.545 \wedge R_{05} > 2.045 \rightarrow NB$	1.000
	$R_{12} \leq 5.005 \wedge R_{04} > 0.545 \wedge R_{05} \leq 2.045 \rightarrow B$	0.294
2007	$R_{11} > 5.265 \rightarrow NB$	1.000
	$R_{11} \leq 5.265 \wedge R_{12} > 7.745 \rightarrow NB$	1.000
	$R_{11} \leq 5.265 \wedge R_{12} \leq 7.745 \wedge R_{03} > 0.615 \rightarrow NB$	1.000
	$R_{11} \leq 5.265 \wedge R_{12} \leq 7.745 \wedge R_{03} \leq 0.615 \wedge R_{11} \leq 2.650 \rightarrow NB$	1.000
	$R_{11} \leq 5.265 \wedge R_{12} \leq 7.745 \wedge R_{03} \leq 0.615 \wedge R_{11} > 2.650 \rightarrow B$	0.156
2008	$R_{11} > 4.930 \rightarrow NB$	1.000
	$R_{11} \leq 4.930 \wedge R_{02} \leq 1.325 \rightarrow NB$	1.000
	$R_{11} \leq 4.930 \wedge R_{02} > 1.325 \wedge R_{07} > 10.165 \rightarrow NB$	1.000
	$R_{11} \leq 4.930 \wedge R_{02} > 1.325 \wedge R_{07} \leq 10.165 \wedge R_{12} > 10.700 \rightarrow NB$	1.000
	$R_{11} \leq 4.930 \wedge R_{02} > 1.325 \wedge R_{07} \leq 10.165 \wedge R_{12} \leq 10.700 \wedge R_{11} \leq 2.740 \rightarrow NB$	1.000
	$R_{11} \leq 4.930 \wedge R_{02} > 1.325 \wedge R_{07} \leq 10.165 \wedge R_{12} \leq 10.700 \wedge R_{11} > 2.740 \rightarrow B$	0.238
2009	$R_{12} > 4.125 \rightarrow NB$	1.000
	$R_{12} \leq 4.125 \wedge R_{03} > 0.455 \rightarrow NB$	1.000
	$R_{12} \leq 4.125 \wedge R_{03} \leq 0.455 \wedge R_{01} \leq 1.160 \rightarrow NB$	1.000
	$R_{12} \leq 4.125 \wedge R_{03} \leq 0.455 \wedge R_{01} > 1.160 \wedge R_{07} > 4.920 \rightarrow NB$	1.000
	$R_{12} \leq 4.125 \wedge R_{03} \leq 0.455 \wedge R_{01} > 1.160 \wedge R_{07} \leq 4.920 \rightarrow B$	0.294

Probability—the estimated probability for a terminal node

there are no significant differences between the trees constructed for a given year on the basis of a set cleaned and not cleaned of atypical financially sound companies. Due to the above-mentioned similarity of results, it was decided to discuss at length classification trees constructed on sets not cleaned of atypical financially sound companies.

Table 4 presents decision rules resulting from the construction of classification trees for subsequent years on the basis of sets not cleaned of atypical financially sound companies.

Based on the results presented in Table 4, it can be concluded that the estimated probability for terminal node, which allows to extract the class of companies that had problems with solvency before 2005, decreases from 0.667 in 2005 to 0.156 in 2007 and then slightly increases to 0.294 in 2009. It may indicate positive effects of the remedial process conducted in the companies concerned. The smallest estimated probability for a terminal node was recorded in 2007. This may be associated with the slowdown of the Polish economy in this period, resulting from the global economic crisis, and—consequently—difficult conditions for the operation of companies, including financially sound ones.

5 Summary

The results of the conducted empirical studies confirm the usefulness of classification methods applied to predicting bankruptcy of companies for the evaluation of the financial standing of companies after the declaration of bankruptcy in comparison with the situation of financially sound companies. The logit model enabled the authors to single out, from among companies declared bankrupt, those bankrupt enterprises whose financial standing in the five consecutive years after the declaration of bankruptcy had improved to such a degree that it could be regarded as similar to the situation typical of financially sound companies. The classification trees, in turn, provided decision rules indicating those areas of operations of companies which require particular attention in the remedial process of bankrupt companies.

The logit models estimated on the basis of sets cleaned of atypical financially sound companies indicated the importance in the evaluation of the financial standing of companies after the declaration of bankruptcy:

- 1 year—only of the liability ratios
- 2 years—of the liability and profitability ratios
- 3–5 years—of the liability, profitability, and productivity ratios

The classification trees provided decision rules based on:

- 1–2 years after the bankruptcy—the liquidity, liability and productivity ratios
- 3 years after the bankruptcy—the liquidity and productivity ratios
- 4 years after the bankruptcy—the liquidity, profitability and productivity ratios
- 5 years after the bankruptcy—the liquidity and productivity ratios

In light of the results obtained, it can be concluded that the development paths of companies B_3 and B_5 can be a source of valuable guidance on selecting appropriate remedial programmes for companies having a problem with solvency. On the other hand, the analysis of decisions taken in companies B_1, B_2 and B_4 after the declaration of bankruptcy can provide information useful for avoiding the deterioration of their financial standing. In the authors' opinion, it is worth continuing the undertaken studies and attempting to test the results achieved and confront them with reality, depending on the availability of data.

In further research, authors are planning to incorporate other classification methods. They also intend to consider other methods of detecting outliers, and other approaches (e.g. V-fold cross-validation) in the verification of the obtained results.

Acknowledgements Publication was financed from the funds granted to the Faculty of Management at Cracow University of Economics, within the framework of the subsidy for the maintenance of research potential.

References

1. Altman, E.I.: Financial ratios, discriminant analysis and prediction of corporate bankruptcy. J. Financ. **23**(4), 589–609 (1968)
2. Altman, E.I., Branch, B.: The bankruptcy system's chapter 22 recidivism problem: how serious is it? Financ. Rev. **50**, 1–26 (2015)
3. Barnett, V., Lewis, T.: Outliers in Statistical Data. Wiley, New York (1994)
4. Birdsall, T.G.: The theory of signal detectability: ROC curves and their character, Cooley electronics laboratory. Technical Report, No. 177. Department of Electrical and Computer Engineering, The University of Michigan, Ann Arbor, Michigan (1973)
5. Breiman, L., Friedman, J., Olshen, R., Stone, C.: Classification and Regression Trees. CRC Press, London (1984)
6. De Andrés, J., Sánchez–Lasheras, F., Lorca, P., De Cos Juez, F.J.: A hybrid device of self organizing maps (SOM) and multivariate adaptive regression splines (MAR) for the forecasting of Firms' bankruptcy. Account. Manage. Inf. Syst. **10**(3), 351–374 (2011)
7. Eberhart, A.C., Altman, E.I., Aggarwal, R.: The equity performance of firms emerging from bankruptcy. J. Financ. **54**(5), 1855–1868 (1999)
8. Frydman, H., Altman, E.I., Kao, D.: Introducing recursive partitioning for financial classification: the case of financial distress. J. Financ. **40**(1), 269–291 (1985)
9. García, V., Marqués, A.I., Sánchez, S.S.: An insight into the experimental design for credit risk and corporate bankruptcy prediction systems. J. Intell. Inf. Syst. **44**(1), 159–189 (2015). doi:10.1007/s10844-014-0333-4
10. Hodge, V.J., Austin, J.: A survey of outlier detection methodologies. Artif. Intell. Rev. **22**(2), 85–126 (2004)
11. Kosiorowski, D.: Robust classification and clustering based on the projection depth function. In: Brito, P. (ed.) COMPSTAT 2008. Proceedings in Computational Statistics, pp. 209–216. Physica-Verlag, Heidelberg (2008)
12. McFadden, D.: Conditional logit analysis of qualitative choice behavior. In: Zarembka, P. (ed.) Frontiers in Econometrics. Academic, New York (1973)
13. Neter, J., Wasserman, W., Kutner, M.H.: Applied Linear Regression Models, 2nd edn., Irwin, Homewood, IL (1989)
14. Odom, M.D., Sharda, R.: A neural network model for bankruptcy prediction. In: Proceedings of IEEE International Conference on Neural Networks, vol. 2, pp. 163–168, San Diego (1990)
15. Ohlson, J.: Financial ratios and the probabilistic prediction of bankruptcy. J. Account. Res. **18**(1), 109–131 (1980)
16. Pawełek, B., Kostrzewska, J., Lipieta, A.: The problem of outliers in the research on the financial standing of construction enterprises in Poland. In: Papież, M., Śmiech, S. (eds.) Proceedings of the 9th Professor Aleksander Zeliaś International Conference on Modelling and Forecasting of Socio-Economic Phenomena, pp. 164–173. Foundation of the Cracow University of Economics, Cracow (2015)
17. Platt, H.D., Platt, M.B.: A re-examination of the effectiveness of the bankruptcy process. J. Bus. Financ. Account. **29**(9–10), 1209–1237 (2002)
18. Shumway, T.: Forecasting bankruptcy more accurately: a simple hazard model. J. Bus. **74**(1), 101–124 (2001)
19. Spicka, J.: The financial condition of the construction companies before bankruptcy. Eur. J. Bus. Manage. **5**(23), 160–169 (2013)

20. Sun, J., Li, H., Huang, Q.-H., HE, K.-Y.: Predicting financial distress and corporate failure: a review from the state-of-the-art definitions, modeling, sampling, and featuring approaches. Knowl.-Based Syst. **57**, 41–56 (2014)
21. Tukey, J.W.: Exploratory Data Analysis. Addison-Wesley, Reading, MA (1977)
22. Wu, Y., Gaunt, C., Gray, S.: A comparison of alternative bankruptcy prediction models. J. Contemp. Account. Econ. **6**, 34–45 (2010)
23. Zuo, Y.: Projections-based depth functions and associated medians. Ann. Stat. **31**(5), 1460–1490 (2003)
24. Zuo, Y., Serfling, R.: General notions of statistical depth functions. Ann. Stat. **28**(2), 461–482 (2000)

On the Identification of Correlated Differential Features for Supervised Classification of High-Dimensional Data

Shu Kay Ng and Geoffrey J. McLachlan

Abstract Many real problems in supervised classification involve high-dimensional feature data measured for individuals of known origin from two or more classes. When the dimension of the feature vector is very large relative to the number of individuals, it presents formidable challenges to construct a discriminant rule (classifier) for assigning an unclassified individual to one of the known classes. One way to handle this high-dimensional problem is to identify highly relevant differential features for constructing a classifier. Here a new approach is considered, where a mixture model with random effects is used firstly to partition the features into clusters and then the relevance of each feature variable for differentiating the classes is formally tested and ranked using cluster-specific contrasts of mixed effects. Finally, a non-parametric clustering approach is adopted to identify networks of differential features that are highly correlated. The method is illustrated using a publicly available data set in cancer research for the discovery of correlated biomarkers relevant to the cancer diagnosis and prognosis.

1 Introduction

In supervised classification, the data are classified with respect to g known classes and the intent is to construct a discriminant rule or classifier on the basis of these classified data for assigning an unclassified individual to one of the g classes on the basis of its feature vector. Many real problems in supervised classification, however, involve high-dimensional feature vectors. While there is a vast literature on dimensional reduction and/or feature selection in supervised classification [4, 8, 13],

S.K. Ng (✉)
School of Medicine and Menzies Health Institute Queensland, Griffith University, Nathan, QLD 4111, Australia
e-mail: s.ng@griffith.edu.au

G.J. McLachlan
Department of Mathematics, University of Queensland, St Lucia, QLD 4072, Australia
e-mail: g.mclachlan@uq.edu.au

F. Palumbo et al. (eds.), *Data Science*, Studies in Classification, Data Analysis, and Knowledge Organization, DOI 10.1007/978-3-319-55723-6_4

some of the methods may become inapplicable or unreliable when the dimension of the feature vector is very large relative to the number of individuals [2, 10, 15, 24]. An example of such an application is the analysis of gene-expression data, where expression levels of genes (features) are available from patients in g known classes of distinct disease stages or outcomes and the aim is to identify a small subset of "marker" genes that characterize the different classes and construct a discriminant rule to predict the class of origin of an unclassified patient [11, 17]. One way to handle this high-dimensional problem is to identify genes that are differentially expressed among the g classes of tissue samples. In this context, multiple hypothesis test-based approaches [27–29] have been proposed to assess statistical significance of differential expression for each gene separately, with control for the false discovery rate (FDR) which is defined as the expected proportion of false positives among the genes declared to be differentially expressed [1]. Clustering-based approaches have also been considered, but these methods either work on gene-specific summary statistics [14, 23] or reduced forms of gene-expression data [6]. Alternatively, clustering methods that can handle full gene-expression data rely on the assumption that pure clusters of null (non-differentially expressed) genes and differentially expressed genes exist [12, 26]; see also [25]. More recently, a mixture model-based approach with random-effects terms was proposed to draw inference on differences between classes using full gene-expression data [22]. This method does not rely on the clusters being pure as to whether all cluster members are differentially expressed or null genes. In this paper, we propose a new three-step method that extends this mixture model-based approach in order to identify networks of correlated differential features (genes) for supervised classification of high-dimensional data.

The rest of the paper is organized as follows. In Sect. 2, we describe the mixture model with random-effects terms [20] that is adopted in the first step to cluster the genes using full gene-expression data. We also present the second step, where the relevance of each feature variable for differentiating the classes is formally tested and ranked on the basis of cluster-specific contrasts of mixed effects. In Sect. 3, we describe the final third step in which a non-parametric clustering approach is used to further explore the group structures of selected highly ranked differential features for each cluster identified in the first step. Section 4 presents the application of the proposed method to a publicly available gene-expression data set in cancer research for the discovery of correlated biomarkers relevant to the cancer prognosis. Discussion is given in Sect. 5.

2 Mixture Model with Random-Effects Terms

With supervised classification, it is supposed that an individual belongs to one of g classes, denoted by $C_1, \ldots, C_g$, and that there is a vector of p feature variables measured on each individual. Based on the observed feature vectors, represented by an $n \times p$ matrix, the intent is to construct a discriminant rule for allocating an unclassified individual to one of the g classes [15]. For applications in the context of supervised classification with gene-expression data, the number of individual tissue samples n is very small relative to the number of genes p. To handle this high-dimensional problem, it is proposed to adopt a mixture model with random-effects terms to firstly cluster the p genes and then identify those genes that are highly differentiated between the g classes of tissue samples.

Let $\boldsymbol{y}_j = (y_{1j}, \ldots, y_{nj})^T$ contain the measurements on the jth gene ($j = 1, \ldots, p$), where the superscript T denotes vector transpose and p is much greater than n. It is assumed that $\boldsymbol{y}_j$ has a h-component mixture distribution with probability π_i of belonging to the ith cluster ($i = 1, \ldots, h$), where the π_i sum to one. We let the h-dimensional vector z_j denote the cluster membership of $\boldsymbol{y}_j$, where $z_{ij} = (z_j)_i = 1$ if $\boldsymbol{y}_j$ belongs to the ith cluster and zero otherwise ($i = 1, \ldots, h$). A mixture model with random-effects terms [20] is required because it is anticipated that repeated measurements of gene expression for a tissue sample and expression levels for a gene are both correlated; see also [19]. Specific random effects are thus considered in the mixture model to capture individual gene effects and the correlation between gene-expression levels among the tissue classes [22]. Conditional on its membership of the ith cluster, the distribution of $\boldsymbol{y}_j$ is specified by the linear mixed model

$$\boldsymbol{y}_j = \boldsymbol{X}\boldsymbol{\eta}_i + \boldsymbol{U}\boldsymbol{b}_{ij} + \boldsymbol{V}\boldsymbol{c}_i + \boldsymbol{\varepsilon}_{ij}, \tag{1}$$

where $\boldsymbol{X}, \boldsymbol{U}$, and $\boldsymbol{V}$ denote the known design matrices corresponding to the fixed effects terms $\boldsymbol{\eta}_i$ and to the random-effects terms $\boldsymbol{b}_{ij}$ and $\boldsymbol{c}_i$ ($i = 1, \ldots, h$; $j = 1, \ldots, p$), respectively. The vector $\boldsymbol{b}_{ij} = (b_{1ij}, \ldots, b_{gij})^T$ contains the unobservable gene-specific random effects for each of the g tissue classes, and $\boldsymbol{c}_i = (c_{1i}, \ldots, c_{ni})^T$ contains the random effects common to all genes from the ith cluster. The measurement error vector $\boldsymbol{\epsilon}_{ij}$ is taken to be multivariate normal $N_n(\boldsymbol{0}, \boldsymbol{A}_i)$, where $\boldsymbol{A}_i$ is a diagonal matrix. The vectors $\boldsymbol{b}_{ij}$ and $\boldsymbol{c}_i$ of random-effects terms are taken to be multivariate normal $N_g(\boldsymbol{0}, \boldsymbol{B}_i)$ and $N_n(\boldsymbol{0}, \boldsymbol{C}_i)$, respectively, where the variance component $\boldsymbol{C}_i$ is assumed to be diagonal and $\boldsymbol{B}_i$ is a non-diagonal $g \times g$ matrix, where the correlation between gene-specific random effects b_{lij} ($l = 1, \ldots, g$) is modelled via the off-diagonal elements in $\boldsymbol{B}_i$; see, for example, [22]. The assignment of the p genes into h clusters is implemented using the estimated conditional posterior probabilities of cluster membership given $\boldsymbol{y}_j$ and $\hat{\mathbf{c}}_l$($j = 1, \ldots, p$; $l = 1, \ldots, g$):

$$\tau_i(\boldsymbol{y}_j; \hat{\boldsymbol{\Psi}}, \hat{\mathbf{c}}) = \mathrm{pr}(Z_{ij} = 1 | \boldsymbol{y}_j, \hat{\mathbf{c}}) = \frac{\hat{\boldsymbol{\pi}}_i f(\boldsymbol{y}_j | z_{ij} = 1; \hat{\boldsymbol{\psi}}_i, \hat{\mathbf{c}}_i)}{\sum_{m=1}^{h} \hat{\boldsymbol{\pi}}_m f(\boldsymbol{y}_j | z_{mj} = 1; \hat{\boldsymbol{\psi}}_m, \hat{\mathbf{c}}_m)}, \tag{2}$$

where $\boldsymbol{\psi}_i$ is the parameter vector for the ith component density containing the unknown parameters $\boldsymbol{\eta}_i$ and distinct elements in $\boldsymbol{A}_i$, $\boldsymbol{B}_i$, and $\boldsymbol{C}_i (i = 1, \ldots, h)$, and

$$\log f(\boldsymbol{y}_j|z_{ij} = 1; \hat{\boldsymbol{\psi}}_i, \hat{\mathbf{c}}_i) = -\frac{1}{2}\left\{ \log|\hat{\mathbf{D}}_i| + (\boldsymbol{y}_j - \boldsymbol{X}\hat{\boldsymbol{\eta}}_i - \boldsymbol{V}\hat{\mathbf{c}}_i)^T \hat{\mathbf{D}}_i^{-1} (\boldsymbol{y}_j - \boldsymbol{X}\hat{\boldsymbol{\eta}}_i - \boldsymbol{V}\hat{\mathbf{c}}_i) \right\}$$

is the log density of $\boldsymbol{y}_j$ conditioned on $\hat{\mathbf{c}}_i$ and the membership of the ith cluster, apart from an additive constant, and where $\hat{\mathbf{D}}_i = \hat{\mathbf{A}}_i + \boldsymbol{U}\hat{\mathbf{B}}_i\boldsymbol{U}^T$; see [20].

To quantify the relevance of each gene for differentiating the g classes, we consider an individual observation-specific contrast in the estimates of the fixed and random effects weighted by the estimated posterior probabilities (2) of cluster membership:

$$W_j = \sum_{i=1}^{h} \tau_i(\boldsymbol{y}_j; \hat{\boldsymbol{\Psi}}, \hat{\mathbf{c}})\, \hat{S}_{ij} \qquad (j = 1, \ldots, p), \tag{3}$$

where

$$\hat{S}_{ij} = \boldsymbol{d}_j^T (\hat{\boldsymbol{\eta}}_i^T, \hat{\mathbf{b}}_{G_i}^T, \hat{\mathbf{c}}_i^T)^T \Big/ \sqrt{\boldsymbol{d}_j^T \hat{\boldsymbol{\Omega}}_i \boldsymbol{d}_j} \tag{4}$$

is the cluster-specific normalized contrast with the BLUP estimator of the mixed effects, and where $\boldsymbol{d}_j$ is a vector whose elements sum to zero, $\boldsymbol{b}_{G_i} = (\boldsymbol{b}_{i_1}^T, \ldots, \boldsymbol{b}_{i_{p_i}}^T)^T$ contains the gene-specific random-effects terms for the p_i genes belonging to the ith cluster G_i $(i = 1, \ldots, h)$, and $\hat{\boldsymbol{\Omega}}_i$ is the covariance matrix of the BLUP estimator of the mixed effects, which can be partitioned conformally corresponding to $\boldsymbol{\eta}_i|\boldsymbol{b}_{G_i}|\boldsymbol{c}_i$, respectively, as described in [22].

Based on the weighted contrast W_j $(j = 1, \ldots, p)$ given in (3), the p genes can be ranked in the order of their relevance for differentiating the g classes (with respect to the defined form of d_j for the normalized contrast (4)). In the final step of the proposed method to be described in the next section, we intend to explore the group structure of top-ranked differentially expressed genes in each identified cluster $G_i (i = 1, \ldots, h)$, say, for those genes with contrast W_j more extreme than thresholds w_{0u} or w_{0d} for upregulated and downregulated genes, respectively. A guide to plausible values of w_{0u} and w_{0d} can be obtained using the percentile rank of $W_j (j = 1, \ldots, p)$, whereby the percentiles are taken to be the mixing proportions of the non-central portions of W_j fitted by a three-component mixture of t-distributions (these two components are considered as representing the distribution of W_j for upregulated and downregulated differentially expressed genes).

3 A Non-parametric Clustering Approach for Identification of Correlated Features

We consider the r_i top-ranked genes with W_j more extreme than either w_{0u} or w_{0d} in Cluster $G_i(i = 1, \ldots, h)$ and adopt a non-parametric method to cluster the r_i genes into networks of differentially expressed genes that are highly correlated. The method starts with the calculation of pairwise correlation coefficients for each pair of the r_i genes in $G_i(i = 1, \ldots, h)$. Significance of the pairwise correlation coefficients is then assessed with the use of a permutation method [21] to determine the null distribution of correlation coefficients. Precisely, the n class labels of tissue samples are randomly permuted separately for each gene. We pool the permutations for all $N_{r_i} = r_i(r_1 - 1)/2$ pairs of genes to determine the null distribution of correlation coefficients. In this paper, we consider the use of $S = 100$ repetitions of permutations and estimate the P-value for each pair of genes by

$$P_l = \sum_{s=1}^{S} \frac{\#\{m : R_{0m}^{(s)} \geq R_l, m = 1, \ldots, N_{r_i}\}}{N_{r_i} S} \quad (l = 1, \ldots, N_{r_i}), \tag{5}$$

where $R_{0m}^{(s)}$ is the null version of correlation coefficient for the mth pair of genes after the sth repetition of permutations ($m = 1, \ldots, N_{r_i}$; $s = 1, \ldots, S$). Let $P_{(1)} \leq \cdots \leq P_{(N_{r_i})}$ be the ordered observed P-values obtained from (5). The Benjamini–Hochberg procedure [1] is adopted to determine the cut-off $\hat{k}$, where

$$\hat{k} = \arg\max\{k : P_{(k)} \leq \alpha k / N_{r_i}\}, \tag{6}$$

with control of the FDR at level α. Pairwise correlation coefficients corresponding to P-values $P_{(1)} \leq \cdots \leq P_{(\hat{k})}$ are identified to be significant. Significance of the pairwise correlation coefficients is represented by an $r_i \times r_i$ symmetric binary matrix M with elements of one or zero indicating that the corresponding correlation coefficients are significance or not. Finally, we search in M to identify networks of differentially expressed genes in which all members in a group significantly correlate with one another [21]. This non-parametric clustering approach obtains overlapping groups (networks) of correlated differentially expressed genes.

4 Real Example

We consider the colorectal cancer gene-expression data set [5], which comprised expression values of 15,552 genes for plasma samples from 12 colorectal cancer patients and 8 healthy donors. The original study aims to validate the power of four randomly selected markers (from a list of 40 genes differentially upregulated in cancer patients) in enabling differentiation of the tumour from the healthy condition [5]. With the proposed three-step approach, we first fitted a mixture model with random-effects terms to the column-normalized gene-expression data set with h=3 to h=20 clusters, taking $\boldsymbol{X} = \boldsymbol{U}$ to be a 20×2 zero-one matrix (the first 12 rows are $(1, 0)$ and the next 8 rows are $(0, 1)$) and taking $\boldsymbol{V}$ to be $\boldsymbol{I}_{20}$. Based on the Bayesian information criterion (BIC) for model selection, we identified that there are 15 clusters of genes. The ML estimates of the unknown parameters are presented in Table 1. The ranking of differentially expressed genes is then implemented on the basis of the weighted estimates of a contrast in the mixed effects (3). For the case of g=2 classes of tissue samples (tumour versus healthy), we consider $\boldsymbol{d}_j$ of the form as

$$\boldsymbol{d}_j^T = (1\ -1 \ \vdots\ 0\ 0,\ \ldots,\ 0\ 0,\ 1\ -1,\ 0\ 0,\ \ldots\ \vdots\ 0\ \ldots\ 0), \tag{7}$$

where only one pair of $(1\ \ -1)$ exists in the second partition corresponding to $\boldsymbol{b}_{G_i}$; see Eq. (4). We then fitted a three-component mixture of t-distributions [16] to W_j and obtained the mixing proportions of the components corresponding to the non-central portion of W_j, which are 11.5 and 7.2% for upregulated and downregulated genes in the tumour tissues, respectively. Thus we selected $w_{0u} = 1.661$ (the 88.5th percentile of W_j) and $w_{0d} = -2.236$ (the 7.2th percentile of $W_j (j = 1, \ldots, p)$). There are a total of 2907 differentially expressed genes with W_j more extreme than w_{0u} or w_{0d} ($W_j > 1.661$ or $W_j < -2.236$). Among them, 1581 genes have valid identifiers (1073 upregulated

Table 1 Estimates of the mixture model with random-effects terms for the colorectal cancer data set (15 clusters)

		η_i	A_i	B_i	C_i
i	π_i	(η_{1i}, η_{2i})	$(\sigma_{1i}, \sigma_{2i})$	$(\sigma_{b1i}, \sigma_{b2i}, \sigma_{b12i})$	(σ_{ci})
1	0.024	−0.601, −0.591	0.643, 0.919	0.144, 0.085, 0.105	0.020
2	0.031	0.144, 0.239	0.954, 2.790	0.033, 0.095, 0.039	0.006
3	0.109	−0.232, −0.030	0.578, 0.511	0.024, 0.029, 0.023	0.011
4	0.035	0.032, 0.059	1.721, 0.336	0.054, 0.035, 0.013	0.001
5	0.114	0.070, 0.126	0.646, 0.235	0.055, 0.053, 0.031	0.004
6	0.036	0.217, 0.466	0.382, 0.567	0.035, 0.060, 0.037	0.004
7	0.092	−0.043, −0.299	0.265, 0.480	0.050, 0.037, 0.036	0.024
8	0.092	−0.004, −0.067	0.664, 1.304	0.028, 0.066, 0.030	0.000
9	0.026	−0.052, −0.535	1.622, 2.926	0.215, 0.156, 0.165	0.068
10	0.104	0.034, 0.094	1.487, 1.590	0.039, 0.088, 0.039	0.008
11	0.034	0.151, 0.037	2.770, 1.602	0.087, 0.119, 0.048	0.014
12	0.069	−0.418, −0.173	0.905, 0.715	0.022, 0.023, 0.018	0.034
13	0.130	0.211, 0.151	1.314, 0.859	0.036, 0.094, 0.036	0.002
14	0.034	0.454, −0.016	0.680, 0.691	0.019, 0.053, -0.002	0.020
15	0.070	0.012, 0.081	0.179, 0.196	0.049, 0.071, 0.048	0.004

Table 2 Descriptive statistics of W_j for the differentially expressed genes with valid gene identifiers and W_j more extreme than either w_{0u} or w_{0d} (15 clusters)

i	r_i	Mean (SD)	Median (IQR)	(Minimum, maximum)
1	1	−2.269 (n.a.)	−2.269 (n.a.)	N.a.
2	0	N.a.	N.a.	N.a.
3	173	−2.412 (0.127)	−2.378 (0.195)	(−2.782, −2.240)
4	0	N.a.	N.a.	N.a.
5	27	−0.124 (2.167)	1.679 (4.232)	(−2.486, 2.436)
6	44	−2.471 (0.194)	−2.458 (0.255)	(−3.052, −2.239)
7	714	2.391 (0.406)	2.429 (0.639)	(1.662, 3.635)
8	2	1.772 (0.005)	1.772 (n.a.)	(1.768, 1.776)
9	101	2.231 (0.415)	2.160 (0.567)	(1.669, 3.244)
10	1	1.775 (n.a.)	1.775 (n.a.)	N.a.
11	4	1.996 (0.142)	2.019 (0.262)	(1.803, 2.142)
12	264	−2.607 (0.161)	−2.725 (0.221)	(−2.940, −2.237)
13	10	1.856 (0.186)	1.696 (0.277)	(1.667, 2.185)
14	224	2.339 (0.468)	2.249 (0.719)	(1.667, 3.873)
15	16	−1.660 (1.706)	−2.362 (0.242)	(−2.772, 1.876)

Notation: *SD* standard deviation, *IQR* interquartile range, *n.a.* not appropriate

and 508 downregulated). Descriptive statistics of W_j for these 1581 differentially expressed genes are provided in Table 2. It can be seen that Clusters 7–11 and 13–14 contain upregulated differentially expressed genes, Clusters 1, 3, 6, and 12 contain downregulated differentially expressed genes, and Clusters 5 and 15 contain both upregulated and downregulated differentially expressed genes.

In the final step, we applied the non-parametric method to identify networks of correlated differentially expressed genes from the r_i genes in Cluster G_i. We set α to be between 0.1 and 0.00005 such that the expected number of false positives among the pairs of genes identified to be significantly correlated is smaller than one; see [21]. With the matrix M, networks of differentially expressed genes were displayed using UCINET6 for Windows [3]. Figure 1 presents the identified networks of upregulated differentially expressed genes in Clusters 7, 9, 13, and 14, where the nodal size of a gene is proportional to the degree of the node (the number of genes that are significantly correlated with the gene). Networks of downregulated differentially expressed genes (Clusters 3, 6, and 12) were provided in Fig. 2. Clusters 5 and 15 had networks of up- and down regulated differentially expressed genes (Fig. 3).

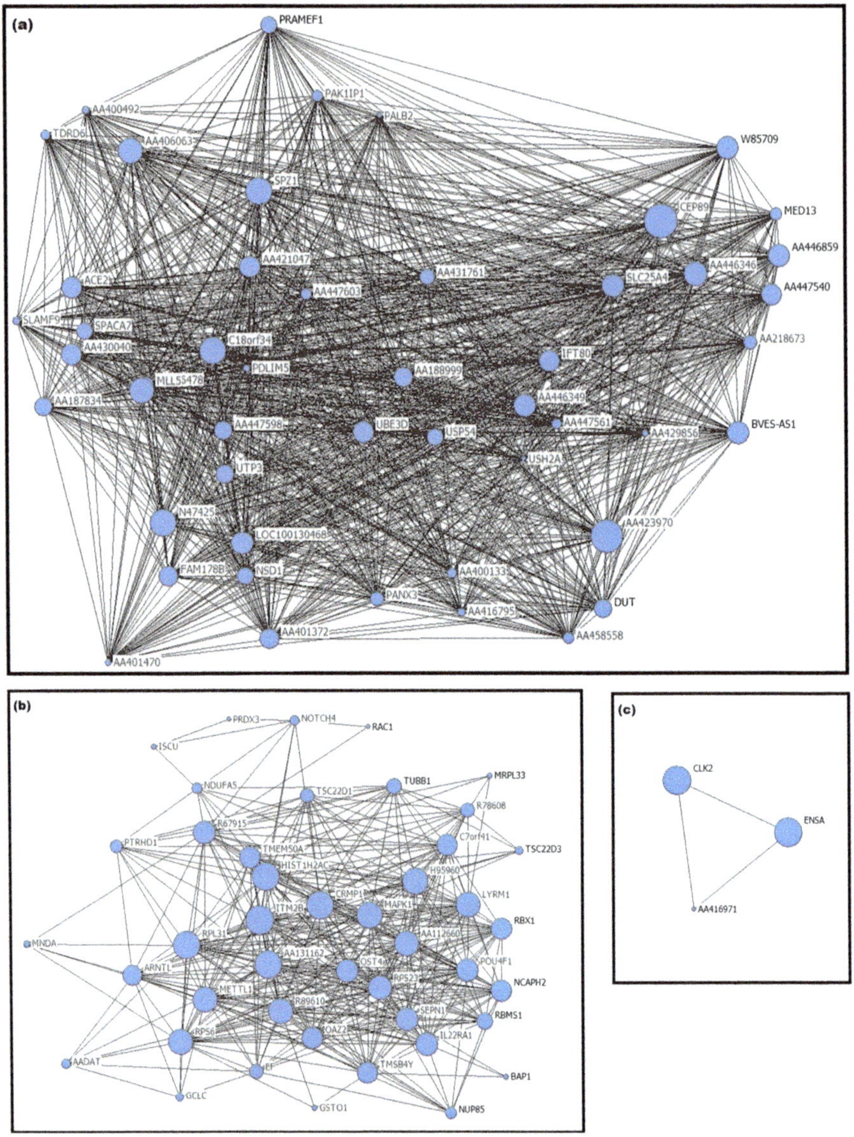

Fig. 1 Network of upregulated differentially expressed genes in (**a**) Cluster 7; (**b**) Cluster 9; (**c**) Cluster 13; and (**d**) Cluster 14. Nodal size is proportional to the degree (the number of genes that are significantly correlated with the gene). For Clusters 7, 9, and 13, only genes with the top 50 degrees were displayed

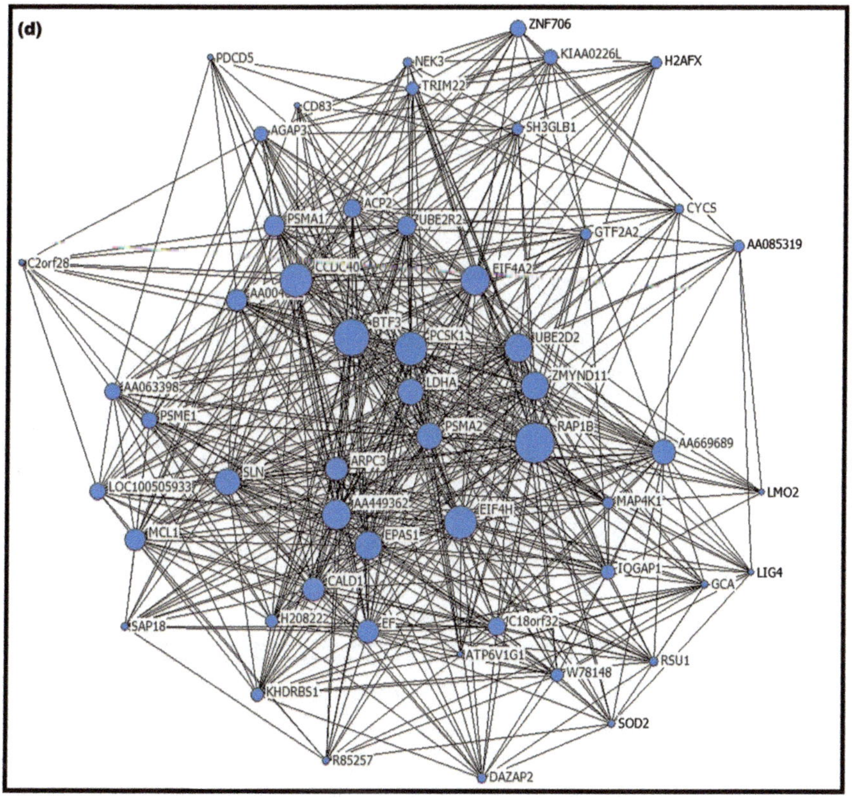

Fig. 1 (continued)

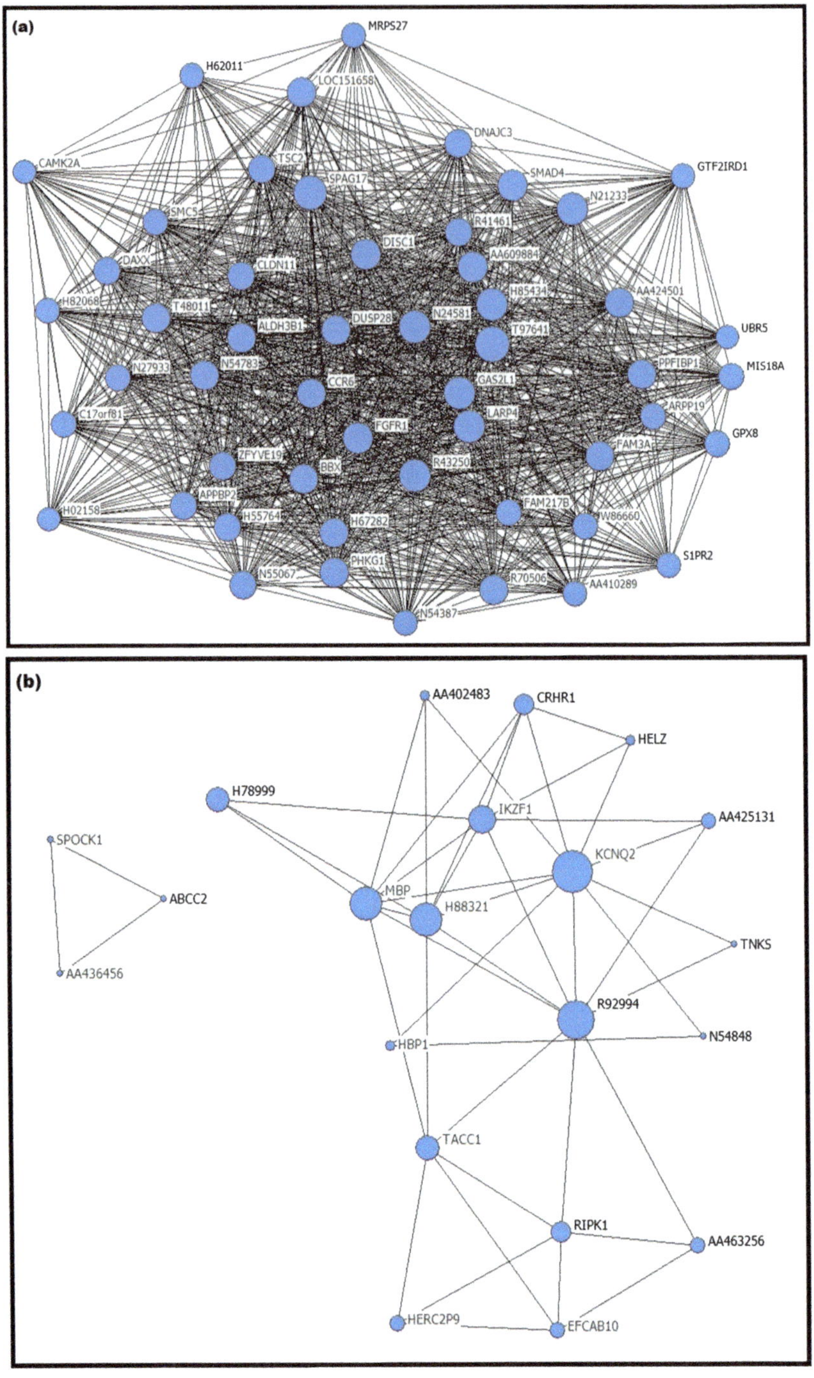

Fig. 2 Network of downregulated differentially expressed genes in (**a**) Cluster 3; (**b**) Cluster 6; and (**c**) Cluster 12. Nodal size is proportional to the degree (the number of genes that are significantly correlated with the gene). For Clusters 3 and 12, only genes with the top 50 degrees were displayed

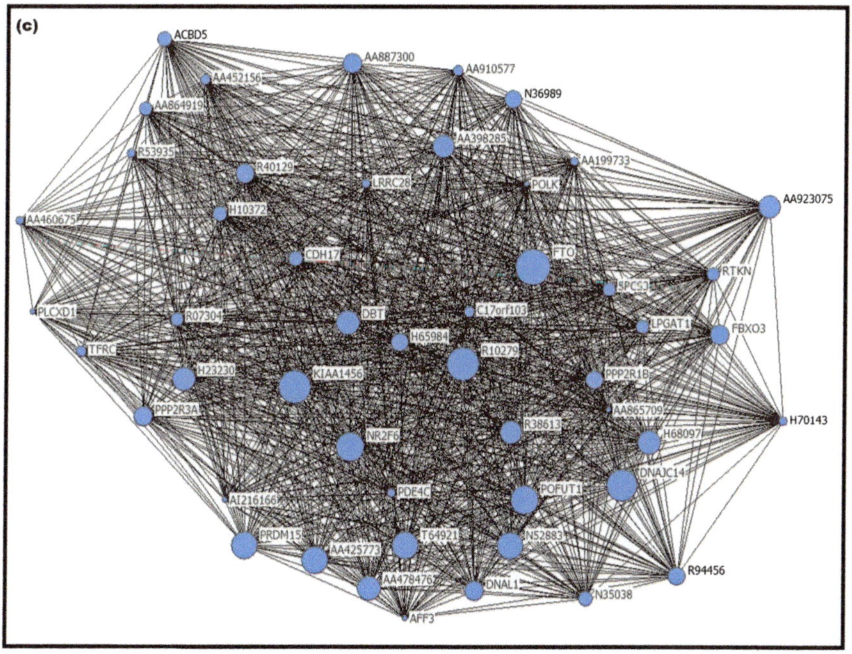

Fig. 2 (continued)

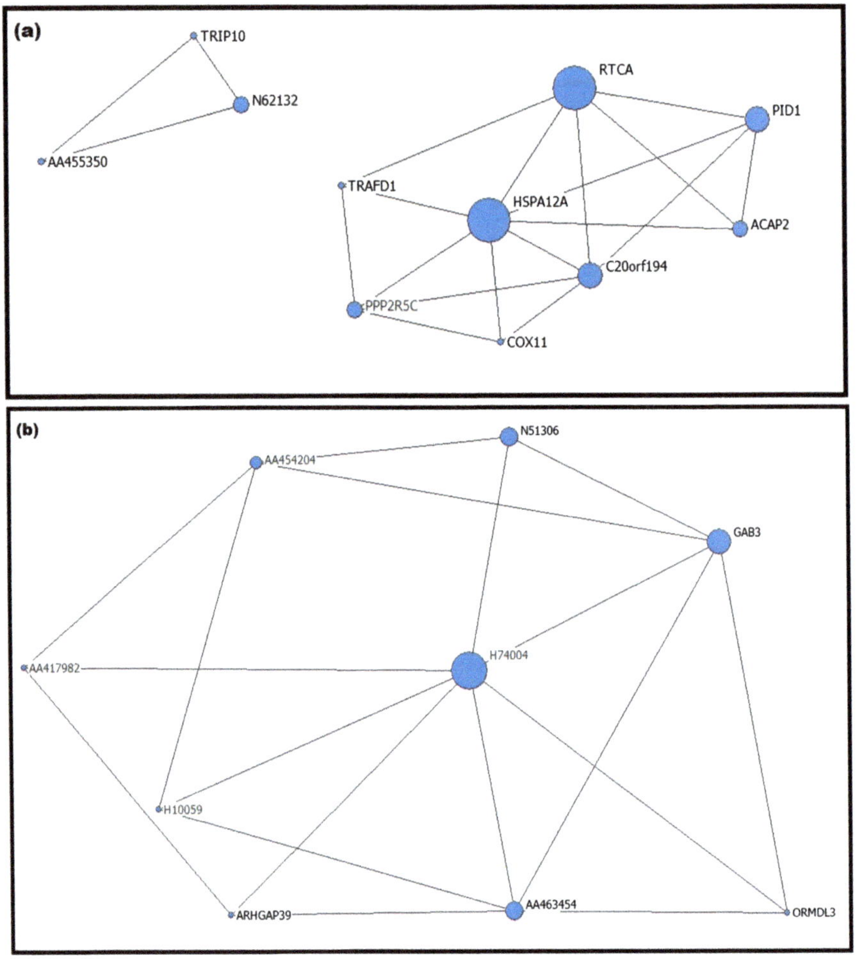

Fig. 3 Network of upregulated and downregulated differentially expressed genes in (**a**) Cluster 5 and (**b**) Cluster 15. Nodal size is proportional to the degree (the number of genes that are significantly correlated with the gene). From (**a**), it can be seen that two separate networks were identified for genes belonging to Cluster 5. One of them contains upregulated genes HSPA12A, RTCA, PID1, C20orf194, ACAP2, PPP2R5C, COX11, and TRAFD1. Another one contains downregulated genes N62132, AA455350, and TRIP10. For Cluster 15 (**b**), the network contains genes that are downregulated except ARHGAP39 (upregulated), which significantly correlated with downregulated genes {H74004, AA417982} and {H74004, AA463454}

A summary of the identified networks of correlated differentially expressed genes for each cluster is given in Table 3. Two isolated networks of differentially expressed genes were identified: {N62132, TRIP10, AA455350} downregulated genes network from Cluster 5 and {CLK2, ENSA, AA416971} upregulated genes network from Cluster 13. It is noted that four upregulated genes were considered

Table 3 A summary of networks of highly correlated differentially expressed genes

i	r_i	Highly correlated differentially expressed genes
1	1	NRGN
3	173[a]	T97641, SPAG17, N24581, LARP4, N21233, R43250, GAS2L1, H85434, DISC1, FGFR1
5	27[b]	(N62132, TRIP10, AA455350), HSPA12A, RTCA, PID1, C20orf194, ACAP2, PPP2R5C, COX11, TRAFD1
6	44[a]	KCNQ2, R92994, H88321, MBP, IKZF1, H78999, TACC1, RIPK1, CRHR1, AA463256, AA425131, EFCAB10, HERC2P9
7	714[a]	CEP89, AA423970, C18orf34, N47425, SPZ1, MLL5, AA406063, AA446346, AA446349, AA446859, W85709
8	2	SELT, PIP4K2A
9	101[a]	ITM2B, CRMP1, AA131162, HIST1H2AC, RPL31, MAPK1, H95960, R89610, LYRM1, AA112660, RPS6, METTL1
10	1	RGS2
11	4	MTG1, QK1, EF, SPARCL1
12	264[a]	FTO, KIAA1456, R10279, DNAJC14, POFUT1, NR2F6, PRDM15, T64921, N52883, AA425773
13	10[b]	(CLK2, ENSA, AA416971)
14	224[a]	RAP1B, BTF3, CCDC40, PCSK1, EIF4H, EIF4A2, AA449362, ZMYND11, EPAS1, UBE2D2
15	16	H74004, GAB3, AA463454, N51306, AA454204, ORMDL3, CPSF6, H10059, AA417982, ARHGAP39

[a]For large networks, only genes with the top ten degrees were listed
[b]Genes that form an isolated network are grouped within a bracket (in Clusters 5 and 13)

in the original study and three of them (EPAS1, UBE2D3, KIAA0101) were validated to be significantly increased in cancer compared to healthy donors [5]. Our clustering results confirmed the same findings; these three genes were identified as differentially expressed genes in Cluster 14 (with contrast W_j = 3.7, 3.3, and 2.0, respectively, and ranked the 2nd, 8th, and 156th among the 224 differentially expressed genes in Cluster 14). The original study could not validate the remaining upregulated gene DDX46. However, our method has sufficient power to identify DDX46 as a differentially expressed gene in Cluster 5, with $W_j = 2.4$ and ranked the 1st among the 14 upregulated differentially expressed genes in Cluster 5.

5 Discussion

We have presented a new approach to identify correlated differential features for supervised classification of high-dimensional data. The method adopts a mixture model with random-effects terms to cluster the feature variables and then ranks them in terms of their cluster-specific contrasts of mixed effects that quantify the evidence of differentiation between the known classes. The final step of the method adopts a non-parametric clustering approach to identify networks of differential features that are highly correlated in each identified cluster.

The proposed method is illustrated using an application on the analysis of gene-expression cancer data. The identified differentially expressed genes and their correlation structures can have significant contribution in the discovery of novel biomarkers relevant to the cancer diagnosis and prognosis; see also [7, 9] for the benefit of using the covariance information among genes for feature selection. Moreover, these differentially expressed genes can be included in a model to construct a classifier with a smaller subset of marker genes, using methods such as mixtures of factor analysers [15, 16] or mixtures of multivariate generalized Bernoulli distributions [18]. This work will be pursued in future research.

Acknowledgements Part of this work has been presented in the Conference of the International Federation of Classification Societies, Bologna, July 2015. This work was supported by a grant from the Australian Research Council.

References

1. Benjamini, Y., Hochberg, Y.: Controlling the false discovery rate: a practical and powerful approach to multiple testing. J. R. Stat. Soc. B **57**, 259–300 (1995)
2. Bickel, P.J., Levina, E.: Some theory for Fisher's linear discriminant function, 'naive Bayes', and some alternatives when there are many more variables than observations. Bernoulli **10**, 989–1010 (2004)
3. Borgatti, S.P., Everett, M.G., Freeman, L.C.: Ucinet for Windows: Software for Social Network Analysis. Analytic Technologies, Harvard, MA (2002). Available via http://www.analytictech.com/. Accessed 8 Dec 2015
4. Cai, T., Liu, W.: A direct estimation approach to sparse linear discriminant analysis. J. Am. Stat. Assoc. **106**, 1566–1577 (2011)
5. Collado, M., Garcia, V., Garcia, J.M., Alonso, I., Lombardia, L., et al.: Genomic profiling of circulating plasma RNA for the analysis of cancer. Clin. Chem. **53**, 1860–1863 (2007)
6. Dahl, D.B., Newton, M.A.: Multiple hypothesis testing by clustering treatment effects. J. Am. Stat. Assoc. **102**, 517–526 (2007)
7. Donoho, D., Jin, J.: Higher criticism for large-scale inference, especially for rare and weak effects. Stat. Sci. **30**, 1–25 (2015)
8. Fan, J., Lv, J.: A selective review of variable selection in high dimensional feature space. Stat. Sin. **20**, 101–148 (2010)
9. Fan, J., Feng, Y., Tong, X.: A road to classification in high dimensional space: the regularized optimal affine discriminant. J. R. Stat. Soc. B **74**, 745–771 (2012)
10. Hall, P., Pittelkow, Y., Ghosh, M.: Theoretic measures of relative performance of classifiers for high-dimensional data with small sample sizes. J. R. Stat. Soc. B **70**, 158–173 (2008)
11. Hall, P., Jin, J., Miller, H.: Feature selection when there are many influential features. Bernoulli **20**, 1647–1671 (2014)
12. He, Y., Pan, W., Lin, J.: Cluster analysis using multivariate normal mixture models to detect differential gene expression with microarray data. Comput. Stat. Data Anal. **51**, 641–658 (2006)
13. Kersten, J.: Simultaneous feature selection and Gaussian mixture model estimation for supervised classification problems. Pattern Recogn. **47**, 2582–2595 (2014)

14. Matsui, S., Noma, H.: Estimating effect sizes of differentially expressed genes for power and sample-size assessments in microarray experiments. Biometrics **67**, 1225–1235 (2011)
15. McLachlan, G.J.: Discriminant analysis. WIREs Comput. Stat. **4**, 421–431 (2012)
16. McLachlan, G.J., Peel, D.: Finite Mixture Models. Wiley, New York (2000)
17. McLachlan, G.J., Do, K.A., Ambroise, C.: Analyzing Microarray Gene Expression Data. Wiley, New York (2004)
18. Ng, S.K.: A two-way clustering framework to identify disparities in multimorbidity patterns of mental and physical health conditions among Australians. Stat. Med. **34**, 3444–3460 (2015)
19. Ng, S.K., McLachlan, G.J.: Mixture models for clustering multilevel growth trajectories. Comput. Stat. Data Anal. **71**, 43–51 (2014)
20. Ng, S.K., McLachlan, G.J., Wang, K., Ben-Tovim, L., Ng, S.-W.: A mixture model with random-effects components for clustering correlated gene-expression profiles. Bioinformatics **22**, 1745–1752 (2006)
21. Ng, S.K., Holden, L., Sun, J.: Identifying comorbidity patterns of health conditions via cluster analysis of pairwise concordance statistics. Stat. Med. **31**, 3393–3405 (2012)
22. Ng, S.K., McLachlan, G.J., Wang, K., Nagymanyoki, Z., Liu, S., Ng, S.-W.: Inference on differences between classes using cluster-specific contrasts of mixed effects. Biostatistics **16**, 98–112 (2015)
23. Pan, W., Lin, J., Le, C.T.: Model-based cluster analysis of microarray gene-expression data. Genome Biol. **3**, 0009.1–0009.8 (2002)
24. Pyne, S., Lee, S.X., Wang, K., Irish, J., Tamayo, P., et al.: Joint modeling and registration of cell populations in cohorts of high-dimensional flow cytometric data. PLoS One **9**, e100334 (2014)
25. Qi, Y., Sun, H., Sun, Q., Pan, L.: Ranking analysis for identifying differentially expressed genes. Genomics **97**, 326–329 (2011)
26. Qiu, W., He, W., Wang, X., Lazarus, R.: A marginal mixture model for selecting differentially expressed genes across two types of tissue samples. Int. J. Biostat. **4**, Article 20 (2008)
27. Smyth, G.: Linear models and empirical Bayes methods for assessing differential expression in microarray experiments. Stat. Appl. Genet. Mol. Biol. **3**, Article 3 (2004)
28. Storey, J.D.: The optimal discovery procedure: a new approach to simultaneous significance testing. J. R. Stat. Soc. B **69**, 347–368 (2007)
29. Zhao, Y.: Posterior probability of discovery and expected rate of discovery for multiple hypothesis testing and high throughput assays. J. Am. Stat. Assoc. **106**, 984–996 (2011)

Part II
Clustering Methods and Applications

T-Sharper Images and T-Level Cuts of Fuzzy Partitions

Slavka Bodjanova

Abstract A fuzzy partition is usually described by a matrix U with elements from the unit interval. For a variety of reasons, U is often approximated by simpler matrices. The most popular are the maximum membership approximation and approximations based on α-level cuts of U. Because they create crisp matrices with membership grades either zero or one, most of the detailed information from U is lost. Two methods of gradual non-crisp approximations of U based on a set of parameters T derived from U are proposed. The first produces T-sharper images of U with gradually decreasing amount of fuzziness. The final image is a convex combination of all possible maximum membership approximations of U. The second method generates T-level cuts of U. They represent a gradual transformation from the lower crisp approximation of U by its core (α-cut at level $\alpha = 1$) to the upper crisp approximation of U by its support (strong α-cut at level $\alpha = 0$). Properties of both methods are discussed and illustrated with examples.

1 Introduction

The relevance of theory of fuzzy sets to cluster analysis has been well documented by many research works and applications [1, 8, 10–12, 14, 16]. A fuzzy partition U of n objects into c clusters is usually described by a matrix of the size $c \times n$ with elements from the unit interval. The use of the unit interval allows to express membership grades of U with infinitely many values from 0 to 1. However, this detailed information is not always needed. A large number of different membership grades complicate interpretation and increase computational cost in further applications. In decision-making, an approximation of U often provides sufficient information. There has been extensive research related to approximation of fuzzy sets [3–5, 9, 13, 15]. However, approximation of fuzzy partitions has not received considerable attention. Some work in this direction can be found in [1, 2, 6–8].

S. Bodjanova
Texas A&M University-Kingsville, Kingsville, TX 78363, USA
e-mail: kfsb000@tamuk.edu

F. Palumbo et al. (eds.), *Data Science*, Studies in Classification, Data Analysis, and Knowledge Organization, DOI 10.1007/978-3-319-55723-6_5

In practice, an approximation of a fuzzy partition U is usually performed by the maximum membership method [1] or by an α-cut [5, 15] of U, where α is a real number from the unit interval. Both methods produce crisp matrices with elements from the set $\{0, 1\}$ and therefore most of the detailed information provided by the original fuzzy partition is lost. The aim of this contribution is to explore some methods of gradual approximations of U by non-crisp matrices. Two approaches, based on two different interpretations of the notion of α-cut of U, are considered. First, the α-cut can be interpreted as a sharper approximation of U in which the large membership grades (at least as large as α) are elevated to 1, while the small membership grades (smaller than α) are reduced to 0. Second, it may be viewed as a coarser approximation of U in which the total number of distinct membership grades of U is reduced to only two values, namely to 0 and 1. In this case, 0 and 1 are the labels for the image of U in the partition of the unit interval into subintervals $[0, \alpha)$ and $[\alpha, 1]$, respectively.

The maximum membership approximation of U is a generalization of the α-cut of U, where the parameter α is replaced by an n-tuple of parameters $T = (t_1, \ldots, t_n)$ derived from U. Therefore, the concept of α-sharpness can be extended to T-sharpness. Also, the maximum membership approximation of U can be obtained from a sequence of its minimum membership approximations [2]. T-sharper approximations and the minimum membership approximation based on T-sharpness will be explored in Sect. 3. Analogously, when the set $\mathbf{T} = \{\alpha\}$ is extended to the set $\mathbf{T} = \{t_1, \ldots, t_m\}$ of parameters from $(0, 1)$, the α-cut of U is generalized to the $\mathbf{T}$-level cut of U, which is a coarser image of U in the partition of the unit interval into at most $m + 1$ subintervals. $\mathbf{T}$-level cuts will be studied in Sect. 4. T-sharper approximations and $\mathbf{T}$-level cuts of a fuzzy partition U represent a compromise between the original matrix U (too fuzzy) and its crisp approximation (too coarse). The parameters in the n-tuple T or in the set $\mathbf{T}$ can be derived from U or can be given by a researcher. The rest of this paper is organized as follows. Section 2 reviews some basic concepts of fuzzy partitions. Section 5 provides some concluding remarks.

2 Preliminaries

We assume that X is a finite universal set with the cardinality $|X| = n$ and $2 \leq c \leq n$. Let V_{cn} denote the set of all real $c \times n$ matrices. Then the hard (crisp) c-partition space for X is the set [1]

$$M_{hcn} = \{U \in V_{cn} : u_{ij} \in \{0, 1\}, \text{ for all } i, j; \ \sum_{i=1}^{c} u_{ij} = 1, \text{ for all } j; \ 0 < \sum_{j}^{n} u_{ij} < n, \text{ for all } i\}, \tag{1}$$

the probabilistic fuzzy c-partition space for X is the set

$$M_{fcn} = \{U \in V_{cn} : u_{ij} \in [0,1], \text{ for all } i,j;\ \sum_{i=1}^{c} u_{ij} = 1, \text{ for all } j;\ 0 < \sum_{j}^{n} u_{ij} < n, \text{ for all } i\}, \quad (2)$$

and the possibilistic fuzzzy c-partition space for X is the set

$$M_{pcn} = \{U \in V_{cn} : u_{ij} \in [0,1], \text{ for all } i,j;\ \sum_{i=1}^{c} u_{ij} > 0, \text{ for all } j;\ 0 < \sum_{j}^{n} u_{ij} < n, \text{ for all } i\}. \quad (3)$$

The degenerated partition spaces M_{hcno}, M_{fcno}, and M_{pcno} are supersets of M_{hcn}, M_{fcn}, and M_{pcn}, respectively, obtained by the condition

$$0 \leq \sum_{j}^{n} u_{ij} < n, \text{ for all } i. \quad (4)$$

The notation I will be used for the set of integers $\{1, \ldots, c\}$ and the notation J will be used for the set of integers $\{1, \ldots, n\}$. For $U, W \in V_{cn}$,

$$U \leq W \text{ if and only if } u_{ij} \leq w_{ij} \text{ for all } (i,j) \in I \times J. \quad (5)$$

Fuzziness of a fuzzy partition $U \in M_{fco}$ can be evaluated by the partition entropy defined by

$$H(U) = \frac{1}{n} \sum_{i \in I} \sum_{j \in J} h(u_{ij}), \quad (6)$$

where $h(u_{ij}) = u_{ij} log_{1/c}(u_{ij})$, when $u_{ij} > 0$ and $h(0) = 0$. Then

1. $0 \leq H(U) \leq 1$
2. $H(U) = 0$ if and only if $U \in M_{hco}$
3. $H(U) = 1$ if and only if U has all elements equal to $1/c$

For $\alpha \in [0,1]$, the α-cut of $U \in M_{fco}$ is the matrix $U_\alpha \in V_{cn}$ with elements

$$(u_\alpha)_{ij} = \begin{cases} 1 & \text{if } u_{ij} \geq \alpha, \\ 0 & \text{otherwise.} \end{cases} \quad (7)$$

The strong α-cut of U is the matrix $U_{\alpha+} \in V_{cn}$ with elements

$$(u_{\alpha+})_{ij} = \begin{cases} 1 & \text{if } u_{ij} > \alpha, \\ 0 & \text{otherwise .} \end{cases} \quad (8)$$

When $\alpha = 1$, then U_α is the core of U, denoted by *cor* U, and when $\alpha = 0$, then $U_{\alpha+}$ is the support of U, denoted by *supp* U.

The maximum membership approximation of $U \in M_{fco}$ is the partition $MM(U) \in M_{fco}$ defined for all $(i,j) \in I \times J$ by

$$MM(u_{ij}) = \begin{cases} 1 & \text{if } u_{ij} = \max\{u_{kj};\ k \in I\}, \\ 0 & \text{otherwise.} \end{cases} \tag{9}$$

Note that U may have several different maximum membership approximations if the maximum membership of a classified object in U is not unique.

3 T-Sharper Approximations of Fuzzy Partitions

Sharpening of membership grades of a fuzzy set is obtained by transformation of large membership grades closer to 1 and small membership grades closer to zero. The notions of large and small are understood with respect to a given threshold $\alpha \in [0, 1]$. The following generalization of α-sharpening can be considered in the case of fuzzy partitions.

Definition 1 Consider $U, W \in M_{fco}$ and n-tuple $T = (t_1, \ldots, t_n)$, $t_j \in [0, 1]$ for all $j \in J$. Then W is T-sharper than U, denoted by $W \prec_T U$, if the membership grades of W satisfy the following properties: for all $(i,j) \in I \times J$,

$$w_{ij} \leq u_{ij} \leq t_j \quad \text{or} \quad t_j \leq u_{ij} \leq w_{ij}. \tag{10}$$

Further in this paper, the notation $T = (t_j)$ will be used instead of $T = (t_1, \ldots, t_n)$.

Remark 1 If $T = (\alpha)$, $\alpha \in [0, 1]$, then T-sharpness is reduced to α-sharpness. In order to obtain a T-sharper approximation of U which is a fuzzy partition different from U, the n-tuple T must contain at least one parameter t_j such that

$$\min\{u_{ij} | u_{ij} > 0, i \in I\} \leq t_j \leq \max\{u_{ij} | u_{ij} < 1, i \in I\}. \tag{11}$$

Proposition 1 *Given $T = (t_1, \ldots, t_n)$ and $U, V, W \in M_{fco}$, the following properties hold:*

1. *$U \prec_T U$*
2. *If $U \prec_T V$ and $V \prec_T U$, then $U = V$*
3. *If $U \prec_T V$ and $V \prec_T W$, then $U \prec_T W$*

Corollary 1 *M_{fco} is partially ordered by the relation $\prec_T$.*

T-sharpening of fuzzy partitions reduces their fuzziness and therefore it can be considered as a method of partial defuzzification.

Proposition 2 *Let $U, W \in M_{fco}$ and $W \prec_T U$. Then $H(W) \leq H(U)$.*

Note that if $H(W) \leq H(U)$, then W does not need to be a T-sharper approximation of U.

The notion of T-sharpness can be used in characterization of the minimum membership approximation of a fuzzy partition. Recall that the method of the maximum membership generates a crisp approximation of U, where the maximum membership grade in each column of U is replaced by 1. On the contrary, the minimum membership grade approximation should be a fuzzy partition where the minimum membership grade in each column of U is replaced by 0. Since this approximation should reduce the partition entropy of U, we request that the minimum membership approximation of U be T-sharper than U.

Definition 2 Assume $U \in M_{fco}$. For each $j \in J$, let $s_j = \max\{u_{ij}|, i \in I\}$ and $r_j = \min\{u_{ij}|u_{ij} > 0, i \in I\}$. Create matrix $W \in V_{cn}$ as follows: for all $(i,j) \in I \times J$

$$w_{ij} = \begin{cases} u_{ij} & \text{if } r_j = s_j, \\ 0 & \text{if } u_{ij} \leq r_j < s_j, \\ \delta_{ij} > 0 & \text{otherwise,} \end{cases} \tag{12}$$

where $\sum_{i=1}^{c} w_{ij} = 1$ for all $j \in J$ and $W \prec_T U$ for some n-tuple T. Then W is a minimum membership approximation of U denoted by $mm(U)$.

Note that minimum membership approximation of U is not unique.

Proposition 3 *Assume $U \in M_{fco}$. For each $j \in J$, let $s_j = \max\{u_{ij}|, i \in I\}$, $r_j = \min\{u_{ij}|u_{ij} > 0, i \in I\}$, $I_j = \{i \in I : u_{ij} = r_j\}$, and $I_j^* = \{i \in I : u_{ij} > r_j\}$. Consider mapping $\varphi_1 : M_{fco} \to M_{fco}$ which assigns to each $u_{ij} \in U$ element $\varphi_1(u_{ij}) \in \varphi_1(U)$ as follows:*

$$\varphi_1(u_{ij}) = \begin{cases} u_{ij} & \text{if } r_j = s_j, \\ 0 & \text{if } u_{ij} \leq r_j < s_j, \\ u_{ij} + \gamma_{ij} & \text{otherwise,} \end{cases} \tag{13}$$

where $\gamma_j = r_j|I_j|/|I_j^*|$. Then $\varphi_1(U) = mm(U)$ and $\varphi_1(U) \prec_T U$, where $T = (r_j)$.

Another way of obtaining a minimum membership approximation of U is based on linear intensification [2].

Proposition 4 *Assume $U \in M_{fco}$. For each $j \in J$, let $s_j = \max\{u_{ij}|, i \in I\}$, $r_j = \min\{u_{ij}|u_{ij} > 0, i \in I\}$, $I_j = \{i \in I|u_{ij} > 0\}$, and $m_j = |I_j|$. Consider the mapping (linear intensification) $\varphi_2 : M_{fco} \to M_{fco}$ which assigns to each $u_{ij} \in U$ element*

$\varphi_2(u_{ij}) \in \varphi_2(U)$ *as follows:*

$$\varphi_2(u_{ij}) = \begin{cases} \frac{1}{m_j} + \frac{1}{1-m_j r_j}(u_{ij} - \frac{1}{m_j}) & \text{if } r_j < s_j, \\ u_{ij} & \text{otherwise}. \end{cases} \tag{14}$$

Then $\varphi_2(U) = mm(U)$ *and* $\varphi_2(U) \prec_T U$, *where* $T = (1/m_j)$.

A slight modification of φ_1 from Proposition 3 also generates $mm(U)$.

Proposition 5 *Assume* $U \in M_{fco}$. *For each* $j \in J$, $r_j = \min\{u_{ij} | u_{ij} > 0, i \in I\}$, $I_j = \{i \in I | \ u_{ij} = r_j\}$, $s_j = \max\{u_{ij}, i \in I\}$, *and* $I_j^{**} = \{i \in I | \ u_{ij} = s_j\}$. *Consider the mapping* $\varphi_3 : M_{fco} \to M_{fco}$ *which assigns to each* $u_{ij} \in U$ *element* $\varphi_3(u_{ij}) \in \varphi_3(U)$ *as follows:*

$$\varphi_3(u_{ij}) = \begin{cases} u_{ij} & \text{if } r_j = s_j \text{ or } r_j < u_{ij} < s_j, \\ 0 & \text{if } u_{ij} \le r_j < s_j, \\ u_{ij} + \gamma_{ij} & \text{otherwise}, \end{cases} \tag{15}$$

where $\gamma_j = r_j |I_j| / |I_j^{**}|$. *Then* $\varphi_3(U) = mm(U)$ *and* $\varphi_3(U) \prec_T U$, *where* $T = (r_j)$.

A sequence of gradual approximations of U by $mm(U)$ can be created according to Algorithm 1 below.

Algorithm 1 Let $U \in M_{fco}$ and $\varphi : M_{fco} \to M_{fco}$ such that $\varphi(U) = mm(U)$.

Step 1: Put $k := 0$ and $mm^k(U) := U$.
Step 2: Create $mm^{k+1}(U) = \varphi(mm^k(U))$.
Step 3: If $mm^{k+1}(U) = mm^k(U)$ or $k = c - 2$, stop. Else put $k := k + 1$ and go to Step 2.

Proposition 6 *Assume a sequence* $mm^k(U)$, $k = 0, \ldots, c-2$ *of approximations of* U *generated by Algorithm 1. Then* $mm^{k+1}(U) \prec_{T_k} mm^k(U)$ *and* $T_{k+1} \ge T_k$.

Corollary 2 *Assume a sequence* $mm^k(U)$, $k = 0, \ldots, c-2$ *of approximations of* U *generated by Algorithm 1. Then* $H(mm^{k+1}(U)) \le H(mm^k(U))$.

Remark 2 When U has in each column only non-repeated positive membership grades, then

- Fuzzy partition $mm^k(U)$, $k = 1, \ldots, c-1$ created by Algorithm 1 is the kth minimum membership approximation of U. Consequently, $mm^{c-1}(U) = MM(U)$,
- If φ_2 from Proposition 4 is used in Algorithm 1, then $mm^{k+1}(U) \prec_{T_k} mm^k(U)$, $k = 0, \ldots, c-2$, where $T_k = (1/(c-k))$ and hence $T_0 = (1/c)$ and $T_{c-1} = (1/2)$.

Each mapping $\varphi : M_{fco} \to M_{fco}$ which creates a minimum membership approximation of a fuzzy partition U reduces the number m_j of clusters to which object $x_j \in X$

is classified ($2 \leq m_j \leq c$). Then, in decision-making, the following guidelines can be used for the choice of φ.

- If there is no other condition placed on $mm(U)$, use the mapping φ_1 from Proposition 6. In this case, membership grades of x_j greater than the minimal one are uniformly elevated.
- If the goal is to create a fuzzy partition $mm(U)$ such that the membership grades of x_j distributed to m_j clusters of U are sharpened with respect to the threshold $1/m_j$, use the mapping φ_2 from Proposition 4.
- If the goal is to create a fuzzy partition $mm(U)$ with more distinct maximum membership grades, use the mapping φ_3 from Proposition 5. Note that φ_3 can be considered as a combination of the minimum membership and the maximum membership methods, because the minimal membership grade of x_j is reduced to zero and only its maximal membership grade is elevated.

Example 1 Let U be the fuzzy partition of elements from the set $X = \{x_1, x_2, x_3, x_4, x_5, x_6\}$ represented by the matrix

$$U = \begin{pmatrix} x_1 & x_2 & x_3 & x_4 & x_5 & x_6 \\ 0.27 & 0.55 & 0.1 & 0 & 0.25 & 0.40 \\ 0.03 & 0.15 & 0.1 & 0.60 & 0.25 & 0.30 \\ 0.35 & 0.30 & 0.2 & 0.28 & 0.25 & 0.09 \\ 0.35 & 0 & 0.6 & 0.12 & 0.25 & 0.21 \end{pmatrix}.$$

Due to repeated maximum membership grades in column 1 and column 5, there are 8 different maximum membership approximations of U. When Algorithm 1 with the mapping φ_1 from Proposition 6 is applied to U, the following non-crisp approximations are obtained. First,

$$mm^1(U) = \begin{pmatrix} 0.12 & 0.625 & 0 & 0 & 0.25 & 0.43 \\ 0 & 0 & 0 & 0.66 & 0.25 & 0.33 \\ 0.44 & 0.375 & 0.3 & 0.34 & 0.25 & 0 \\ 0.44 & 0 & 0.7 & 0 & 0,25 & 0.24 \end{pmatrix},$$

where $mm^1(U) \prec_{T_0} (U)$ and $T_0 = (0.03, 0.15, 0.1, 0.12, 0.25, 0.09)$.
Partition entropy $H(U) = 0.821$ was reduced to $H(mm^1(U)) = 0.643$.
The second and the third approximations are

$$mm^2(U) = \begin{pmatrix} 0 & 1 & 0 & 0 & 0.25 & 0.55 \\ 0 & 0 & 0 & 1 & 0.25 & 0.45 \\ 0.5 & 0 & 0 & 0 & 0.25 & 0 \\ 0.5 & 0 & 1 & 0 & 0.25 & 0 \end{pmatrix}, \quad mm^3(U) = \begin{pmatrix} 0 & 1 & 0 & 0 & 0.25 & 1 \\ 0 & 0 & 0 & 1 & 0.25 & 0 \\ 0.5 & 0 & 0 & 0 & 0.25 & 0 \\ 0.5 & 0 & 1 & 0 & 0,25 & 0 \end{pmatrix},$$

where $mm^3(U) \prec_{T_2} mm^2(U) \prec_{T_1} mm^1(U)$,
$T_1 = (0.12, 0.375, 0.3, 0.34, 0.25, 0.24)$ and $T_2 = (0.5, 1, 1, 1, 0.25, 0.45)$,
$H(mm^2(U)) = 0.333$ and $H(mm^3(U)) = 0.250$.

When Algorithm 1 with the mapping φ_2 from Proposition 4 is applied to U, the following non-crisp approximations are obtained. First,

$$mm^1(U) = \begin{pmatrix} 0.2728 & 0.7273 & 0 & 0 & 0.25 & 0.4844 \\ 0 & 0 & 0 & 0.75 & 0.25 & 0.3281 \\ 0.3636 & 0.2727 & 0.1667 & 0.25 & 0.25 & 0 \\ 0.3636 & 0 & 0.8333 & 0 & 0.25 & 0.1875 \end{pmatrix},$$

where $mm^1(U) \prec_{T_0} (U)$ and $T_0 = (1/4, 1/3, 1/4, 1/3, 1/4, 1/4)$.
Partition entropy $H(U) = 0.821$ was reduced to $H(mm^1(U)) = 0.614$.
The second and the third approximations are

$$mm^2(U) = \begin{pmatrix} 0 & 1 & 0 & 0 & 0.25 & 0.324 \\ 0 & 0 & 0 & 1 & 0.25 & 0.6786 \\ 0.5 & 0 & 0 & 0 & 0.25 & 0 \\ 0.5 & 0 & 1 & 0 & 0.25 & 0 \end{pmatrix}, \quad mm^3(U) = \begin{pmatrix} 0 & 1 & 0 & 0 & 0.25 & 1 \\ 0 & 0 & 0 & 1 & 0.25 & 0 \\ 0.5 & 0 & 0 & 0 & 0.25 & 0 \\ 0.5 & 0 & 1 & 0 & 0.25 & 0 \end{pmatrix}.$$

where $mm^3(U) \prec_{T_2} mm^2(U) \prec_{T_1} mm^1(U)$,
$T_1 = (1/3, 1/2, 1/2, 1/2, 1/4, 1/3)$ and $T_2 = (1/2, 1, 1, 1, 1/4, 1/2)$,
$H(mm^2(U)) = 0.325$ and $H(mm^3(U)) = 0.250$. When Algorithm 1 with the mapping φ_3 from Proposition 5 is applied to U, the following non-crisp approximations are obtained. First,

$$mm^1(U) = \begin{pmatrix} 0.270 & 0.7 & 0 & 0 & 0.25 & 0.49 \\ 0 & 0 & 0 & 0.72 & 0.25 & 0.30 \\ 0.365 & 0.3 & 0.2 & 0.28 & 0.25 & 0 \\ 0.365 & 0 & 0.8 & 0 & 0.25 & 0.21 \end{pmatrix},$$

where $mm^1(U) \prec_{T_0} (U)$ and $T_0 = (0.03, 0.15, 0.1, 0.12, 0.25, 0.09)$.
Partition entropy $H(U) = 0.821$ was reduced to $H(mm^1(U)) = 0.627$.
The second and the third approximations are

$$mm^2(U) = \begin{pmatrix} 0 & 1 & 0 & 0 & 0.25 & 0.55 \\ 0 & 0 & 0 & 1 & 0.25 & 0.45 \\ 0.5 & 0 & 0 & 0 & 0.25 & 0 \\ 0.5 & 0 & 1 & 0 & 0.25 & 0 \end{pmatrix}, \quad mm^3(U) = \begin{pmatrix} 0 & 1 & 0 & 0 & 0.25 & 1 \\ 0 & 0 & 0 & 1 & 0.25 & 0 \\ 0.5 & 0 & 0 & 0 & 0.25 & 0 \\ 0.5 & 0 & 1 & 0 & 0.25 & 0 \end{pmatrix}.$$

where $mm^3(U) \prec_{T_2} mm^2(U) \prec_{T_1} mm^1(U)$,
$T_1 = (0.27, 0.3, 0.2, 0.28, 0.25, 0.21)$ and $T_2 = (0.5, 1, 1, 1, 0.25, 0.45)$,
$H(mm^2(U)) = 0.333$ and $H(mm^3(U)) = 0.250$.

4 T-Level Cuts of Fuzzy Partitions

Another way of approximation of a fuzzy partition U is transformation of U to a matrix with reduced number of distinct membership grades from interval $(0, 1)$. This type of transformation is called coarsening. Given $U \in M_{fco}$, the set

$$\Lambda_U = \{\delta \in (0,1) : \delta = u_{ij} \text{ for some } u_{ij} \in U\} \tag{16}$$

is called the level set of U. Obviously, the level set of a crisp partition is empty.

Definition 3 Consider a matrix $W \in V_{cn}$ with elements from $[0, 1]$. The coarseness of W is evaluated by the coefficient

$$\kappa(U) = 1 - \frac{|\Lambda_U|}{n.c}. \tag{17}$$

Proposition 7 *Assume a fuzzy partition $U \in M_{fco}$. Then*

1. $0 \leq \kappa(U) \leq 1$
2. $\kappa(U) = 1$ *if and only if $U \in M_{hco}$ (crisp partition)*
3. $\kappa(U) = 0$ *if and only if all membership grades of U are non-repeated values between zero and one*

An α-cut of U can be considered as a coarser image of U in the partition of the unit interval into subintervals $[0, \alpha)$ and $[\alpha, 1]$, or simply as the level cut of U based on the set of parameters $\mathbf{T} = \{\alpha\}, \alpha \in (0, 1)$. Let $\{\delta_1, \delta_2, \ldots, \delta_k\}$ be the set of all distinct membership grades of U such that $1 > \delta_1 > \delta_2 > \ldots > \delta_k > 0$. Then the level cuts $U_{\delta_1}, U_{\delta_2}, \ldots, U_{\delta_k}$ create a sequence of gradual crisp approximations of U such that

$$cor\, U < U_{\delta_1} < U_{\delta_2} < \ldots < U_{\delta_k} < supp\, U. \tag{18}$$

Our goal is to find a sequence of gradual non-crisp coarser approximations of U satisfying inequality (18). Based on the concept of p-level sets [3], the concept of $\mathbf{T}$-level cuts of a fuzzy partition U is introduced in the next definition.

Definition 4 Let $U \in M_{fco}$ and $\mathbf{T} = \{t_1, t_2, \ldots, t_m\}$ be a set of parameters such that $0 < t_1 \leq t_2 \leq \ldots \leq t_m < 1$. Then the $\mathbf{T}$-level cut of U is the matrix $\mathbf{T}(U) \in V_{cn}$ defined for all $(i, j) \in I \times J$ as follows:

$$\mathbf{T}(u_{ij}) = \frac{|\{t \in \mathbf{T} : t < u_{ij}\}|}{|\mathbf{T}|}. \tag{19}$$

The membership grade $\mathbf{T}(u_{ij})$ evaluates the proportion of parameters from $\mathbf{T}$ which are dominated by u_{ij}. If $\mathbf{T}$ includes r distinct parameters from $(0, 1)$, then $\mathbf{T}(U)$ is an image of U in the partition of the unit interval into $r + 1$ subintervals. In general, $\mathbf{T}(U)$ is a possibilistic fuzzy partition.

Proposition 8 *Let $U \in M_{fco}$. Assume a $\boldsymbol{T}$-level cut $\boldsymbol{T}(U)$ with the membership grades $\mathbf{T}(u_{ij})$. Then*

1. $\mathbf{T}(0) = 0$ *and* $\mathbf{T}(1) = 1$
2. $\mathbf{T}(u_{ij}) \in \{0, 1/m, 2/m, \dots, (m-1)/m, 1\}$, *where* $m = |\mathbf{T}|$
3. *If* $u_{ij} \leq u_{pq}$ *then* $\mathbf{T}(u_{ij}) \leq \mathbf{T}(u_{pq})$
4. *When* $\mathbf{T} = \{t_1, t_2, \dots, t_m\}$, $\boldsymbol{S} = \{s_1, s_2, \dots, s_m\}$, *and* $t_j \leq s_j$ *for all* $j = 1, \dots, m$, *then* $\mathbf{S}(U) \leq \mathbf{T}(U)$
5. *If* $\mathbf{T} = \{\alpha\}$, $\alpha \in (0, 1)$, *then* $\mathbf{T}(U) = U_{\alpha+}$

Corollary 3 *Let $|\Lambda_U| = m$. If the set $\boldsymbol{T}$ has less than m parameters (not necessarily distinct), then $\boldsymbol{T}(U)$ is a coarser approximation of U.*

A sequence of gradual approximations of U by coarser $\mathbf{T}$-level cuts of U can be created according to Algorithm 2 below.

Algorithm 2 Consider $U \in M_{fco}$.

Step 1: Let k be the number of crisp columns in U (i.e., columns containing elements equal only to zero and one). Put $n^* := n - k$.

Step 2: Remove from U all crisp columns and arrange the elements in each remaining column from the largest to the smallest. Denote the resulting matrix by U^*. Let c^* be the total number of distinct rows in U^* with all elements from $(0, 1)$.

Step 3: For $q = 1, \dots c^*$, arrange the elements in the qth row of U^* from the smallest to the largest. Denote the resulting matrix by U^{**}. Assign the qth-row of U^{**} to n^*-tuple $\mathbf{T}_q$ and create $\mathbf{T}_q(U)$.

Proposition 9 *Consider the $\boldsymbol{T}$-level cuts of $U \in M_{pc}$ created by Algorithm 2. Then*

1. *cor* $U < \mathbf{T}_1(U) < \mathbf{T}_2(U) < \mathbf{T}_{c^*}(U) <$ *supp* U
2. $\kappa(\mathbf{T}_q(U)) < \kappa(U)$ *for all* $q = 1, \dots, c^*$
3. $\mathbf{T}_q(U) \in M_{pco}$ *for* $q \geq 2$

Remark 3 $\mathbf{T}_1(U)$ created by Algorithm 2 contains one column with all elements equal to zero (the column, where the smallest from all n maximal membership grades is located). Therefore, $\mathbf{T}_1(U)$ is not a possibilistic fuzzy partition of n objects.

Example 2 Consider the fuzzy partition U from Example 1. Then Algorithm 2 applied to U creates the following matrix U^*

$$U^* = \begin{pmatrix} 0.35 & 0.55 & 0.6 & 0.60 & 0.25 & 0.40 \\ 0.35 & 0.30 & 0.2 & 0.28 & 0.25 & 0.30 \\ 0.27 & 0.15 & 0.1 & 0.12 & 0.25 & 0.21 \\ 0.03 & 0 & 0.1 & 0 & 0.25 & 0.09 \end{pmatrix}.$$

Then $T_1 = \{0.25, 0.35, 0.4, 0.55, 0.6, 0.6\}$ and

$$T_1(U) = \begin{pmatrix} 1/6 & 3/6 & 0 & 0 & 0 & 2/6 \\ 0 & 0 & 0 & 4/6 & 0 & 1/6 \\ 1/6 & 1/6 & 0 & 1/6 & 0 & 0 \\ 1/6 & 0 & 4/6 & 0 & 0 & 0 \end{pmatrix}.$$

Coefficient of coarseness $\kappa(U) = 0.375$ is increased to $\kappa(T_1(U)) = 0.833$.
The second and the third approximations are

$$T_2(U) = \begin{pmatrix} 2/6 & 1 & 0 & 0 & 1/6 & 1 \\ 0 & 0 & 0 & 1 & 1/6 & 3/6 \\ 5/6 & 3/6 & 0 & 2/6 & 1/6 & 0 \\ 5/6 & 0 & 1 & 0 & 1/6 & 1/6 \end{pmatrix}, \quad T_3(U) = \begin{pmatrix} 5/6 & 1 & 0 & 0 & 4/6 & 1 \\ 0 & 2/6 & 0 & 1 & 4/6 & 1 \\ 1 & 1 & 3/6 & 1 & 4/6 & 0 \\ 1 & 0 & 1 & 1/6 & 4/6 & 3/6 \end{pmatrix}.$$

with $T_2 = \{0.2, 0.25, 0.28, 0.3, 0.3, 0.35\}$ and $T_3 = \{0.1, 0.12, 0.15, 0.21, 0.25, 0.27\}$,
and $\kappa(T_2(U)) = \kappa(T_1(U)) = 0.833$. Obviously,

$$core\ U < T_1(U) < T_2(U) < T_3(U) < supp\ U.$$

5 Conclusion

Two methods of a gradual non-crisp approximation of a fuzzy partition U based on parameters which may be derived from U were proposed. While the first method focuses on reduction of fuzziness of U, the second method increases coarseness of U. The relation of T-sharpness and the coefficient of coarseness were introduced. Although the work in this paper refers to a fuzzy partition of n objects into c clusters, it can be applied to any situation where information is represented by a matrix with elements from the unit interval. This includes, e.g., tables of relative frequencies or normalized evaluations of n objects by c experts. In future work, approximations of fuzzy partitions based on combination of sharpening and coarsening will be studied.

References

1. Bezdek, J.C.: Pattern Recognition with Fuzzy Objective Function Algorithms. Plenum Press, New York (1981)
2. Bodjanova, S.: Linear intensification of probabilistic fuzzy partitions. Fuzzy Sets Syst. **141**, 319–332 (2004)
3. Bodjanova, S., Kalina, M.: Coarsening of fuzzy sets. In: Proceedings of IEEE 11th International Symposium on Intelligent Systems and Informatics, pp. 51–56. IEEE, New York (2013)

4. Deng, X., Yao, Y.: Decision-theoretic three-way approximations of fuzzy sets. Inf. Sci. **279**, 702–715 (2014)
5. Dubois, D., Prade, H.: Fundamentals of Fuzzy Sets. Kluwer, Dordrecht (2000)
6. Dumitrescu, D., Pop, H.F.: Degenerate and non-degenerate convex decomposition of finite fuzzy partitions - I. Fuzzy Sets Syst. **73**, 365–376 (1995)
7. Dumitrescu, D., Pop, H.F.: Degenerate and non-degenerate convex decomposition of finite fuzzy partitions - II. Fuzzy Sets Syst. **96**, 111–118 (1998)
8. Dumitrescu, D., Lazzerini, B., Jain, L.C.: Fuzzy Sets and Their Applications to Clustering and Training. CRP Press, Boca Raton (2000)
9. Li, T.J., Zhang, W.X.: Rough fuzzy approximations on two universes of discourse. Inf. Sci. **178**, 892–906 (2008)
10. Maji, P., Pal, S.: Rough set based generalized fuzzy c-means algorithm and quantitative indices. Trans. Syst. Man Cybern. Part B **37**, 1529–1540 (2007)
11. Mitra, S., Banka, H., Pedrycz, W.: Rough-fuzzy collaborative clustering. Trans. Syst. Man Cybern. Part B **37**, 795–805 (2005)
12. Mitra, S., Pedrycz, W., Barman, B.: Shadowed c-means: Integrating fuzzy and rough clustering. Pattern Recogn. **43**, 1282–1291 (2010)
13. Pedrycz, W., Dong, F., Hirota, K.: Finite cut-based approximation of fuzzy sets and its evolutionary optimization. Fuzzy Sets Syst. **160**, 3550–3564 (2009)
14. Peters, G., Crespo, F., Lingras, P., Weber, R.: Soft clustering-fuzzy and rough approaches and their extensions and derivatives. Int. J. Approx. Reason. **54**, 307–322 (2013)
15. Radecki, T.: Level-fuzzy sets. J. Cybern. **7**, 189–198 (1977)
16. Zhou, J., Pedrycz, W., Miao, D.Q.: Shadowed sets in the characterization of rough-fuzzy clustering. Pattern Recogn. **44**, 1738–1749 (2011)

Benchmarking for Clustering Methods Based on Real Data: A Statistical View

Anne-Laure Boulesteix and Myriam Hatz

Abstract In analogy to clinical trials, in a benchmark experiment based on real datasets we can see the considered datasets as playing the role of patients and the compared methods as playing the role of treatments. This view of benchmark experiments, which has already been suggested in the literature, brings to light the importance of statistical concepts such as testing, confidence intervals, power calculation, and sampling procedure for the interpretation of benchmarking results. In this paper we propose an application of these concepts to the special case of benchmark experiments comparing clustering algorithms. We present a simple exemplary benchmarking study comparing two classical clustering algorithms based on 50 high-dimensional gene expression datasets and discuss the interpretation of its results from a critical statistical perspective. The R-codes implementing the analyses presented in this paper are freely available from: http://www.ibe.med.uni-muenchen.de/organisation/mitarbeiter/020_professuren/boulesteix/boulesteixhatz.

1 Introduction

Real data are more complex than simulated data. In practice data never follow well-known distributions. To assess the behavior of data analysis methods in concrete situations of practical relevance, benchmarking on real data is essential.

Whereas supervised learning methods can be evaluated in real data settings based on, e.g., their cross-validation error, there is no obvious criterion to be used to evaluate clustering methods. Quite generally, benchmarking—in particular benchmarking using real data—is a very complex issue in the context of unsupervised learning and to date there still exists no guidance in the literature on how to design and interpret such an experiment. One of the goals of the so-called Task Force on Benchmarking initiated by members of the International Federation of Classification Societies (IFCS) [12] is to provide such guidance.

A.-L. Boulesteix (✉) • M. Hatz
Department of Medical Informatics, Biometry and Epidemiology, University of Munich, München, Germany
e-mail: boulesteix@ibe.med.uni-muenchen.de; myriam.hatz@gmail.com

F. Palumbo et al. (eds.), *Data Science*, Studies in Classification, Data Analysis, and Knowledge Organization, DOI 10.1007/978-3-319-55723-6_6

In the special case where a/the true cluster structure is known, it can be used as a target to be achieved by the clustering method. The agreement between this true cluster structure and the cluster structure output by the method of interest can then be considered as a goodness criterion for evaluating the considered clustering method.

Papers comparing clustering methods typically include simulation studies and an application to a small to moderate number of real datasets. In the present paper, we critically discuss these real data applications from a statistical point of view. In particular, we draw a parallel between benchmark experiments and clinical trials as already suggested by Boulesteix and colleagues [1, 3] for the case of real data and Doove et al. [6] in the context of simulations. In our framework, real datasets play the role of patients, and clustering methods play the role of therapies. With this metaphor in mind, we claim that, in order to make clear statements from real data benchmark experiments, one has to analyze and interpret their results following statistical principles, as illustrated through an exemplary benchmark experiment based on 50 microarray datasets.

Our goal is fourfold: (1) illustrating the variability of benchmarking results across real datasets, (2) propagating statistical thinking in the context of benchmark experiments, where datasets are considered as statistical units, (3) discussing the notion of power in this context, (4) illustrating a possible strategy for the interpretation of benchmark studies based on real datasets through an exemplary study.

The paper is structured as follows. Section 2 briefly presents the clustering methods, data and evaluation criterion used in the exemplary benchmark experiment. The statistical interpretation of the results is given in Sect. 3, including discussions of the concepts of statistical testing, sample size calculation, dependence on datasets' characteristics, and sampling.

2 An Illustrative Benchmark Study: Methods and Data

This section briefly presents the clustering methods, data and evaluation criterion used in the exemplary benchmark experiment.

2.1 Data

The collection of datasets used in our exemplary study was first described by de Souza et al. [5] and used in the context of benchmarking for supervised classification by Boulesteix et al. [4]. It includes 50 clinical gene expression datasets with binary response variable (e.g., diseased vs. healthy), with numbers of patients between $n = 23$ and $n = 286$ and number of variables (genes) between 1098 and 54,680 variables. The datasets can be freely downloaded from the companion website of the paper by Boulesteix et al. [4].

In our study, the interest is in clustering the patients, a task commonly performed in clinical research with the aim, say, to identify typical patient profiles or to discover new disease subtypes. In this context, we would like clustering methods to be able to recover the true cluster structure given by the binary response variable, since it is known to be clinically relevant. Our study includes datasets with binary response variables only to make the comparison of the results across datasets easier.

2.2 Goodness Criterion

As a goodness criterion for clustering methods, we thus simply consider the adjusted Rand index (ARI) [8] measuring the agreement between the true cluster structure (denoted as "partition $\mathscr{C}_{true}$" of $\{1, \ldots, n\}$) defined by the binary response variable and the cluster structure ("partition $\mathscr{C}_M$") output by the clustering method M of interest. The Rand index (RI) can be seen as the proportion of pairs of objects that are either in the same cluster or in different clusters according to both $\mathscr{C}_{true}$ and $\mathscr{C}_M$:

$$RI = \frac{\binom{n_{11}}{2} + \binom{n_{12}}{2} + \binom{n_{21}}{2} + \binom{n_{22}}{2}}{\binom{n}{2}},$$

where $n_{11}, n_{12}, n_{21}, n_{22}$ are the entries of the table showing the numbers of observations from each of the two classes ($Y = 0, 1$) assigned to each of the two clusters ($C = 1, 2$) by clustering method M, see Table 1.

The *adjusted* Rand index (ARI) is an adjusted version of the Rand index accounting for the random agreement and defined as

$$\begin{aligned} ARI &= \frac{RI - mean(RI)}{max(RI) - mean(RI)} \\ &= \frac{\sum_{i,j}\binom{n_{ij}}{2} - \left[\sum_{i=1}^{2}\binom{n_{i.}}{2}\sum_{j=1}^{2}\binom{n_{.j}}{2}\right]/\binom{n}{2}}{\frac{1}{2}\left[\sum_{i=1}^{2}\binom{n_{i.}}{2} + \sum_{j=1}^{2}\binom{n_{.j}}{2}\right] - \left[\sum_{i=1}^{2}\binom{n_{i.}}{2}\sum_{j=1}^{2}\binom{n_{.j}}{2}\right]/\binom{n}{2}}. \end{aligned}$$

Table 1 Partition $\mathscr{C}_M$ output by clustering method M and true clustering $\mathscr{C}_{true}$ given by the binary variable (e.g., diseased vs. healthy)

	$\mathscr{C}_{true}$		
$\mathscr{C}_M$	$Y = 0$	$Y = 1$	Σ
$C = 1$	n_{11}	n_{12}	$n_{1.}$
$C = 2$	n_{21}	n_{22}	$n_{2.}$
Σ	$n_{.1}$	$n_{.2}$	n

2.3 Clustering Methods

In this paper we consider two simple standard clustering methods, since our focus is on issues related to benchmarking and interpretation rather than on the methods themselves. These very widely used methods can be seen as representatives of two important families of clustering methods, namely partitioning methods and hierarchical methods. The first method we consider is partitioning around medoids (PAM) as implemented in the function "pam" of the R package "cluster." The second method is agglomerative hierarchical clustering with euclidean distance as implemented in the function "hclust." These two methods are applied to obtain $K = 2$ clusters, by setting the number of clusters to 2 in "pam" and by cutting the tree in order to obtain two clusters. The choice of $K = 2$ corresponds to the true cluster structure reflected by the binary response variable (note that it would be interesting to also perform analyses with other values of K but this would lead to the problem of the choice of K, which goes beyond the scope of this paper).

3 Statistical Interpretation of Results

In this section, the results of our exemplary benchmark experiment presented in Sect. 2 are discussed from a statistical perspective. Most importantly, we propose to adopt and extend the statistical framework presented by Boulesteix et al. [4] to the context of unsupervised learning.

3.1 Main Results

We obtain the results in the form of a 50×2 matrix containing the ARI-values for all 50 datasets and both methods. A straightforward way to visualize the results is to display the ARI-values and differences in the form of boxplots as depicted in Fig. 1.

Paired tests can be performed to compare the ARI-values of the two methods, as described in Boulesteix et al. [4] in the different case of error rates of classification methods. p-Values of 0.001 and 0.0005 are obtained from the Wilcoxon test and t-test, respectively, whereby the Wilcoxon test seems to be more appropriate considering the skewness of the difference's distribution. In the same vein, one can compute confidence intervals for the median: the bootstrap confidence interval for the median is $(0, 0.053)$ with the percentile method and $(0, 0.046)$ with the bias-corrected accelerated bootstrap method [7].

Beyond statistical tests and the consideration of confidence intervals, further issues related to benchmarking can be advantageously considered from a statistical perspective, in particular in light of clinical trials methodology. They are discussed in the following subsections.

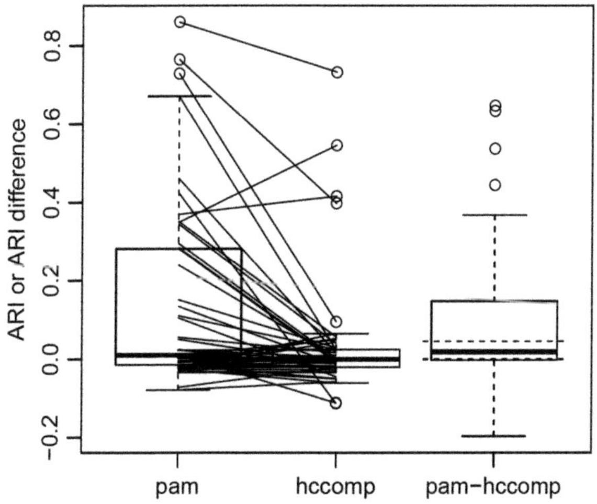

Fig. 1 Adjusted Rand Index (ARI) for the 50 datasets for PAM clustering (*left*) and agglomerative hierarchical clustering with complete linkage (*middle*); difference between these two ARI values (*right*) with the confidence intervals (of the median) obtained by the bias-corrected accelerated bootstrap method represented as *dashed lines*

3.2 Sample Size Calculation

Obviously, the number of datasets included in a benchmark experiment greatly influences the results of the testing procedure. The larger the number of datasets the higher the power to detect differences, and the lower the variance of the estimated difference between the two methods.

To illustrate this issue, we determine the median ARI-difference and the p-value of Wilcoxon's test obtained for 1000 random subsets of datasets drawn out of the 50 considered datasets. The corresponding boxplots are displayed in Fig. 2 for different subset sizes ($J = 3$, $J = 5$, $J = 10$, $J = 25$ datasets). As expected, the more datasets one includes in the benchmark experiment, the higher the stability of the median difference in ARI and the lower the p-values. If one performs the benchmark experiment based on only $J = 3, 5$, or 10 datasets instead of $J = 50$, the result may look completely different from the results with $J = 50$ datasets. Of note, a number of very large differences (> 0.2) are obtained for $J = 3, 5, 10$. Furthermore, most subsets of size $J = 10$ yield p-values> 0.05.

The notion of power of benchmark experiments in relationship with the number of included datasets can be formally addressed within the statistical testing framework. For simplicity, we assume that the paired t-test is used to compare the two methods. Considering the slightly skewed distribution of the differences between ARI-values of the two methods displayed in the right boxplot of Fig. 1, the Wilcoxon is certainly more appropriate. But sample size calculation is essentially an approximative procedure intended to provide orders of magnitude, so considering

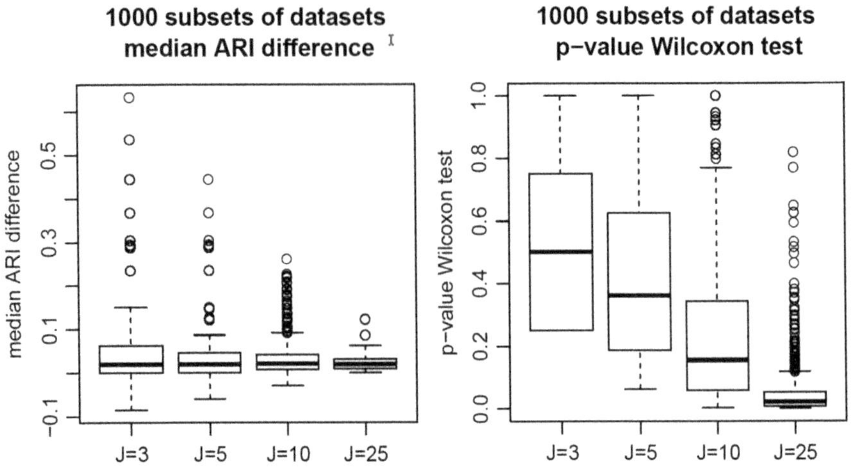

Fig. 2 *Left*: Median difference in *ARI* between the two methods for 1000 subsets of $J = 3, 5, 10, 25$ datasets drawn randomly out of the 50 datasets. *Right*: p-Value of the Wilcoxon test comparing the *ARI*-values of the two methods for 1000 subsets of $J = 3, 5, 10, 25$ datasets drawn randomly out of the 50 datasets

the t-test for the purpose of sample size calculation is acceptable in our context if one keeps in mind that the size has to be slightly increased if the Wilcoxon test is applied instead of the t-test.

The number of observations required to detect a difference of Δ at a significance level of α with a power of $1 - \beta$ using a two-sided paired t-test is approximated by

$$J \approx \frac{(z_{\alpha/2} + z_\beta)^2}{(\Delta/\sigma)^2} \tag{1}$$

where σ denotes the standard deviation of the difference and z_q denotes the q-quantile of the standard normal distribution. Note that this formula is based on the approximation of the Student distribution as standard normal distribution (the exact formula is less easy to apply since it involves the quantiles of the Student distribution, which themselves depend on J).

In our context, σ corresponds to the standard deviation of the difference that is displayed in the right boxplot of Fig. 1 for the 50 datasets. We obtain $\hat{\sigma} = 0.18$. Using Eq. (1), we compute that 25 resp. 102 datasets are required to detect differences of $\Delta = 0.1$ and $\Delta = 0.05$, respectively. Thus, even for a large difference of $\Delta = 0.1$, and for a relatively homogenous set of datasets as considered here (gene expression data, continuous variables, small to moderate sample size), the number of required datasets by far exceeds the size of typical benchmark experiments.

Our results, even if based on a particular example, suggest that it is unrealistic to draw statistically valid conclusions on average superiority of a method over the other based on real datasets without much time and effort. This problem becomes even more pointed if one does not consider average effects but tries to establish relationships between superiority of methods and datasets' characteristics. This issue is discussed and illustrated in the next section.

3.3 Dependence on Datasets' Characteristics

It can be argued that average superiority over a whole area of application is of poor relevance, since it is expected that the behavior of methods varies a lot depending on datasets' characteristics. Investigating the relationship between datasets' characteristics and methods' performance amounts to examining average superiority within a reduced area defined by particular datasets' characteristics. In this perspective, the issues discussed in the previous sections are also relevant when relating performance/superiority to datasets' characteristics based on real data—and certainly even more since the numbers of datasets are smaller.

It is important to investigate average superiority when elaborating guidelines and establish standard approaches. In an ideal world, methods that establish themselves as standard are those which are superior to other "on average"—even if this is not explicitly tested. Similarly, in an ideal world drugs that are routinely prescribed to patients are those that work best on average according to adequate statistical testing within clinical trials.

However, in the same way as the superior drug may not be the same for two different patients, the superior algorithm may not be the same for different datasets. In both cases, part of these differences might be explained by individual characteristics such as, say, age and sex of the patient and size and number of variables of the dataset, to cite only a few trivial examples. In the same way as a doctor wants to know which drug will best help the patient sitting in front of him, the data analyst wants to know which method performs best for the dataset at hand.

In the clinical context, two strategies have been pursued to address this problem: the search for subgroups in which treatment effects are different, on the one hand, and regression analysis for relating treatment effects to patients' characteristics, on the other hand. In a classical clinical trial with two parallel groups receiving a different treatment, regression analysis is usually performed as follows. The regression model relates the outcome of interest (dependent variable) to the treatment group, the patient's characteristic and their interaction (independent variables). In the context of benchmarking considered here, both methods are applied to all datasets, so the regression model simplifies to a model with the difference of performance as dependent variable and the dataset's characteristic as independent variable.

The search for subgroups can be performed using recursive partitioning methods both in clinical settings [11] and benchmarking settings. This is the approach adopted by Doove et al. [6] in the context of simulation-based benchmarking for clustering methods. In analogy to the term "treatment regime" used in the clinical context, Doove et al. [6] aim at deriving the "optimal data-analytic regime" depending on the dataset's characteristic in the context of benchmark studies.

With real data, however, things are more complex [6], not only because of the lack of straightforward goodness criterion. Firstly, the problem of the limited power is even more of an issue when relating performance to datasets' characteristics than when simply testing average superiority as described in Sect. 3.2. That is because the focus is now essentially on subgroups of datasets. Secondly, datasets' characteristics may be highly correlated, making it different to distinguish their respective effects. In simulation-based benchmarking, some of the relevant datasets' characteristics are controlled by design, hence mitigating this problem. The combination of these two problems makes the investigation of relationships between datasets' characteristics and methods' performance very difficult when using real datasets. On the one hand, the independent effects of the datasets' characteristics can only be assessed by including many of them in the model (keeping in mind that they are not all observable in real data settings!). On the other hand, increasing the number of datasets' characteristics in the model also decreases the power to identify individual effects.

To sum up, claims on the relationships between datasets' characteristics and methods' performance based on real datasets should be formulated very cautiously.

3.4 Sampling Issues and Over-Optimism

In clinical trials precise inclusion criteria for patients are defined before starting patient recruitment, for example, "age>18," "male sex," "no diabetes," etc. All patients fulfilling these criteria are considered for inclusion in the study and asked for their consent. After the data have been collected, it is not allowed to exclude patients from the analysis a posteriori based on their response to therapy.

Such sensible rules should ideally also be adopted in real data-based benchmark studies. Obviously, not all datasets are appropriate to be included in the benchmark study. Or the other way around, a method is not appropriate to all datasets. If some criteria that the dataset has to fulfill to be analyzed with the method are known before performing the benchmark study, candidate datasets should be checked for these criteria and included in the benchmark study only if they fulfill them. All datasets allowed to enter the study should be considered when reporting the results, even those yielding very bad results for the authors' "favorite" method.

Removing these bad datasets from the results has two detrimental consequences: (1) potential important relationships between method performance and datasets' characteristics in the vein of Sect. 3.3 may be overlooked; (2) the overall performance of the "favorite" method may be substantially over-estimated, as outlined theoretically [13] and empirically [9, 10] in the case of supervised learning. By eliminating bad datasets from reporting, one violates rule 4 from the "Ten simple rules to avoid over-optimism in computational research" [2]. This kind of "fishing for datasets" makes the results of real data-based benchmarking even less representative of further datasets.

The definition of inclusion criteria for benchmarking could ideally follow similar principles as in clinical trials. Too strict inclusion criteria lead to study results that are very specific to the considered settings and may not be of broad interest. Conversely, including heterogeneous datasets may make interpretation difficult.

An important difference between benchmarking settings and clinical settings is the "recruitment procedure." For a clinical trial one may, for example, recruit consecutive patients presenting to the hospital with some given symptoms. In the context of benchmarking, however, datasets have to be actively looked for (e.g., in databases or from the companion websites of published papers). This active role of the researcher in the recruitment introduces some arbitrariness and complicates the statistical formalization of the sampling procedure.

There is no straightforward sampling procedure for the population of datasets and the datasets can often not be considered as an *i.i.d.* sample drawn from the population of interest. This may induce biases and dependencies between observations that are difficult to avoid. They should be taken into account when interpreting the results of the benchmarking study. Otherwise, the statistical interpretation of benchmarking may give the readers a false sense of security/scientific correctness. Such issues may be devoted more attention in the context of benchmarking research in the future.

4 Conclusion

Applications to "one or few real datasets" are useful and important. However, they should be considered as illustrative and not representative of what we would obtain with further datasets [1] as long as no statistical inference is performed. Statistical inference requires many datasets and raises important challenges. In particular, there is no straightforward sampling procedure for the population of datasets. Bias is difficult to avoid. In conclusion, results of benchmark experiments based on real datasets should be interpreted with highest caution.

Acknowledgements We thank Sarah Tegenfeldt for language correction and the IFCS Task Force on Benchmarking, in particular to Iven van Mechelen, for very fruitful discussions on the topics of our paper.

References

1. Boulesteix, A.-L.: On representative and illustrative comparisons with real data in bioinformatics: response to the letter to the editor by Smith et al. Bioinformatics **29**(20), 2664–2666 (2013)
2. Boulesteix, A.-L.: Ten simple rules for reducing overoptimistic reporting in methodological computational research. PLOS Comput. Biol. **11**, e1004191 (2015)

3. Boulesteix, A.L., Lauer, S., Eugster, M.J.E.: A plea for neutral comparison studies in computational sciences. PLoS One **8**(4), e61562 (2013)
4. Boulesteix, A.-L., Hable, R., Lauer, S., Eugster, M.J.: A statistical framework for hypothesis testing in real data comparison studies. Am. Stat. **69**, 201–212 (2015)
5. de Souza, B., de Carvalho, A., Soares, C.: A comprehensive comparison of ml algorithms for gene expression data classification. In: Neural Networks (IJCNN), The 2010 International Joint Conference on IEEE, pp. 1–8 (2010)
6. Doove, L., Wilderjans, T., Calcagni, A., van Michelen, I.: Deriving optimal data-analytic regimes from benchmarking studies. Comput. Stat. Data Anal. **107**, 81–91 (2017). http://doi.org/10.1016/j.csda.2016.10.016. http://www.sciencedirect.com/science/article/pii/S0167947316302432
7. Efron, B.: Better bootstrap confidence intervals. J. Am. Stat. Assoc. **82**(397), 171–185 (1987)
8. Hubert, L., Arabie, P.: Comparing partitions. J. Classif. **2**(1), 193–218 (1985)
9. Jelizarow, M., Guillemot, V., Tenenhaus, A., Strimmer, K., Boulesteix, A.-L.: Over-optimism in bioinformatics: an illustration. Bioinformatics **26**(16), 1990–1998 (2010)
10. Macià, N., Bernadó-Mansilla, E., Orriols-Puig, A., Ho, T.K.: Learner excellence biased by data set selection: a case for data characterisation and artificial data sets. Pattern Recogn. **46**(3), 1054–1066 (2013)
11. Seibold, H., Zeileis, A., Hothorn, T.: Model-based recursive partitioning for subgroup analyses. Int. J. Biostat. **12**(1), 45–63 (2016)
12. Steinley, D., van Mechelen, I., IFCS Task Force on Benchmarking, 2015: Benchmarking in cluster analysis: preview of a white paper. Abstract. Conference of the International Federation of Classification Society, Bologna, 6th to 8th July 2015
13. Yousefi, M.R., Hua, J., Sima, C., Dougherty, E.R.: Reporting bias when using real data sets to analyze classification performance. Bioinformatics **26**(1), 68–76 (2010)

Representable Hierarchical Clustering Methods for Asymmetric Networks

Gunnar Carlsson, Facundo Mémoli, Alejandro Ribeiro, and Santiago Segarra

Abstract This paper introduces the generative model of representability for hierarchical clustering methods in asymmetric networks, i.e., the possibility to describe a method through its action on a collection of networks called representers. We characterize the necessary and sufficient structural conditions needed on these representers in order to generate a method which is scale preserving and admissible with respect to two known axioms and, based on this result, we construct the family of cyclic clustering methods.

1 Introduction

The relevance of clustering in modern data analysis is indubitable given its usage in multiple fields of knowledge from, e.g., medicine [22] to marketing [16]. There are literally hundreds of methods that can be applied to the determination of hierarchical [8, 11] and non-hierarchical clusters in finite metric (thus symmetric) spaces—see, e.g., [17]. Even in the case of asymmetric networks [18], multiple methods have been developed to extend the notion of clustering into this less intuitive domain [1, 7, 14, 15, 19, 20]. Although not as developed as its practice [6], the theoretical framework for clustering has been developed over the last decade for non-hierarchical [2, 3, 10, 12, 13, 21, 24] and hierarchical clustering [2, 4, 5]. Of special interest to us is this last direction, where two axioms were used to determine a bounded family of admissible clustering methods for asymmetric

G. Carlsson
Department of Mathematics, Stanford University, Stanford, CA, USA
e-mail: gunnar@math.stanford.edu

F. Mémoli (✉)
Department of Mathematics, The Ohio State University, Columbus, OH, USA

Department of Computer Science, The Ohio State University, Columbus, OH, USA
e-mail: memoli@math.osu.edu

A. Ribeiro • S. Segarra
Department of Electrical and Systems Engineering, University of Pennsylvania, Philadelphia, PA, USA
e-mail: aribeiro@seas.upenn.edu; ssegarra@seas.upenn.edu

F. Palumbo et al. (eds.), *Data Science*, Studies in Classification, Data Analysis, and Knowledge Organization, DOI 10.1007/978-3-319-55723-6_7

networks. However, we consider that admissibility is an insufficient characterization criterion to identify clustering methods of practical importance, thus, we introduce additional features with the objective to further winnow the space of admissible methods.

In the current paper, we introduce and analyze the notion of representability for hierarchical clustering methods. A representable method is one that can be specified by determining its action on a collection of networks called representers. Our main characterization result [cf. Theorem 2] states the conditions needed on these representers to obtain an associated clustering method with desirable practical properties. In Sect. 3, we present the notion of scale preservation and highlight its practical utility. In Sect. 4 we introduce the concept of representability through a reinterpretation of reciprocal clustering and build upon this to develop a more general theory. We also connect representability to the more practical concepts of scale preservation and admissibility (Sect. 4.2), and we introduce the family of cyclic clustering methods (Sect. 4.3).

Proofs, algorithmic details, and computational results are omitted due to space constraints.

2 Preliminaries

We define a network N as a pair (X, A_X) where X is a set of n points or nodes and $A_X : X \times X \to \mathbb{R}_+$ is a dissimilarity function. Dissimilarities $A_X(x, x')$ from x to x' are nonnegative, null if and only if $x = x'$, need not satisfy the triangle inequality and may be asymmetric, i.e., $A_X(x, x') \neq A_X(x', x)$ for some $x, x' \in X$. We focus our study on asymmetric networks since these general structures include, as particular cases, symmetric networks and finite metric spaces.

The output of hierarchically clustering a network $N = (X, A_X)$ is a dendrogram D_X, i.e., a nested collection of partitions $D_X(\delta)$ indexed by a resolution parameter $\delta \geq 0$ [9]. Dendrograms can be represented as trees. The interpretation of a dendrogram is that of a structure which yields different clusterings at different resolutions.

Given a network (X, A_X) and $x, x' \in X$, a chain $C(x, x')$ is an ordered sequence of nodes, $C(x, x') = [x = x_0, x_1, \ldots, x_{l-1}, x_l = x']$, which starts at x and finishes at x'. The *cost* of chain $C(x, x')$ is defined as the maximum dissimilarity $\max_{i|x_i \in C(x,x')} A_X(x_i, x_{i+1})$ encountered when traversing its links in order. We define the directed minimum chain cost $\tilde{u}^*_X(x, x')$ between x and x' as the minimum cost among all the chains connecting x to x',

$$\tilde{u}^*_X(x, x') = \min_{C(x,x')} \max_{i|x_i \in C(x,x')} A_X(x_i, x_{i+1}). \tag{1}$$

An ultrametric u_X on the set X is a function $u_X : X \times X \to \mathbb{R}_+$ that satisfies symmetry $u_X(x, x') = u_X(x', x)$, identity $u_X(x, x') = 0 \iff x = x'$, and the strong triangle inequality

$$u_X(x, x') \leq \max\big(u_X(x, x''), u_X(x'', x')\big), \tag{2}$$

for all $x, x', x'' \in X$. It can be shown that dendrograms and finite ultrametric spaces are equivalent [2], where the ultrametric value between two points is given by the resolution at which they first merge in thc dendrogram. However, ultrametrics are more convenient than dendrograms for the results developed in this paper.

Remark 1 For the particular case when N is a finite metric space, the minimum chain cost (1) is an ultrametric and coincides with the ultrametric output of the single linkage clustering method [5].

A hierarchical clustering method is defined as a map $\mathcal{H} : \mathcal{N} \to \mathcal{D}$ from the space of networks $\mathcal{N}$ to the space of dendrograms $\mathcal{D}$, or, equivalently, as a map $\mathcal{H} : \mathcal{N} \to \mathcal{U}$ from $\mathcal{N}$ to the space $\mathcal{U}$ of networks with ultrametrics as dissimilarity functions. This loose definition permits the existence of an ample diversity of methods, many of them carrying limited practical utility. Thus, we recall an axiomatic construction formulated to determine a subclass of admissible clustering methods.

2.1 Admissible Hierarchical Clustering Methods

In [4, 5], the authors impose the two following requirements on clustering methods:

(A1) Axiom of Value

Consider a two-node network $N = (X, A_X)$ with $X = \{p, q\}$, $A_X(p, q) = \alpha$, and $A_X(q, p) = \beta$. The ultrametric $(X, u_X) = \mathcal{H}(N)$ output by $\mathcal{H}$ satisfies $u_X(p, q) = \max(\alpha, \beta)$.

(A2) Axiom of Transformation

Given networks $N_X = (X, A_X)$ and $N_Y = (Y, A_Y)$ and a dissimilarity reducing map $\phi : X \to Y$, that is a map ϕ such that for all $x, x' \in X$ it holds $A_X(x, x') \geq A_Y(\phi(x), \phi(x'))$, the output ultrametrics $(X, u_X) = \mathcal{H}(X, A_X)$ and $(Y, u_Y) = \mathcal{H}(Y, A_Y)$ satisfy $u_X(x, x') \geq u_Y(\phi(x), \phi(x'))$.

Axiom (A1) states that in a network with two nodes p and q, the dendrogram D_X has them merging at the maximum value of the two dissimilarities α and β. This is reasonable because at resolutions $\delta < \max(\alpha, \beta)$ one node can influence

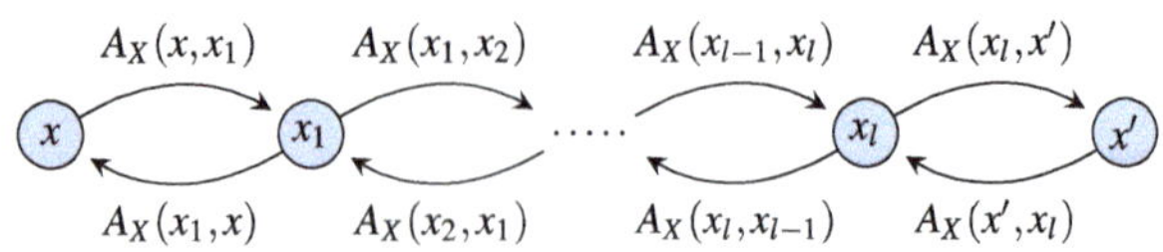

Fig. 1 Reciprocal clustering. x, x' cluster at resolution δ if they are joined by a bidirectional chain of max. dissimilarity δ [cf. (3)]

the other but not vice versa, which in most situations means that the nodes are not alike. Axiom (A2) states that a contraction of the dissimilarity function A_X entails a contraction of the ultrametric u_X.

A hierarchical clustering method $\mathscr{H}$ is *admissible* if it satisfies axioms (A1) and (A2). Two admissible methods of interest are reciprocal and nonreciprocal clustering. The *reciprocal* clustering method $\mathscr{H}^{\mathrm{R}}$ with output $(X, u_X^{\mathrm{R}}) = \mathscr{H}^{\mathrm{R}}(X, A_X)$ is the one for which the ultrametric $u_X^{\mathrm{R}}(x, x')$ between points x and x' is given by

$$u_X^{\mathrm{R}}(x, x') = \min_{C(x,x')} \max_{i | x_i \in C(x,x')} \bar{A}_X(x_i, x_{i+1}), \tag{3}$$

where $\bar{A}_X(x, x') = \max(A_X(x, x'), A_X(x', x))$ for all $x, x' \in X$. Intuitively, in (3) we search for chains $C(x, x')$ linking nodes x and x'. Then, for a given chain, walk from x to x' and determine the maximum dissimilarity, in either the forward or backward direction, across all links in the chain. The reciprocal ultrametric $u_X^{\mathrm{R}}(x, x')$ is the minimum of this value across all possible chains; see Fig. 1. We define the *nonreciprocal* clustering method $\mathscr{H}^{\mathrm{NR}}$ as one where the ultrametric between x and x' is given by the maximum of the directed minimum chain costs (1) in both directions

$$u_X^{\mathrm{NR}}(x, x') = \max\left(\tilde{u}_X^*(x, x'),\ \tilde{u}_X^*(x', x)\right). \tag{4}$$

In (4) we implicitly consider forward chains $C(x, x')$ going from x to x' and backward chains $C(x', x)$ from x' to x. We then determine the respective maximum dissimilarities and search independently for the forward and backward chains that minimize the respective maximum dissimilarities. The nonreciprocal ultrametric $u_X^{\mathrm{NR}}(x, x')$ is the maximum of these two minimum values; see Fig. 2. Reciprocal and nonreciprocal clustering bound the range of ultrametrics generated by any other admissible method $\mathscr{H}$.

Theorem 1 [4]) *Given any network* $N = (X, A_X)$ *and the associated reciprocal* u_X^{R} *and nonreciprocal* u_X^{NR} *ultrametrics as defined in* (3) *and* (4)*. Then, for any admissible method* $\mathscr{H}$*, the output ultrametric* $(X, u_X) = \mathscr{H}(X, A_X)$ *is such that* $u_X^{\mathrm{NR}}(x, x') \leq u_X(x, x') \leq u_X^{\mathrm{R}}(x, x')$ *for all* x, x'.

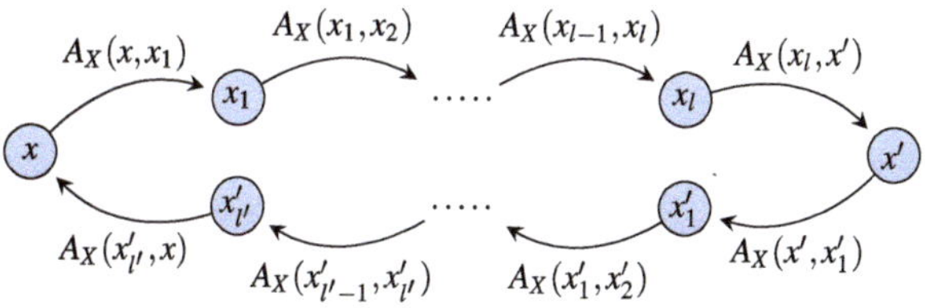

Fig. 2 Nonreciprocal clustering. x, x' cluster at resolution δ if they are joined in both directions by chains of max. dissimilarity δ [cf. (4)]

3 Scale Preservation

In this section we begin to explore additional conditions to winnow the space of admissible methods. One such condition is the concept of a scale preserving hierarchical clustering method.

(P1) Scale Preservation

Consider a network $N_X = (X, A_X)$ and a nondecreasing function $\psi : \mathbb{R}_+ \to \mathbb{R}_+$ with $\psi(0) = 0$, $\psi(z) > 0$ for all $z > 0$ and $\lim_{z\to\infty} \psi(z) = \infty$. Define the network $N_X^\psi = (X, \psi(A_X))$ with the same set of nodes and dissimilarities transformed by ψ. A hierarchical clustering method $\mathscr{H}$ is said to be *scale preserving* if for an arbitrary network $N_X = (X, A_X)$ and a function ψ satisfying the above requirements, the outputs $(X, u_X) = \mathscr{H}(N_X)$ and $(X, u_X^\psi) = \mathscr{H}(N_X^\psi)$ satisfy $u_X^\psi = \psi \circ u_X$.

Scale preservation entails two consequences of practical relevance. First, since ultrametric outcomes vary according to the same function that transforms the dissimilarities, scale preserving methods are invariant with respect to units. Second, in some applications we may be interested in clustering a network of which we are given a similarity function instead of dissimilarities. In such case, in order to apply a dissimilarity based method, we must apply a decreasing function to the similarity information to transform it into dissimilarities. Scale preservation ensures that the structure of the clustering hierarchy does not depend on the decreasing function used to turn similarities into dissimilarities. This property is helpful in many practical instances. In, e.g., trust networks it is more natural for subjects to express their trust of neighbors rather than their distrust. In social networks, proximity indicators—like number of exchanged messages—are more common than distance indicators.

4 Representability

We build upon the notion of representable methods—introduced for non-hierarchical clustering in [3]—to specify the hierarchical clustering of arbitrary networks through the clustering of particular examples that we call *representers*. To explain the concept of a representable method we first present an alternative definition for the reciprocal ultrametric (3). Start by considering a given asymmetric network $N = (X, A_X)$ and define $\mho_2 = (\{p, q\}, A_{p,q})$ as a two-node network with both dissimilarities equal to 1. Define the λ-multiple of the network $\mho_2$ as the network $\lambda * \mho_2 = (\{p, q\}, \lambda A_{p,q})$ whose underlying set is the same and its dissimilarities are linearly scaled by a given $\lambda > 0$. Further define the Lipschitz constant of a map $\phi : \{p, q\} \to X$ from $\mho_2$ to N as

$$L(\phi; \mho_2, N) = \max\left(A_X(\phi(p), \phi(q)), A_X(\phi(q), \phi(p))\right), \tag{5}$$

i.e., the maximum dissimilarity into which one of the unit dissimilarities in $\mho_2$ is mapped. For any nodes $x, x' \in X$ we define the optimal multiple $\lambda_X^{\mho_2}(x, x')$ between x and x' with respect to $\mho_2$ as

$$\lambda_X^{\mho_2}(x, x') = \{L(\phi; \mho_2, N) \mid \phi : \{p, q\} \to X,\ x, x' \in \operatorname{Im}(\phi)\}. \tag{6}$$

Notice that $\lambda_X^{\mho_2}(x, x')$ is the minimum multiple needed for the existence of a dissimilarity reducing map with x and x' in its image between a multiple of $\mho_2$ and N.

We define the representable clustering method $\mathcal{H}^{\mho_2}$ associated with the representer network $\mho_2$ as the method with ultrametric output $(X, u_X^{\mho_2}) = \mathcal{H}^{\mho_2}(N)$ given by

$$u_X^{\mho_2}(x, x') = \min_{C(x,x')} \max_{i | x_i \in C(x,x')} \lambda_X^{\mho_2}(x_i, x_{i+1}), \tag{7}$$

for all $x, x' \in X$. Definition (7) is illustrated in Fig. 3.

It is immediate that the method $\mathcal{H}^{\mho_2}$ with output ultrametrics as in (7) is equivalent to reciprocal clustering $\mathcal{H}^{\mathrm{R}}$ with output ultrametrics as in (3). Indeed, given a network $N = (X, A_X)$ and points $x, x' \in X$, there are only two possible maps from $\lambda * \mho_2$ to N containing nodes x and x' in their images. One map takes p to x and q to x' and the other reverses the images and takes p to x' and q to x. However, either maps have the same Lipschitz constant as defined in (5), ensuring that the optimal multiple in (6) is well defined. Consequently, we obtain that

$$\lambda_X^{\mho_2}(x, x') = \max\left(A_X(x, x'), A_X(x', x)\right) = \bar{A}_X(x, x'). \tag{8}$$

Comparing (7) with (3) and using the observation in (8) the equivalence $\mathcal{H}^{\mathrm{R}} \equiv \mathcal{H}^{\mho_2}$ follows, i.e., $\mathcal{H}^{\mathrm{R}}(N) = \mathcal{H}^{\mho_2}(N)$ for all networks N. Thus, we say that

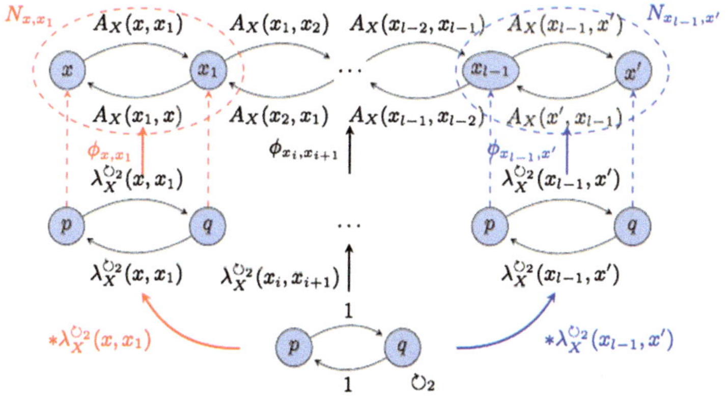

Fig. 3 Representable method $\mathscr{H}^{\circlearrowright_2}$ with ultrametric output as in (7). For every pair x_i, x_{i+1} of consecutive nodes in the chain $C(x, x')$ we multiply the network $\circlearrowright_2$ by the minimum multiple $\lambda_X^{\circlearrowright_2}(x_i, x_{i+1})$ that allows the existence of a dissimilarity reducing map $\phi_{x_i,x_{i+1}}$ containing nodes x_i and x_{i+1} in its image. The maximum among all these multiples determines the cost of the chain $C(x, x')$. The value $u_X^{\circlearrowright_2}(x, x')$ is the minimum chain cost over all possible chains linking x to x'

the reciprocal clustering method is represented by the network $\circlearrowright_2$. That is, if we consider $\circlearrowright_2$ as an interaction modality defining a basic clustering unit, the hierarchical clustering of a generic network follows from application of (7).

The definition in (7) is certainly more cumbersome than (3). However, the former can be generalized to cases in which we consider arbitrary representers in lieu of $\circlearrowright_2$ as we explain next.

4.1 Representable Hierarchical Clustering Methods

Generalizing $\mathscr{H}^{\circlearrowright_2}$ entails redefining the Lipschitz constant of a map and the optimal multiples so that they are calculated with respect to an arbitrary representer network $\omega = (X_\omega, A_\omega)$ instead of $\circlearrowright_2$. In representer networks ω, we allow the domain $\mathrm{dom}(A_\omega)$ of the dissimilarity function A_ω to be a proper subset of the product space, i.e., we may have $\mathrm{dom}(A_\omega) \neq X_\omega \times X_\omega$. This is a technical modification that allows representer networks to have some dissimilarities that can be interpreted as arbitrarily large. Generalizing (5), given an arbitrary network $N = (X, A_X)$ we define the *Lipschitz constant* of a map $\phi : X_\omega \to X$ from ω to N as

$$L(\phi; \omega, N) = \max_{\substack{(z,z') \in \mathrm{dom}(A_\omega) \\ z \neq z'}} \frac{A_X(\phi(z), \phi(z'))}{A_\omega(z, z')}. \quad (9)$$

Notice that $L(\phi; \omega, N)$ is the minimum multiple of the network ω such that the considered map ϕ is dissimilarity reducing from $L(\phi; \omega, N) * \omega$ to N. Further, observe that (9) reduces to (5) when $\omega = \mho_2$. Notice as well that the maximum in (9) is computed for pairs (z, z') in the domain of A_ω. Pairs not belonging to the domain could be mapped to any dissimilarity without modifying the value of the Lipschitz constant. Mimicking (6), for arbitrary nodes $x, x' \in X$ we define the optimal multiple $\lambda_X^\omega(x, x')$ between x and x' with respect to ω as

$$\lambda_X^\omega(x, x') = \min\{L(\phi; \omega, N) \mid \phi : X_\omega \to X,\ x, x' \in \mathrm{Im}(\phi)\}. \tag{10}$$

This means that $\lambda_X^\omega(x, x')$ is the minimum Lipschitz constant among those maps that have x and x' in its image. Observe that (10) reduces to (6) when $\omega = \mho_2$.

Representable methods are generalized to cases in which we are given a nonempty set Ω of representer networks ω. In such case we define the function λ_X^Ω by considering the infimum across all representers $\omega \in \Omega$,

$$\lambda_X^\Omega(x, x') = \inf_{\omega \in \Omega} \lambda_X^\omega(x, x'), \tag{11}$$

for all $x, x' \in X$. The value $\lambda_X^\Omega(x, x')$ is the infimum across all multiples $\lambda > 0$ such that there exists a dissimilarity reducing map from $\lambda * \omega$ to the network containing x and x' in its image for some representer $\omega \in \Omega$. For a given network $N = (X, A_X)$, the representable clustering method $\mathscr{H}^\Omega$ associated with the collection of representers Ω is the one with outputs $(X, u_X^\Omega) = \mathscr{H}^\Omega(X, A_X)$ such that the ultrametric u_X^Ω is given by

$$u_X^\Omega(x, x') = \min_{C(x,x')} \max_{i | x_i \in C(x,x')} \lambda_X^\Omega(x_i, x_{i+1}), \tag{12}$$

for all $x, x' \in X$. The definition in (12) is interpreted in Fig. 4.

We say that Ω is *uniformly bounded* if there exists a finite M such that for all $\omega = (X_\omega, A_\omega) \in \Omega$ we have that $\max_{(z,z') \in \mathrm{dom}(A_\omega)} A_\omega(z, z') \leq M$. We now formally define the notion of representability.

(P2) Representability

We say that a clustering method $\mathscr{H}$ is representable if there exists a uniformly bounded collection Ω of weakly connected representers each with finite number of nodes such that $\mathscr{H} \equiv \mathscr{H}^\Omega$ where $\mathscr{H}^\Omega$ has output ultrametrics as in (12). If the collection Ω is finite, we say that $\mathscr{H}$ is finitely representable.

For every collection of representers Ω satisfying the conditions in property (P2), (12) defines a valid ultrametric. Moreover, every representable clustering method abides by axiom (A2), as stated next.

Proposition 1 *Given a collection of representers Ω satisfying the conditions in (P2), the representable method $\mathscr{H}^\Omega$ is valid, i.e., u_X^Ω defined in* (12) *is an ultrametric for all networks $N = (X, A_X)$, and satisfies the Axiom of Transformation (A2)*

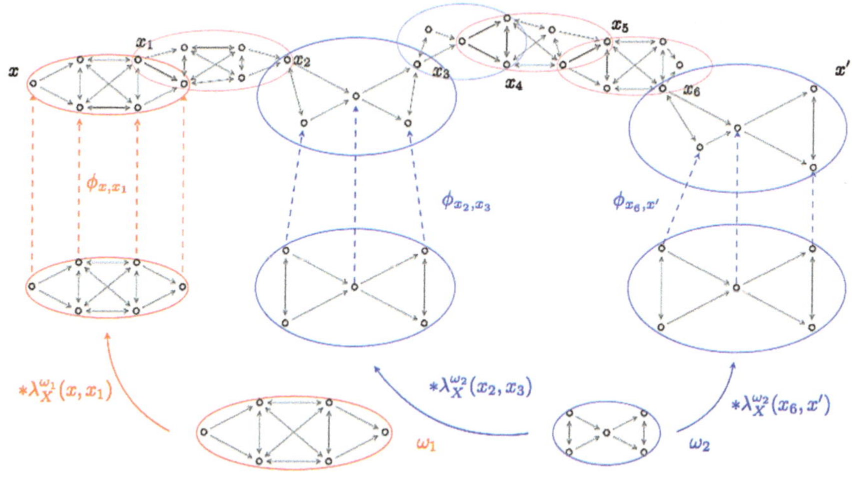

Fig. 4 A representable clustering method $\mathscr{H}^{\Omega}$. The collection of representers Ω is composed by two representers ω_1 and ω_2 shown at the *bottom* of the figure. In order to compute the ultrametric value $u_X^{\Omega}(x, x')$ we link x and x' through a chain, e.g., $[x, x_1, \ldots, x_6, x']$ in the figure, and link pairs of consecutive nodes with multiples of the representers. We depict these multiples for pairs (x, x_1), (x_2, x_3), and (x_6, x') and the corresponding dissimilarity reducing maps ϕ_{x,x_1}, ϕ_{x_2,x_3}, $\phi_{x_6,x'}$ from the multiple of the representers to the network, containing the corresponding pair of nodes in their images. The ultrametric value $u_X^{\Omega}(x, x')$ is given by minimizing over all chains joining x and x' the maximum multiple of a representer used to link consecutive nodes in the chain (12)

The condition in (P2) that a valid representable method is defined by a set of weakly connected [23] representers is necessary and sufficient. However, the condition in (P2) that Ω be uniformly bounded is sufficient but not necessary for $\mathscr{H}^{\Omega}$ to output a valid ultrametric. Although (A2) is guaranteed for every representable method, the Axiom of Value (A1) need not be satisfied. Thus, admissibility and representability are independent properties.

Remark 2 Representability is a mechanism for defining universal hierarchical clustering methods from given representative examples. Each representer $\omega \in \Omega$ can be interpreted as defining a particular structure that is to be considered a cluster unit. The scaling of this unit structure [cf. (10), (11)] and its replication through the network [cf. (12)] indicate the resolution at which nodes become part of a cluster. The interest in representability is that it is easier to state desirable clustering structures for particular networks rather than for arbitrary ones. We refer the reader to Sect. 4.3 for particular examples of representer networks that give rise to intuitively appealing clustering methods.

4.2 Representability, Scale Preservation, and Admissibility

Are all representable clustering methods relevant in practice? To answer this question we seek to characterize methods that satisfy some desired properties that we deem reasonable. In particular, we consider methods that are admissible with respect to the axioms of value and transformation (A1) and (A2) as well as scale preserving in the sense of (P1).

In characterizing admissible, representable, and scale preserving methods, the concept of structure representer appears naturally. We say that a representer $\omega = (X_\omega, A_\omega)$ is a *structure representer* if and only if $|X_\omega| \geq 2$ and

$$A_\omega(z, z') = 1, \text{ for all } z \neq z' \text{ s.t. } (z, z') \in \text{dom}(A_\omega). \tag{13}$$

The requirement in (13) implies that structure representers define the relationships that are necessary in a cluster unit but do not distinguish between different levels of influence. In the following theorem we claim that admissible, representable, and scale preserving hierarchical clustering methods are those represented by a collection Ω of strongly connected, structure representers.

Theorem 2 *A representable clustering method $\mathcal{H}$ satisfies axioms (A1)–(A2) and scale preservation (P1) if and only if $\mathcal{H} \equiv \mathcal{H}^{\Omega}$ where Ω is a collection of strongly connected, structure representers as defined by the condition in* (13).

Recalling the interpretation of representability as the extension of clustering defined for particular cases, Theorem 2 entails that the definitions of particular cases cannot present dissimilarity degrees if we require scale preservation. That is, the dissimilarity between every pair of distinct nodes in the representers must be either 1 or undefined. The edges with value 1 imply that the corresponding influence relations are required for the formation of a cluster whereas the influence relations associated with undefined edges are not required. Conversely, Theorem 2 states that encoding different degrees of required influence for different pairs of nodes within the representers is impossible if we want the resulting clustering method to be scale preserving.

4.3 Cyclic Clustering Methods

Let $\circlearrowright_t = (\{1, \ldots, t\}, A_t)$ denote a cycle network with t nodes such that the domain of the dissimilarity function $\text{dom}(A_t) = \{(i, i+1)\}_{i=1}^{t-1} \cup (t, 1)$ and every defined dissimilarity is equal to 1. In this section we study representable methods where the representer collections contain cycle networks. We first note that the method defined by a representer collection that contains a finite number of cycle networks is equivalent to the method represented by the longest cycle in the collection.

Proposition 2 *Given a finite collection $\Omega = \{\circlearrowleft_t | t \in \mathcal{T}\}$ of cyclic representers, we have that $\mathcal{H}^{\Omega} \equiv \mathcal{H}^{\circlearrowleft_{t_{\max}}}$, where $t_{\max} = \max \mathcal{T}$.*

The method $\mathcal{H}^{\circlearrowleft_t}$ is referred to as the tth cyclic method. Cyclic methods $\mathcal{H}^{\circlearrowleft_t}$ for all $t \geq 2$ are admissible and scale preserving as stated in the following corollary of Theorem 2.

Corollary 1 *Cyclic methods $\mathcal{H}^{\circlearrowleft_t}$ satisfy axioms (A1)–(A2) and the scale preservation property (P1).*

The corollary follows from the fact that networks $\circlearrowleft_t$ are strongly connected and structure representers. The second cyclic method $\mathcal{H}^{\circlearrowleft_2}$ was used to introduce the concept of representable clustering in (5)–(7) and shown to coincide with the reciprocal clustering method $\mathcal{H}^{\mathrm{R}}$ in (8). Interpreting $\circlearrowleft_2$ as a basic cluster unit we can then think of reciprocal clustering $\mathcal{H}^{\mathrm{R}} \equiv \mathcal{H}^{\circlearrowleft_2}$ as a method that allows propagation of influence through cycles that contain at most two nodes. Likewise, the method $\mathcal{H}^{\circlearrowleft_3}$ can be interpreted as a method that allows propagation of influence through cycles that contain at most three nodes, and so on.

As we increase t, the output ultrametrics of methods $\mathcal{H}^{\circlearrowleft_t}$ become smaller, in particular smaller than those output by $\mathcal{H}^{\circlearrowleft_2} \equiv \mathcal{H}^{\mathrm{R}}$. This is consistent with Theorem 1 and is indicative of the status of these methods as relaxations of the condition of direct mutual influence. As we increase the length of the cycles, the question arises of whether we recover nonreciprocal clustering. This is not true for any $\circlearrowleft_t$ where t is finite. However, if we define $\mathcal{C}_\infty = \{\circlearrowleft_t\}_{t=1}^{\infty}$ the following result holds.

Proposition 3 *The clustering method $\mathcal{H}^{\mathcal{C}_\infty}$ represented by the family of all cycle networks $\mathcal{C}_\infty$ is equivalent to the nonreciprocal clustering method $\mathcal{H}^{\mathrm{NR}}$ with output ultrametrics as defined in* (4).

Combining the results in Propositions 2 and 3, it follows that by considering methods $\mathcal{H}^{\circlearrowleft_t}$ for finite t and method $\mathcal{H}^{\mathcal{C}_\infty}$ we are considering every method that can be represented by a countable collection of cyclic representers. The reformulation in Proposition 3 expresses nonreciprocal clustering through the consideration of particular cases, namely cycles of arbitrary length. This not only uncovers a drawback of nonreciprocal clustering—propagating influence through cycles of arbitrary length is perhaps unrealistic—but also offers alternative formulations that mitigate this limitation—restrict the propagation of influence to cycles of certain length. In that sense, cyclic methods of length t can be interpreted as a tightening of nonreciprocal clustering. This interpretation is complementary of their interpretation as relaxations of reciprocal clustering that we discussed above. Given this dual interpretation, cyclic clustering methods are of practical importance.

5 Conclusion

The notion of representability was introduced as the possibility of specifying a hierarchical clustering method through its action on a collection of representers. Moreover, the characteristics needed on the representers to obtain an admissible and scale preserving method were detailed. We then focused our attention on cyclic methods, a particular family within the representable methods.

References

1. Boyd, J.: Asymmetric clusters of internal migration regions of France. IEEE Trans. Syst. Man Cybern. **10**(2), 101–104 (1980)
2. Carlsson, G., Mémoli, F.: Characterization, stability and convergence of hierarchical clustering methods. J. Mach. Learn. Res. **11**, 1425–1470 (2010)
3. Carlsson, G., Mémoli, F.: Classifying clustering schemes. Found. Comput. Math. **13**(2), 221–252 (2013)
4. Carlsson, G., Memoli, F., Ribeiro, A., Segarra, S.: Axiomatic construction of hierarchical clustering in asymmetric networks. In: IEEE International Conference on Acoustics, Speech and Signal Process (ICASSP), pp. 5219–5223 (2013)
5. Carlsson, G., Memoli, F., Ribeiro, A., Segarra, S.: Axiomatic construction of hierarchical clustering in asymmetric networks (2014). arXiv:1301.7724v2
6. Guyon, I., von Luxburg, U., Williamson, R.: Clustering: science or art? Tech. rep. Paper presented at the NIPS 2009 Workshop Clustering: Science or Art? (2009)
7. Hubert, L.: Min and max hierarchical clustering using asymmetric similarity measures. Psychometrika **38**(1), 63–72 (1973)
8. Jain, A.K., Dubes, R.C.: Algorithms for Clustering Data. Prentice-Hall, Upper Saddle River, NJ (1988)
9. Jardine, N., Sibson, R.: Mathematical Taxonomy. Wiley Series in Probability and Mathematical Statistics. Wiley, London (1971)
10. Kleinberg, J.M.: An impossibility theorem for clustering. In: NIPS, pp. 446–453 (2002)
11. Lance, G.N., Williams, W.T.: A general theory of classificatory sorting strategies 1. Hierarchical systems. Comput. J. **9**(4), 373–380 (1967)
12. Meila, M.: Comparing clusterings: an axiomatic view. In: Proceedings of the 22nd International Conference on Machine Learning, pp. 577–584. ACM, New York (2005)
13. Meila, M.: Comparing clusterings – an information based distance. J. Multivar. Anal. **98**(5), 873–895 (2007)
14. Murtagh, F.: Multidimensional clustering algorithms. In: Compstat Lectures, vol. 1. Physika Verlag, Vienna (1985)
15. Pentney, W., Meila, M.: Spectral clustering of biological sequence data. In: Proceedings of National Conference on Artificial Intelligence (2005)
16. Punj, G., Stewart, D.W.: Cluster analysis in marketing research: review and suggestions for application. J. Market. Res. **20**(2), 134–148 (1983)
17. Rui, X., Wunsch-II, D.: Survey of clustering algorithms. IEEE Trans. Neural Netw. **16**(3), 645–678 (2005)
18. Saito, T., Yadohisa, H.: Data Analysis of Asymmetric Structures: Advanced Approaches in Computational Statistics. CRC Press, Boca Raton, FL (2004)
19. Slater, P.: Hierarchical internal migration regions of France. IEEE Trans. Syst. Man Cybern. **6**(4), 321–324 (1976)

20. Tarjan, R.E.: An improved algorithm for hierarchical clustering using strong components. Inf. Process. Lett. **17**(1), 37–41 (1983)
21. Van Laarhoven, T., Marchiori, E.: Axioms for graph clustering quality functions. J. Mach. Learn. Res. **15**(1), 193–215 (2014)
22. Walsh, D., Rybicki, L.: Symptom clustering in advanced cancer. Support. Care Cancer **14**(8), 831–836 (2006)
23. West, D.B., et al.: Introduction to Graph Theory, vol. 2. Prentice Hall, Upper Saddle River (2001)
24. Zadeh, R., Ben-David, S.: A uniqueness theorem for clustering. In: Proceedings of Uncertainty in Artificial Intelligence (2009)

A Median-Based Consensus Rule for Distance Exponent Selection in the Framework of Intelligent and Weighted Minkowski Clustering

Renato Cordeiro de Amorim, Nadia Tahiri, Boris Mirkin, and Vladimir Makarenkov

Abstract The intelligent Minkowski and weighted Minkowski K-means are recently developed effective clustering algorithms capable of computing feature weights. Their cluster-specific weights follow the intuitive idea that a feature with a low dispersion in a specific cluster should have a greater weight in this cluster than a feature with a high dispersion. The final clustering provided by these techniques obviously depends on the selection of the Minkowski exponent. The median-based central consensus rule we introduce in this paper allows one to select an optimal value of the Minkowski exponent. Our rule takes into account the values of the Adjusted Rand Index (ARI) between clustering solutions obtained for different Minkowski exponents and selects the clustering that provides the highest average value of ARI. Our simulations, carried out with real and synthetic data, show that the proposed median-based consensus procedure usually outperforms clustering strategies based on the selection of the highest value of the Silhouette or Calinski–Harabasz cluster validity indices.

R.C. de Amorim
School of Computer Science, University of Hertfordshire, College Lane, Hatfield AL10 9AB, UK
e-mail: r.amorim@herts.ac.uk

N. Tahiri • V. Makarenkov (✉)
Département d'informatique, Université du Québec à Montréal, C.P. 8888 succ. Centre-Ville, Montreal, QC, Canada H3C 3P8
e-mail: tahiri.nadia@courrier.uqam.ca; makarenkov.vladimir@uqam.ca

B. Mirkin
Department of Data Analysis and Machine Intelligence, National Research University, Higher School of Economics, Moscow, Russia

Department of Computer Science and Information Systems, Birkbeck University of London, Malet Street, London WC1E 7HX, UK
e-mail: bmirkin@hse.ru

F. Palumbo et al. (eds.), *Data Science*, Studies in Classification, Data Analysis, and Knowledge Organization, DOI 10.1007/978-3-319-55723-6_8

1 Introduction

Clustering algorithms are aiming at revealing the structure of a given dataset Y by partitioning it into a set of K clusters $S = \{S_1, S_2, \ldots, S_K\}$. Each cluster $S_k \in S$ should contain homogeneous entities, according to a given similarity measure. Clustering algorithms have been used to address the most diverse problems, including those in the fields of bioinformatics, business, computer vision, data mining, and security [9, 14, 17, 24].

Over the years many clustering algorithms have been developed, thanks to the considerable effort put by the scientific community. These algorithms can be easily categorized by the clustering approach: hierarchical or partitioning. Algorithms following the former principle produce a tree-like relationship between the entities, which can be visualized using a dendrogram [18, 19]. Partitioning algorithms produce clustering solutions in which each entity is assigned either to a single cluster, i.e., in the case of crisp clustering, or to multiple clusters, i.e., in the case of overlapping clustering [17].

The K-means algorithm [2, 13] is one of the most popular clustering methods. Its implementations can be found in various software frequently used in data analysis, such as MATLAB [15], R [22], SPSS [6], and SciPy [11]. Given a dataset Y composed of N entities y_i, each described over the same set of V features, K-means produces a set of disjoint clusters $S = \{S_1, S_2, \ldots, S_K\}$. K-means also provides a set of centroids $C = \{c_1, c_2, \ldots, c_K\}$ of these clusters by iteratively minimizing the following objective function:

$$W(S, C) = \sum_{k=1}^{K} \sum_{i \in S_k} \sum_{v=1}^{V} (y_{iv} - c_{kv})^2. \quad (1)$$

Despite its popularity, the K-means algorithm has several important weaknesses that are as follows: (1) it requires the number of clusters, K, to be known beforehand; (2) the final clustering, S, is heavily dependent of the starting random partition; (3) the provided solution is usually a local minimum; (4) K-means treats all features equally, regardless of their degree of relevance.

Here, we are particularly interested in issue (4). We have addressed this issue in our recent work, where we have introduced the Minkowski weighted K-means (MWK-means) algorithm [5] (more details are given in Sect. 2). However, our algorithm requires the selection of a suitable Minkowski exponent p. This selection is subjective to the data structure of Y, making it impossible to suggest a value of p that will produce good results in terms of cluster recovery in all cases. In this paper, we present a median-based central consensus rule allowing one to select a suitable value of the Minkowski exponent p in the framework of the intelligent Minkowski and weighted Minkowski clustering [5]. Our new technique will be tested empirically on real-world and synthetic datasets with different numbers of clusters and features.

2 Minkowski Weighted K-Means

We have recently introduced the MWK-means algorithm [5], which allows one to calculate optimal cluster-based feature weights. This calculation follows the intuitive idea that the weight of feature v in cluster k, denoted here as w_{kv}, can vary from one cluster to another. In fact, the more a feature is dispersed in a cluster, the lower will be its weight in this cluster. The Minkowski distance between entity y_i and centroid c_k is given by $\sqrt[p]{\sum_{v=1}^{V} |y_{iv} - c_{kv}|^p}$.

We added to this distance the feature weights and used its pth power. The latter is analogous to the use of the squared Euclidean distance instead of the Euclidean distance:

$$d_p(y_i, c_k) = \sum_{v=1}^{V} w_{kv}^p |y_{iv} - c_{kv}|^p. \tag{2}$$

This weighted distance measure leads to the MWK-means objective function to be minimized:

$$W(S, C, w) = \sum_{k=1}^{K} \sum_{i \in S_k} \sum_{v=1}^{V} w_{kv}^p |y_{iv} - c_{kv}|^p, \tag{3}$$

subject to $\sum_{v=1}^{V} w_{kv} = 1$ and $w_{kv} \geq 0$ for $k = 1, 2, \ldots, K$ and $v = 1, 2, \ldots, V$. The computation is carried out in the framework of crisp clustering in which every entity y_i is assigned to a single cluster S_k.

The criterion (3) has some interesting implications. As both the distances and the weights use the same exponent p, one can see the feature weights as feature re-scaling factors. This interpretation is not possible in other feature weighting algorithms, such as Weighted K-means [7], Attribute weighting K-means [4], and Improved K-prototypes [10]. The criterion (3) supports cluster-dependent feature weights, and perhaps most importantly, a p-dependent cluster shape bias.

Any distance measure introduces a bias to the shapes of the clusters. Assuming a two-dimensional space for easier visualization, the squared Euclidean distance used in (1) makes K-means biased towards circular clusters. At values of p equal to one, two, and tending to infinity, the distance (2) is equivalent to the weighted versions of the Manhattan, squared Euclidean, and Tchebychev distances, respectively. For instance, a value of p between one and two leads to a bias towards a shape situated between a rhombus and a circle.

Within MWK-means, each weight w_{kv} is inversely proportional to the dispersion of v in the cluster S_k, which is given by $D_{kv} = \sum_{i \in S_k} |y_{iv} - c_{kv}|^p$. Thus, the feature weight is computed as follows:

$$w_{kv} = \frac{1}{\sum_{u \in V} [D_{kv}/D_{ku}]^{1/(p-1)}}. \tag{4}$$

This computation of weights follows the principle that feature v having the lowest dispersion in cluster S_k should have the highest weight in this cluster. The Minkowski weighted K-means iteratively minimizes the criterion (3) following the steps below:

1. *Initial settings.* Choose the number of clusters, K, and the Minkowski exponent, p. Set $S \leftarrow \emptyset$, and $w_{kv} = 1/V$ for $k = 1, 2, \ldots, K$ and $v = 1, 2, \ldots, V$.
2. *Centroids setting.* Assign the values of K entities selected at random from Y to be the initial cluster centroids $c_1, c_2, \ldots, c_K$.
3. *Clusters update.* Assign each entity $y_i \in Y$ to the cluster S_k represented by the nearest c_k as per (2), generating the clustering $S' = \{S'_1, S'_2, \ldots, S'_K\}$. If $S' = S$, then go to Step 6 (i.e., computation finishes).
4. *Centroids update.* Update each $c_k \in C$ to the component-wise Minkowski center of $y_i \in S_k$.
5. *Weights update.* Update each feature weight w_{kv} using Eq. (4). Set $S \leftarrow S'$, then go to Step 3.
6. *Final step.* Output the clustering $S = \{S_1, S_2, \ldots, S_K\}$, centroids $C = \{c_1, c_2, \ldots, c_K\}$, and feature weights w.

As the K-means algorithm, the MWK-means algorithm is a heuristic method that cannot guarantee to provide an optimal clustering solution. When using K-means, this issue is often addressed by running the algorithm thousands of times and selecting the clustering S that provides the optimal value of the objective function (1). This strategy can be followed within MWK-means for a given value of p. However, it cannot be used for finding an optimal value of p within MWK-means because the values of the objective function (3) are not comparable at different values of p.

To address the problem of the solution dependence from multiple random starts of MWK-means, we proposed the intelligent Minkowski weighted K-means (iMWK-means) algorithm [5], applying the concept of anomalous patterns to provide good initial centroids and weights to MWK-means. It is worth noting that the concept of anomalous patterns was originally introduced by Mirkin in the framework of the intelligent K-means algorithm [17]. Below, we present the main steps of this algorithm, including our modifications to allow the use of the Minkowski distance:

1. *Initial settings.* Choose a value for the threshold θ and the Minkowski exponent p. Set c_c to the component-wise Minkowski center of Y. Set $C_t \leftarrow \emptyset$, and $w_{kv} = 1/V$ for $k = 1, 2, \ldots, K$ and $v = 1, 2, \ldots, V$.
2. *Define centroid.* Set a tentative centroid, c_t, as the entity farthest away from c_c according to the distance measure defined by (2).
3. *Clustering procedure.* Carry out the MWK-means algorithm using two centroids, c_t and c_c, in order to generate the clustering $S = \{S_t, S_c\}$ (without allowing c_c to change during the centroids update step).
4. *Remove small clusters.* If $|S_t| \geq \theta$, then add c_t to C_t and the corresponding weights to w. In all the cases, remove S_t from the dataset.

5. *Check entities to be clustered.* If some entities to be clustered remain, then go to Step 2.
6. *Final step.* Carry out MWK-means, initialized with the centroids from C_t and the weights from w.

The intelligent K-means algorithm, which serves as our foundation for iMWK-means, can be used for both estimating the correct number of clusters, K, in a dataset and providing good initial centroids. In this study, as in all our previous experiments [5], we were not particularly interested in estimating K. In our simulations, we assumed that the correct number of clusters was known.

3 The Median-Based Central Consensus Rule

The final clustering generated by MWK-means and iMWK-means depends on the value of the Minkowski exponent p, which is also the exponent of feature weights w_{kv}, and defines the shape towards which the clustering will be biased. With this in mind, we devised a median-based central consensus rule that allows us to select a good value of this important clustering parameter.

Selecting the best clustering from a set of clustering solutions is indeed a problem that has attracted considerable attention [17]. A common approach is to apply a cluster validity index (CVI) to each of the available solutions, hoping this index is capable of sensing the structure of the data, and then output the clustering that better optimizes the selected CVI. There are indeed many cluster validity indices (for a recent and comprehensive review see [1]), but no sole CVI is clearly superior to all the others in all cases. However, the Silhouette width (SW) [23] and the Calinski–Harabasz (CH) [3] index tend to be among the top performers [1, 16, 21].

Our median-based central consensus rule can be used with the algorithms that should be carried out only once, like iMWK-means, as well as with those that are based on multiple random initializations, like MWK-means. In the latter case we also employ a CVI, here using either SW or CH. In both cases, we first need to select a range of values of p for our search. We decided to use the values of p from the range 1.0–5.0 with the step of 0.1. We believe that there is no need in using large values of p, as in our previous comprehensive simulations [5] the best partitions have very rarely appeared at the values of p greater than 5.

The median-based central consensus rule for MWK-means is defined as follows:

1. For each value of p, carry out MWK-means 100 times saving only the run with the optimal CVI value (i.e., the highest value of SW or CH in our case). This generates 41 clustering solutions, one for each value of the Minkowski exponent p (i.e., the values of p vary from 1.0 to 5.0 with the step of 0.1).
2. Calculate the Adjusted Rand Index (ARI) [8] between each pair of optimal clustering solutions found in Step 1.

3. Output the clustering solution with the highest average ARI value over all optimal clustering solutions.

In the case of the iMWK-means algorithm, which does not rely on a random start strategy, there is no need in applying a CVI. Here, a given value of p will provide a single clustering solution.

The median-based central consensus rule for iMWK-means is defined as follows:

1. For each value of p, carry out iMWK-means. This generates 41 clustering solutions, one for each value of the Minkowski exponent p (i.e., the values of p vary from 1.0 to 5.0 with the step of 0.1).
2. Calculate the ARI between each pair of clustering solutions found in Step 1 (here each p generates one clustering).
3. Output the clustering solution with the highest average ARI value over all available clustering solutions.

As in [5], in all our experiments we first considered clustering solutions that generated the expected number of clusters, discarding all the others. When no such correct solutions were provided, we accepted the solutions generated by our algorithms regardless of the number of clusters.

4 Setting of the Experiments

In this study, we carried out simulations with real-world benchmark datasets as well as with synthetic data, with and without noise features in both cases. We have examined 12 real-world datasets; 11 of them come from the popular UCI machine learning repository [12] and one from the study of Pal and Majumder [20].

Our synthetic data were composed of spherical Gaussian clusters so that the covariance matrices were diagonal, with the same diagonal value σ^2 generated at each cluster and varying between 0.5 and 1.5 (see also [5]). All centroid components were generated independently using the standard normal distribution. Cluster cardinalities were generated using a uniform distribution, with a constraint that each generated cluster should have included at least 20 entities. In this study, we tested six GMMs configurations that were different in terms of the number of features and clusters. They were as follows: 1000 entities over four features, partitioned into two clusters (1000×4-2); 1000 entities over six features, partitioned into three clusters (1000×6-3); 1000 entities over eight features, partitioned into four clusters (1000×8-4); 1000 entities over ten features, partitioned into five clusters (1000×10-5); 1000 entities over 20 features, partitioned into six clusters (1000×20-6); 1000 entities over 40 features, partitioned into eight clusters (1000×40-8). We generated 100 different datasets for each of these configurations.

Moreover, we also analyzed datasets containing noise features. Here, a noise feature is one composed entirely of uniformly random values. In these experiments,

50% of the number of original features were added to the original datasets. Thus, if an original dataset had eight features, its "noisy" version had a total of 12 features (eight original and four composed of noise).

We standardized each dataset by its range using the following formula:

$$y_{iv} = \frac{y_{iv} - \bar{y_v}}{max(y_v) - min(y_v)}. \tag{5}$$

Often clustering experiments are run with datasets standardized using the popular z-score method. We believe that the above-presented equation is a good alternative normalization option. Imagine, for instance, a dataset with two features, a unimodal feature v_1 and a multimodal feature v_2. The standard deviation of v_2 will be higher than that of v_1, leading to lower z-score values of v_2 in comparison with v_1. This means that the feature v_1 will have a higher contribution to the clustering despite the fact that the feature v_2 has a clearer cluster structure.

5 Results

In this section, we compare the results provided by the introduced median-based consensus MWK-means and iMWK-means algorithms to those given by the traditional K-means, MWK-means, and iMWK-means approaches. All the tables in this section present the ARI of the generated clusterings in relation to the known ground labels. Our tables consist of two major columns. Under "Original," we present the results provided by the traditional algorithms by simply applying a CVI (SW or CH) to all generated clustering solutions, and then selecting the solution corresponding to the highest CVI value. The ARI between this solution and the known ground truth is then reported. Under "Consensus," we present the ARI results obtained after carrying out the central consensus versions of MWK-means and iMWK-means algorithms. In this case, the results of iMWK-means algorithm are not presented for SW and CH indices because our iMWK-means consensus strategy is not based on multiple random starts of the algorithm (see Sect. 3).

Table 1 reports the ARI results for 12 real-world datasets without noise features. Here, our consensus method applied in the framework of MWK-means, using the Silhouette index as CVI, was equal or outperformed the original MWK-means for 10 of the 12 considered datasets. For the two datasets where our method was unable to perform equally or better, i.e., Vehicle Silhouette and Wine data, the ARI difference was very small (0.011 and 0.015, respectively). In the same table, the experiments with MWK-means applied with the CH index provided equal or better results for the consensus strategy for 11 of the 12 datasets. The Wine dataset was the only one for which the original MWK-means strategy provided a higher ARI value. However, the difference in the ARI values in this case was 0.03, only. The results generated by our iMWK-means consensus strategy were rather mixed. Here, the consensus algorithm

Table 1 ARI results for 12 real-world datasets without noise features

	Original						Consensus		
	KM		MWK		iMWK		MWK		
	SW	CH	SW	CH	SW	CH	SW	CH	iMWK
AustraCC	0.499	0.499	0.001	0.504	0.001	0.504	0.504	0.504	0.409
BreastOriginal	0.839	0.839	0.797	0.828	0.823	0.828	0.828	0.855	0.839
Ecoli	0.693	0.454	0.707	0.424	0.004	0.038	0.760	0.564	0.038
Glass	0.560	0.483	0.274	0.507	0.306	0.306	0.292	0.557	0.231
Hepatitis	0.190	0.141	0.396	0.122	0.138	0.149	0.396	0.122	0.149
Ionosphere	0.168	0.168	0.004	0.173	0.209	0.168	0.004	0.173	0.209
Iris	0.716	0.716	0.716	0.716	0.802	0.802	0.745	0.745	0.886
PimaIndians	0.011	0.102	0.008	0.096	0.082	0.094	0.100	0.100	0.077
TuluguVowels	0.534	0.377	0.467	0.356	0.428	0.496	0.467	0.358	0.496
VehicleSil	0.078	0.076	0.085	0.081	0.074	0.075	0.074	0.083	0.086
Wine	0.868	0.868	0.850	0.867	0.819	0.819	0.835	0.837	0.772
Yeast	0.168	0.157	0.172	0.168	0.003	0.012	0.182	0.175	0.003

Results under "Original" are the ARIs calculated between the clustering solutions obtained using the original partitioning algorithms, based on the highest CVI approach, and the known ground truth labels. Results under "Consensus" are the ARIs calculated between the clustering solutions obtained using the consensus versions of partitioning algorithms and the known ground truth labels

Table 2 ARI results for 12 real-world datasets with 50% of noise features added to the data

	Original						Consensus		
	KM		MWK		iMWK		MWK		
	SW	CH	SW	CH	SW	CH	SW	CH	iMWK
AustraCC	0.499	0.499	0.001	0.504	0.504	0.216	0.504	0.504	0.504
BreastOriginal	0.839	0.839	0.855	0.872	0.855	0.872	0.866	0.861	0.866
Ecoli	0.626	0.394	0.737	0.444	0.004	0.038	0.737	0.403	0.038
Glass	0.304	0.408	0.250	0.491	0.303	0.303	0.292	0.555	0.217
Hepatitis	0.243	0.122	0.036	0.122	0.396	0.150	0.407	0.122	0.113
Ionosphere	0.168	0.168	0.004	0.173	0.209	0.096	0.004	0.183	0.209
Iris	0.730	0.730	0.445	0.730	0.757	0.851	0.445	0.730	0.851
PimaIndians	0.011	0.103	0.002	0.100	0.100	0.087	0.099	0.100	0.101
TuluguVowels	0.402	0.388	0.417	0.402	0.478	0.488	0.439	0.403	0.479
VehicleSil	0.079	0.075	0.088	0.080	0.075	0.074	0.073	0.082	0.088
Wine	0.869	0.847	0.867	0.819	0.833	0.833	0.867	0.867	0.785
Yeast	0.069	0.069	0.109	0.174	0.003	0.012	0.103	0.099	0.012

Table 1 column description applies here

was able to provide equal or better results than the SW or CH-based original iMWK-means strategies for 7 of the 12 datasets.

Table 2 presents the results of our experiments with benchmark data to which we added 50% of noise features. Again, we measured the cluster recovery in terms of ARI in relation to the known ground truth. Our consensus strategy carried out

Table 3 Average ARI results provided by the K-means, MWK-means, and iMWK-means algorithms for our synthetic data composed of spherical Gaussian clusters without noise features

	Original						Consensus		
	KM		MWK		iMWK		MWK		
	SW	CH	SW	CH	SW	CH	SW	CH	iMWK
1000×									
4-2	0.470/0.33	0.458/0.33	0.472/0.32	0.457/0.33	0.426/0.32	0.423/0.33	0.463/0.32	0.454/0.32	0.434/0.31
6-3	0.586/0.23	0.552/0.23	0.620/0.21	0.568/0.23	0.527/0.23	0.537/0.22	0.613/0.21	0.561/0.22	0.549/0.23
8-4	0.649/0.19	0.596/0.20	0.675/0.19	0.607/0.20	0.537/0.17	0.559/0.20	0.665/0.19	0.606/0.20	0.605/0.21
10-5	0.685/0.18	0.650/0.19	0.712/0.16	0.660/0.18	0.594/0.17	0.575/0.15	0.706/0.17	0.664/0.18	0.625/0.18
20-6	0.933/0.09	0.913/0.12	0.930/0.07	0.926/0.11	0.740/0.16	0.767/0.17	0.933/0.08	0.934/0.08	0.883/0.16
40-8	0.992/0.01	0.980/0.07	0.988/0.02	0.995/0.02	0.780/0.16	0.811/0.15	0.996/0.01	0.994/0.02	0.977/0.03

The standard deviation of each ARI is indicated after a slash. Table 1 column description applies here

Table 4 Average ARI results provided by the K-means, MWK-means, and iMWK-means algorithms for our synthetic data composed of spherical Gaussian clusters with 50% of added noise features

	Original						Consensus		
	KM		MWK		iMWK		MWK		
	SW	CH	SW	CH	SW	CH	SW	CH	iMWK
1000×									
4-2	0.001/0.00	0.001/0.00	0.001/0.00	0.000/0.00	0.150/0.31	0.141/0.31	0.000/0.00	0.000/0.00	0.067/0.23
6-3	0.050/0.11	0.010/0.06	0.117/0.20	0.010/0.06	0.142/0.20	0.176/0.24	0.100/0.20	0.010/0.06	0.180/0.26
8-4	0.053/0.13	0.053/0.12	0.117/0.20	0.072/0.15	0.320/0.26	0.283/0.24	0.152/0.21	0.105/0.17	0.371/0.26
10-5	0.078/0.11	0.069/0.10	0.246/0.25	0.114/0.15	0.379/0.23	0.375/0.22	0.331/0.25	0.183/0.17	0.432/0.25
20-6	0.445/0.25	0.424/0.22	0.865/0.14	0.729/0.19	0.793/0.17	0.790/0.17	0.893/0.13	0.914/0.12	0.825/0.15
40-8	0.859/0.18	0.789/0.19	0.983/0.03	0.993/0.01	0.859/0.09	0.862/0.09	0.970/0.04	0.981/0.03	0.966/0.15

The standard deviation of each ARI is indicated after a slash. Table 1 column description applies here in the framework of MWK-means, using the Silhouette index as CVI, was equal or outperformed the original MWK-means for 10 of the 12 real datasets affected by noise. It is worth noting that the ARI differences for the two remaining datasets were really minor, i.e., 0.015 for Vehicle Silhouette and 0.003 for Yeast. On the other hand, the experiments with MWK-means used with the CH index provided equal or better results for the consensus strategy for 9 of the 12 datasets. The results generated by the iMWK-means consensus algorithm were again mixed. Here, the consensus technique was able to provide equal or better results than the SW or CH-based original iMWK-means algorithms for 7 of the 12 datasets.

Table 3 reports the results of our simulations with synthetic data composed of spherical Gaussian clusters without noise features. The presented results are the averages taken over 100 datasets for each considered parameter combination. The standard deviations of the obtained ARIs are also reported. Here, our median-based consensus strategy used in the framework of MWK-means algorithm with either SW or CH validity indices provided no improvement compared to the original MWK-means algorithm. The corresponding differences between the average ARI results are often negligible, being most of the times below 0.01. However, the application of our central consensus strategy in the framework of iMWK-means allowed us to improve the results of the original iMWK-means used with either SW or CH for all the six parameter configurations considered in our simulations. The greater the number of features and clusters, the greater this improvement was. For instance, under the configuration 1000×40-8 the improvement in the ARI results provided by consensus iMWK-means was 0.196 in the case of SW and 0.165 in the case of CH.

Finally, Table 4 illustrates the results of our experiments with synthetic data with 50% of added noise features. Here, all of the presented original and consensus MWK and iMWK strategies clearly outperformed the traditional K-means algorithms which did not cope well with the noise features added to the data. We can also observe an important improvement that the consensus MWK-means strategy provided over original MWK-means in the case of both SW and CH optimization for the 1000×8-4, 1000×10-5 and 1000×20-6 data configurations. In the case of iMWK-means, a clear improvement yielded by the consensus strategy can be observed for five of the six data configurations.

6 Conclusion

In this article, we presented an original median-based consensus strategy that can be used in the framework of the weighted Minkowski clustering for selecting an optimal value of the Minkowski exponent p and determining a good consensus partition. Figures 1 and 2 summarize our simulations results obtained for synthetic data consisting of spherical Gaussian clusters without and with noise features. These graphics confirm that the proposed median-based central consensus rule is effective and should be used in the framework of the weighted Minkowski clustering [5] with both MWK-means and iMWK-means algorithms. The advantages provided by

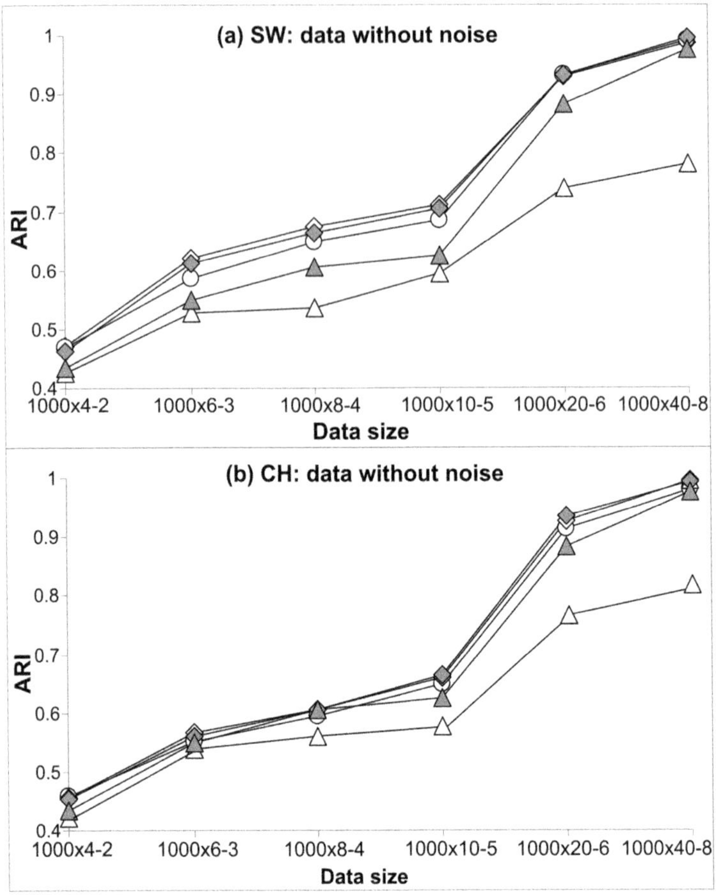

Fig. 1 Average ARI results provided by the K-means, MWK-means, and iMWK-means algorithms for our original synthetic data without noise, used with the Silhouette width (SW, panel **a**) and Calinski-Harabasz (CH, panel **b**) cluster validity indices. The results of the original K-means algorithm are represented by *open circles*, of the original MWK-means algorithm by *open diamonds*, and of the original iMWK-means algorithm by *open triangles*. The results of our median-based central consensus algorithms are represented by *gray diamonds* (MWK-means) and *gray triangles* (iMWK-means)

the introduced consensus strategy are noticeable with the noisy data in the case of MWK-means (see Fig. 2) and with both original and noisy data in the case of iMWK-means (see Figs. 1 and 2). Moreover, both MWK-means and iMWK-means clearly outperform the traditional K-means algorithm when noise features are present in the data. Mention that the Silhouette width generally outperforms the Calinski–Harabasz index in the framework of our consensus MWK-means analysis. These trends have been also confirmed by the analysis of real-world benchmark data (see Tables 1 and 2). It would be interesting to see in the future if the presented

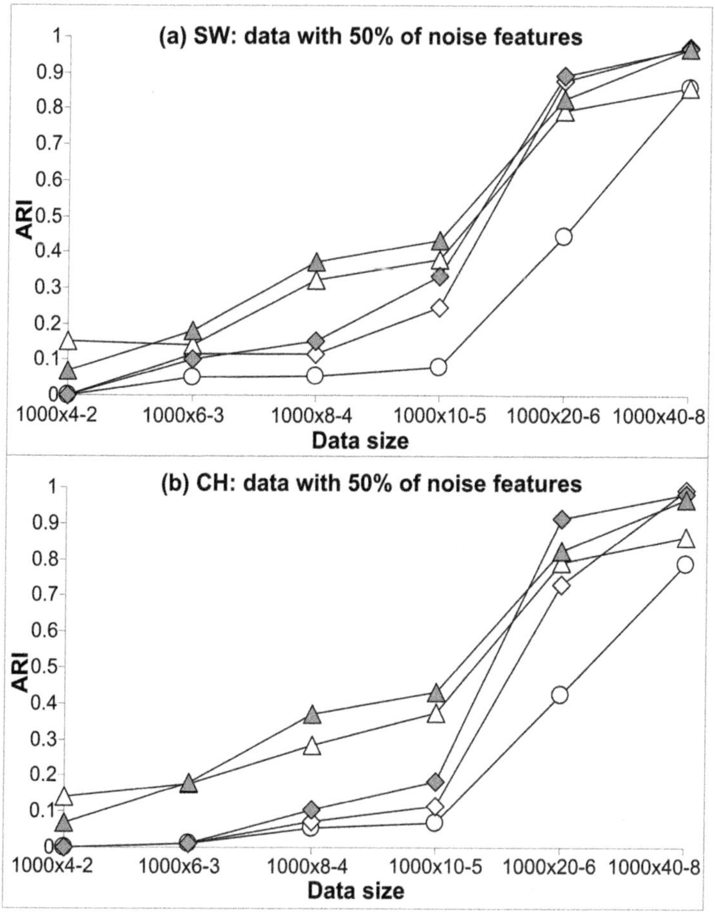

Fig. 2 Average ARI results provided by the K-means, MWK-means, and iMWK-means algorithms for our synthetic data composed of spherical Gaussian clusters with 50% of added noise features. Figure 1 panel description applies here

central consensus rule could be also used as a cluster validity index to allow one to select the correct number of clusters in the data.

References

1. Arbelaitz, O., Gurrutxaga, I., Muguerza, J., Pérez, J.M., Perona, I.: An extensive comparative study of cluster validity indices. Pattern Recogn. **46**, 243–256 (2012)
2. Ball, G.H., Hall, D.J.: A clustering technique for summarizing multivariate data. Behav. Sci. **12**, 153–155 (1967)

3. Caliński, T., Harabasz, J.: A dendrite method for cluster analysis. Commun. Stat.-Theory Methods **3**, 1–27 (1974)
4. Chan, E.Y., Ching, W.K., Ng, M.K., Huang, J.Z.: An optimization algorithm for clustering using weighted dissimilarity measures. Pattern Recogn. **37**, 943–952 (2004)
5. de Amorim, R.C., Mirkin, B.: Minkowski metric, feature weighting and anomalous cluster initializing in K-means clustering. Pattern Recogn. **45**, 1061–1075 (2012)
6. Field, A.: Discovering Statistics Using SPSS. SAGE Publications, New Delhi (2005)
7. Huang, J.Z., Ng, M.K., Rong, H., Li, Z.: Automated variable weighting in k-means type clustering. IEEE Trans. Pattern Anal. Mach. Intell. **27**, 657–668 (2005)
8. Hubert, L., Arabie, P.: Comparing partitions. J. Classif. **2**, 193–218 (1985)
9. Jain, A.K.: 50 years beyond K-means. Pattern Recogn. Lett. **31**, 651–666 (2010)
10. Ji, J., Bai, T., Zhou, C., Ma, C., Wang, Z.: An improved k-prototypes clustering algorithm for mixed numeric and categorical data. Neurocomputing **120**, 590–596 (2013)
11. Jones, E., Oliphant, T., Peterson, P., et al.: SciPy: Open Source Scientific Tools for Python. R Foundation for Statistical Computing, Vienna (2011). Available via DIALOG
12. Lichman, M.: UCI Machine Learning Repository. School of Information and Computer Sciences, University of California, Irvine (2013). Available via DIALOG
13. MacQueen, J.: Some methods for classification and analysis of multivariate observations. In: Broy, M. (ed.) Proceedings of the Fifth Berkeley Symposium on Mathematical Statistics and Probability, pp. 281–297. University of California Press, Berkeley, CA (1967)
14. Makarenkov, V., Legendre, P.: Optimal variable weighting for ultrametric and additive trees and K-Means partitioning. J. Classif. **169**, 245–271 (2001)
15. MATLAB: MATLAB:2010. The MathWorks Inc., Natick, MA (2010)
16. Milligan, G.W., Cooper, M.C.: An examination of procedures for determining the number of clusters in a data set. Psychometrika **50**, 159–179 (1985)
17. Mirkin, B.: Clustering: A Data Recovery Approach. CRC Press, London (2012)
18. Murtagh, F.: Complexities of hierarchic clustering algorithms: state of the art. Comput. Stat. **1**, 101–113 (1984)
19. Murtagh, F., Contreras, P.: Methods of hierarchical clustering (2011). arXiv preprint arXiv:1105.0121
20. Pal, S.K., Majumder, D.D.: Fuzzy sets and decision making approaches in vowel and speaker recognition. IEEE Trans. Syst. Man Cyber. **7**, 625–629 (1977)
21. Pollard, K.S., Van Der Laan, M.J.: A method to identify significant clusters in gene expression data. Bepress, pp. 318–325 (2002)
22. R Core Team: R: A Language and Environment for Statistical Computing. R Foundation for Statistical Computing, Vienna (2013). Available via DIALOG
23. Rousseeuw, P.J.: Silhouettes: a graphical aid to the interpretation and validation of cluster analysis. J. Comput. Appl. Math. **20**, 53–65 (1987)
24. Steinley, D.: K-means: a half-century synthesis. Br. J. Math. Stat. Psychol. **59**, 1–34 (2006)

Finding Prototypes Through a Two-Step Fuzzy Approach

Mario Fordellone and Francesco Palumbo

Abstract Usually, the aim of cluster analysis is to build prototypes, i.e., typologies of units that present similar characteristics. In this paper, an alternative approach based on consensus clustering between two different clustering methods is proposed to obtain homogeneous prototypes. The clustering methods used are fuzzy c-means (that minimizes the objective function with respect to centers of the groups) and archetypal analysis (that minimizes the objective function with respect to extremes of the groups). The consensus clustering is used to assess the correspondence between the clustering solutions obtained and to find the prototypes as a compromise between the two clustering methods.

1 Introduction

According to Rosch [24, 25], prototypes are those elements that represent a category better than others. The degree of representativeness can be measured using a distance function to a salient entity of the category, i.e., a *prototype* [13]. Prototypes can be observed or unobserved (abstract), and they can be represented by a single value or by interval-valued variables. Several numerical techniques to find prototypes in a given multivariate dataset have been proposed in the statistical literature based on different criteria. The most widely used techniques are generally based on non-hierarchical clustering algorithms [11, 18, 27], although many other approaches can be adopted [19, 20, 22].

M. Fordellone (✉)
Sapienza University of Rome, Rome, Italy
e-mail: mario.fordellone@uniroma1.it

F. Palumbo
Federico II University of Naples, Napoli, Italy
e-mail: fpalumbo@unina.it

F. Palumbo et al. (eds.), *Data Science*, Studies in Classification, Data Analysis, and Knowledge Organization, DOI 10.1007/978-3-319-55723-6_9

This paper proposes a two-step procedure based on consensus clustering (CC) to find prototypes within multidimensional data. The first step aims to define two partitions of the $N \times J$ data matrix $\mathbf{X}$ in K groups, where K is assumed to be known; the second step aims to find the correspondence between these two partitions and define the partition solution as the compromise between the two partitions [1, 7, 17, 29].

This work aims to represent a new approach to finding a set of K prototypes through CC to pair the partitions obtained via two different partitioning methods: fuzzy c-means (FCM) [5] and archetypal analysis (AA) [2, 8]. The former seeks K homogeneous groups vis-à-vis their barycenters, while the latter identifies a set of K extreme points, called archetypes, and creates a group around each archetype. Formally, AA minimizes the sum of distances between each point and a set of K archetypes, as defined by a convex combination of extreme points. K is given or can be selected by running the algorithm for different values of K and choosing the desired value according to the most commonly used methods. Such methods are generally based on graphic displays.

The paper is structured as follows: Sect. 2 presents a brief background on CC, FCM, and AA; Sect. 3 then demonstrates a simulation study in order to examine the reliability of the method. Finally, Sect. 4 reports on the results of an application using real data.

2 Methodology

Given a multivariate dataset and two or more partitioning criteria, consensus analysis aims to find a compromise in the set of the partitions [17]. Consensus clustering aims for the same goal among two or more partitions obtained via cluster analysis approaches. Taking into account the final aim of the analysis, the researcher chooses a consensus measure. The present proposal considers two fuzzy approaches as partitioning methods that optimize two different criteria, given the number of groups K: fuzzy c-means (FCM) and archetypal analysis (AA). Proposed by Bezdek et al. [6], the former aims to maximize the homogeneity within the K groups, while the latter, proposed by Cutler and Breiman [8], identifies the K groups with respect to a set of K extreme points, called archetypes, and aims to maximize the heterogeneity among the K groups.

FCM and AA can be defined in terms of a factorization problem of the data matrix $\mathbf{X}$ under different constraints. Formally, let $\mathbf{X}$ be a generic $N \times J$ data matrix, and let $\mathbf{P}$ be an unknown $K \times J$ prototypes matrix; FCM and AA are based on the solution of the following non-negative factorization problem [4]:

$$f(\mathbf{Y}, \mathbf{P}) = \arg\min_{\mathbf{Y},\mathbf{P}} \|\mathbf{X} - \mathbf{Y}\mathbf{P}\|_2^2, \tag{1}$$

where the notation $\|.\|_2^2$ denotes the quadratic norm, $\mathbf{Y}$ is the generic $N \times K$ memberships matrix, and $\mathbf{P}$ refers to the matrix of the centers in the FCM context and to the archetypes matrix in the AA context. In order to avoid any confusion in the remainder of this paper, the **Y** matrix will be referred to as $\boldsymbol{\Delta}$ and $\boldsymbol{\Gamma}$, and the **P** matrix as **C** and **A** when we refer to the FCM or to the AA, respectively. The generic elements y_{ik} vary in [0,1] and represent the membership degree of the generic unit x_i' to the generic element p_k.

2.1 Fuzzy c-Means

Both the fuzzy c-means clustering method [5, 6] and the traditional k-means method minimize the sum of the weighted squared distances between the N units and the K centers. Formally, given an $N \times J$ data matrix **X** FCM minimizes the objective function shown in Eq. (2).

$$f(\Gamma, \mathbf{C}) = \|\mathbf{X} - \Gamma\mathbf{C}\|_2^2, \tag{2}$$

where Γ represents the memberships matrix with elements γ_{ik}. The function in Eq. (2) is minimized under the constraints $\sum_{k=1}^{K} \gamma_{ik} = 1$ and $\gamma_{ik} \geq 0$. The elements γ_{ik} of the Γ matrix are defined according to Eq. (3), while the **C** matrix is defined according to Eq. (4).

$$\gamma_{ik} = \left(\sum_{k'=1}^{K} \left(\frac{\|\mathbf{x}_i - \mathbf{c}_k\|_2}{\|\mathbf{x}_i - \mathbf{c}_{k'}\|_2} \right)^{\frac{2}{m-1}} \right)^{-1}, \tag{3}$$

$$\mathbf{C} = (\Gamma^T\Gamma)^{-1}\Gamma^T\mathbf{X} \tag{4}$$

Note that m is the fuzzifier parameter, commonly set to 2 [6]. Including Eq. (4) in Eq. (2), the objective function becomes:

$$f(\Gamma) = \left\|\mathbf{X} - \Gamma(\Gamma^T\Gamma)^{-1}\Gamma^T\mathbf{X}\right\|_2^2 \tag{5}$$

Then, once the number of groups K is fixed, the FCM algorithm runs through the following steps [6, 9, 30]:

1. Randomly initialize the cluster centers $\mathbf{C}^{(t)}$ and set $t = 0$;
2. Calculate γ_{ik} using Eq. (3);
3. Calculate $\mathbf{C}^{(t+1)}$ using Eq. (4);
4. If $\left\|\mathbf{C}^{(t)} - \mathbf{C}^{(t+1)}\right\|_2^2 \leq \epsilon$, go to Step 5; otherwise, $\mathbf{C}^{(t)} = \mathbf{C}^{(t+1)}$; set $t = t + 1$ and go to Step 2;
5. Print centers matrix **C** and membership matrix Γ;
6. Stop.

2.2 Archetypal Analysis

The term *archetype* is used in the literature to define different meanings. In the prototyping approach, the challenge is to find a few points (archetypes), not necessarily observed, in a set of multivariate observations such that all the data can be well represented as convex combinations of the archetypes.

Formally, given an $N \times J$ data matrix $\mathbf{X}$, archetypal analysis [8, 10, 12] finds a set of archetypes $\{a_1, \ldots, a_K\}$ that are linear combinations of the data points, as shown in Eq. (6).

$$\mathbf{A} = \mathbf{B}\mathbf{X}, \tag{6}$$

where $\mathbf{B}$ is the $K \times N$ coefficients matrix with $\sum_{k=1}^{K} \beta_{ki} = 1$ and $\beta_{ki} \geq 0$, such that the archetypes resemble the data as a convex mixture. For a given choice of archetypes, AA minimizes the objective function shown in Eq. (7).

$$f(\Delta, \mathbf{A}) = \|\mathbf{X} - \Delta\mathbf{A}\|_2^2, \tag{7}$$

under the constraints $\sum_{k=1}^{K} \delta_{ki} = 1$ and $\delta_{ki} \geq 0$. Including Eq. (6) in Eq. (7), the objective function becomes:

$$f(\Delta, \mathbf{B}) = \|\mathbf{X} - \Delta\mathbf{B}\mathbf{X}\|_2^2. \tag{8}$$

Once the number of groups K is fixed, the AA algorithm then runs through the following steps [2, 8, 12]:

1. Randomly initialize the matrix $\mathbf{B}^{(t)}$ and set $t = 0$;
2. Find coefficient matrix $\Delta^{(t)}$, solving the problem in Eq. (8) under constraints $\delta_{ki} \geq 0$ and $\sum_{k=1}^{K} \delta_{ki} = 1$;
3. Given the coefficients $\delta_{ki}^{(t)}$, compute the intermediate archetypes, solving the equation in (8) for $\mathbf{A}^{(t)}$;
4. Update Eq. (8) over $\mathbf{B}$ under constraints $\sum_{k=1}^{K} \beta_{ki} = 1$ and $\beta_{ki} \geq 0$;
5. Set $t = t + 1$, $\mathbf{B}^{(t)} = \mathbf{B}^{(t+1)}$ and calculate $\mathbf{A}^{(t)} = \mathbf{B}^{(t)}\mathbf{X}$;
6. Compute the objective function and, unless it falls below a threshold, continue with Step 2;
7. Stop.

Note that the matrices $(\boldsymbol{\Gamma}^T\boldsymbol{\Gamma})^{-1}\boldsymbol{\Gamma}^T$ and $\mathbf{B}$ play the same role, i.e., to project the single points in a K-dimensional space. In fact, defining the $(\boldsymbol{\Gamma}^T\boldsymbol{\Gamma})^{-1}\boldsymbol{\Gamma}^T = \mathbf{B}^*$, it is possible to shown that the objective functions of FCM (9) and AA (10) optimize an equivalent criterion.

$$f(\Delta, \mathbf{B}) = \|\mathbf{X} - \Gamma\mathbf{B}^*\mathbf{X}\|_2^2 . \tag{9}$$

$$f(\Delta, \mathbf{B}) = \|\mathbf{X} - \Delta\mathbf{B}\mathbf{X}\|_2^2 . \tag{10}$$

2.3 Consensus Clustering

In this paper, the consensus clustering procedure is structured in two fundamental steps: (1) to find and represent the consensus between the partitions of fuzzy c-means and archetypal analysis through correspondence analysis [3, 15, 16] and (2) to measure the consensus through the principal indices of CC.

Let $\mathbf{X}$ be an $N \times J$ data matrix with $T = \{T_1, \ldots, T_R\}$ and $V = \{V_1, \ldots, V_C\}$ two partitions of $\mathbf{X}$: the consensus between partitions T and V is found by starting from the entries shown in the cross-classifying contingency table (shown in Table 1) and crossing the two partitions [17]. Many proposals have been put forth in the literature for the consensus measurement, including Boulis and Ostendorf [7], Fowlkes and Mallows [14], Hubert and Arabie [17], Steinley [28], Strehl and Ghosh [29], and Yeung and Ruzzo [31]. This paper has used the Adjusted Rand Index (ARI) among these different options [23, 26]. ARI was first proposed by Hubert and Arabie [17] in such a context; the index assumes a generalized hypergeometric distribution as a null hypothesis. The two clusterings are drawn randomly, with a fixed number of clusters and a fixed number of elements in each cluster. ARI is then the normalized difference between the Rand Index and its expected value under the null hypothesis. The ARI is defined as shown in Eq. (11).

$$ARI = \frac{\sum_{r=1}^{R}\sum_{c=1}^{C}\binom{n_{rc}}{2} - \binom{n}{2}^{-1}\sum_{r=1}^{R}\binom{n_{r.}}{2}\sum_{c=1}^{C}\binom{n_{.c}}{2}}{\frac{1}{2}\left[\sum_{r=1}^{R}\binom{n_{r.}}{2} + \sum_{c=1}^{C}\binom{n_{.c}}{2}\right] - \binom{n}{2}^{-1}\sum_{r=1}^{R}\binom{n_{r.}}{2}\sum_{c=1}^{C}\binom{n_{.c}}{2}} . \tag{11}$$

ARI has an expected value of 0 for independent clusterings and a maximum value of 1 for identical clusterings.

Table 1 Cross-table between partition T and partition V

		Partition V				
		v_1	v_2	$\cdots$	v_C	
Partition T	t_1	n_{11}	n_{12}	$\cdots$	n_{1C}	$n_{1\cdot}$
	t_2	n_{21}	n_{22}	$\cdots$	n_{2C}	$n_{2\cdot}$
	$\vdots$	$\vdots$	$\vdots$	$\ddots$	$\vdots$	$\vdots$
	t_R	n_{R1}	n_{R2}	$\cdots$	n_{RC}	$n_{R\cdot}$
		$n_{\cdot 1}$	$n_{\cdot 2}$	$\cdots$	$n_{\cdot C}$	n

3 Simulation Study

This section demonstrates an application of simulated data; in particular, it evaluates the consensus between fuzzy c-means and archetypal analysis in different experimental conditions.

According to Fordellone and Palumbo [13], data were generated from four multivariate Gaussian distributions, each with four dimensions. The first is a multivariate Gaussian distribution with $\mu = [0, 0, 0, 0]^T$ and $\boldsymbol{\Sigma} = \mathbf{I}$ (i.e., noise); the last three are multivariate Gaussian distributions that simulate three groups of units according to the experimental conditions shown in Table 2; the groups follow the scheme shown below:

Group 1: $\mathbf{X} \sim \mathcal{N}(\boldsymbol{\mu} = [\ \text{-}20\ \ 10\ \ 30\ \ 15]^T, \boldsymbol{\Sigma})$
Group 2: $\mathbf{X} \sim \mathcal{N}(\boldsymbol{\mu} = [\ \ 0\ \ 20\ \ 15\ \ \text{-}5]^T, \boldsymbol{\Sigma})$
Group 3: $\mathbf{X} \sim \mathcal{N}(\boldsymbol{\mu} = [\ 15\ \ 5\ \ \text{-}7\ \ 20]^T, \boldsymbol{\Sigma})$

Table 2 also shows the consensus results obtained from the eight experimental conditions; in particular, the Rand and *adjusted* Rand indices are reported, which are measurements of agreement/disagreement (i.e., consensus) between two different partitions (FCM and AA in this case). It is worth noting that the maximum consensus was achieved in the first two experimental conditions, where there was a low correlation between the variables and a normal kurtosis level, whereas the minimum level of consensus was shown in the experimental conditions, which demonstrated platykurtic kurtosis levels (rows 3, 4, 7, and 8 in the table).

The aim of the simulation study was to establish the degree of reliability of consensus prototyping under several different hypotheses. In the eight proposed cases, the lowest levels of the consensus occurred most notably when platykurtic distributions were present; this occurred because a low level of kurtosis stimulated the presence of outlier points, and AA is very sensitive to extreme points.

Table 2 Consensus results from simulated data

Experimental conditions			Prototyping results		Consensus measurement	
N	Cor	Kurt.	Groups	Size	Rand	$Rand_{adj}$
900	0.2–0.4	$G = 3$	3	900 (100.0%)	1.000	1.000
300	0.2–0.4	$G = 3$	3	300 (100.0%)	1.000	1.000
900	0.2–0.4	$G < 3$	3	625 (69.4%)	0.725	0.382
300	0.2–0.4	$G < 3$	3	185 (61.7%)	0.683	0.296
900	0.6–0.8	$G = 3$	3	599 (66.6%)	0.753	0.444
300	$0.6 - 0.8$	$G = 3$	3	202 (67.3%)	0.758	0.460
900	$0.6 - 0.8$	$G < 3$	3	533 (59.2%)	0.698	0.329
300	$0.6 - 0.8$	$G < 3$	3	189 (63.0%)	0.720	0.373

4 Personality Traits Finding

In psychology, trait theory (also called dispositional theory) is an approach that aims to study human personality. Trait theorists are primarily interested in the measurement of different traits that can be defined as habitual patterns of behavior, thought, and emotion [21, 22].

Consensus clustering between fuzzy c-means and archetypal analysis is used in this section to delineate the personality traits (http://personality-testing.info/) from a sample of 898 adults (493 males and 405 females). The dataset has 40 ordinal items (from "Strongly disagree" to "Strongly agree"), with ten items for each measurement scale (numeric). Four different scales were used as part of an experimental "DISC" personality test from the International Personality Item Pool (http://ipip.ori.org/newCPIKey.htm):

- *Assertiveness* (AS): the quality of being self-assured and confident without being aggressive;
- *Social confidence* (SC): generally described as a state of being certain about something;
- *Adventurousness* (AD): represented by a desire to engage in activities with some potential for physical danger;
- *Dominance* (DO): conceptualized as a measure of individual differences in levels of group-based discrimination.

The following subsection shows the results that were obtained by the application of the consensus-prototyping approach on the four quantitative measurement scales. The scales were computed using principal component analysis (PCA) applied to the fifty available items. The number of groups ($K = 3$) was chosen according to the FCM and AA scree-plots shown in Fig. 1. The scree-plots show the reductions in the objective functions at different values of K.

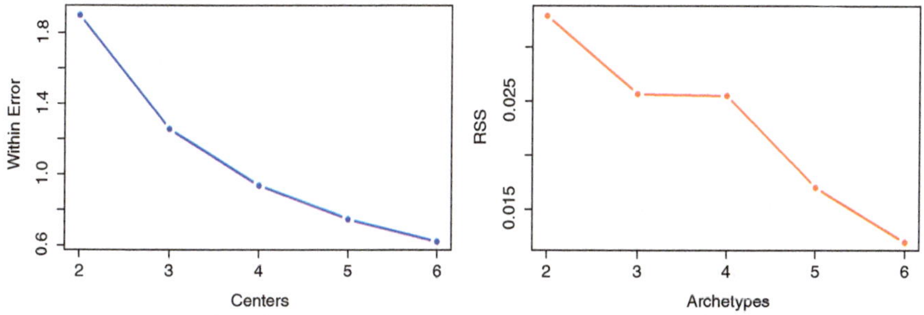

Fig. 1 Scree-plots of the fuzzy c-means and archetypal analysis approaches for different numbers of groups

Table 3 Consensus table between the fuzzy c-means and archetypal analysis approaches

		Archetypal analysis			
		Group 1	Group 2	Group 3	Total
Fuzzy c-means	Group 1	126	35	125	286
	Group 2	53	218	0	271
	Group 3	0	1	340	341
	Total	179	254	465	898

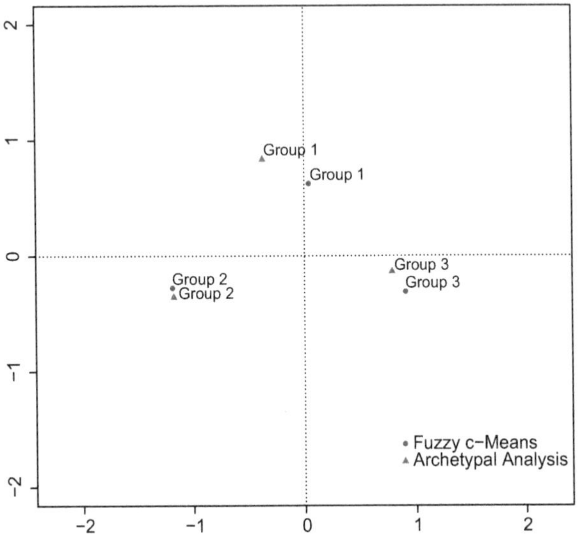

Fig. 2 Correspondence analysis applied to contingency Table 3

The results of CC are shown in consensus Table 3. A correspondence analysis is also applied on the consensus table to graphically illustrate the consensus level. The results are represented in Fig. 2, where we may see that the three groups are very close when they have a maximum level of inertia. In Table 3, in contrast, the consensus prototypes represent roughly 76% of the sample; the Adjusted Rand Index equals 0.499, while the Rand Index equals 0.768.

Figure 3 shows the defined groups and the single distributions of the scales. The different colors (purple, blue, and red) and the different symbols (•, △, and +) of the points represent the three prototypes. The white points in Fig. 3 represent the observation without consensus.

Due to space considerations, we cannot discuss in depth the profiles of the prototypes that were found. Looking at the scatterplot matrix shown in Fig. 3, however, we can state that prototype 1 (the red • symbol) is characterized by a high level of scale *AD*; prototype 2 (the blue △ symbol) presents low values for scales *SC, DO, AS*; and the last prototype (the purple + symbol) is characterized by high levels of scales *SC, AS* and a low level of *AD*.

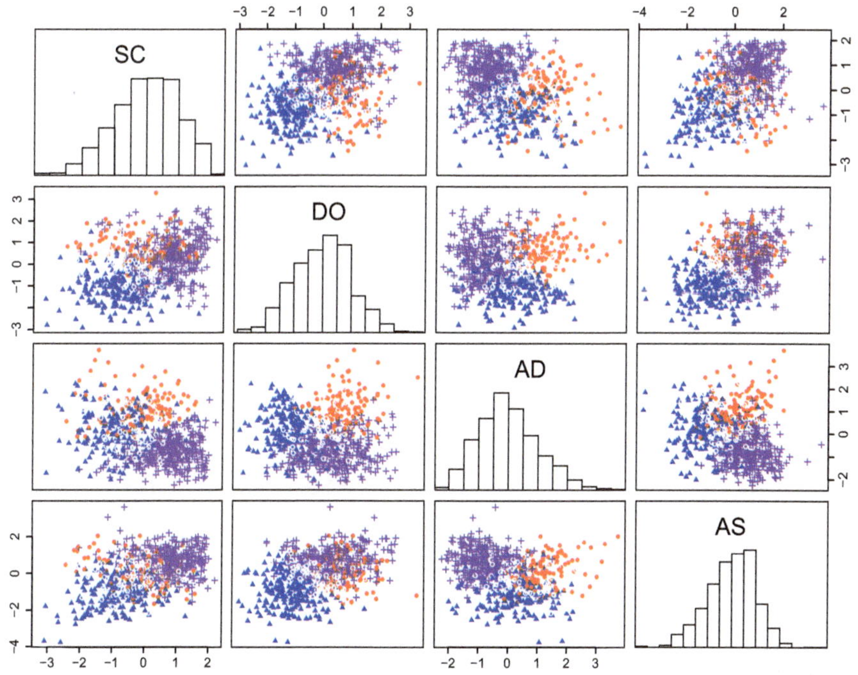

Fig. 3 Scatterplots of the three prototypes computed via consensus clustering

5 Concluding Remarks

The empirical results shown in the present paper lead us to argue that when the groups are well defined (thus avoiding any overlap), the consensus clustering between the two different partitioning methods (i.e., fuzzy c-means and archetypal analysis) clarifies the presence of well-definable prototypes. The simulation study was helpful for appreciating which causes can deeply affect the consensus among the two approaches: the platykurtic level and the presence of multivariate outliers in the data have greatly affected performance of the consensus analysis; the high correlation among variables has worsened performance, but not much in comparison with the previous perturbater effects.

Acknowledgements Dr. Mario Fordellone was a research fellow at the University of Padua at the time when most of the results presented in this chapter were developed. He is grateful for the opportunity to conduct these studies while at the university.

References

1. Arabie, P., Boorman, S.A.: Multidimensional scaling of measures of distance between partitions. J. Math. Psychol. **10**(2), 148–203 (1973)
2. Bauckhage, C., Thurau, C.: Making archetypal analysis practical. In: Pattern Recognition, pp. 272–281. Springer, Heidelberg (2009)
3. Benzécri, J.P.: Correspondence Analysis Handbook. Marcel Dekker, New York (1992)
4. Berry, W.M., Browne, M., Langville, N.A., Pauca, P.V., Plemmons, J.R.: Algorithms and applications for approximate non-negative matrix factorization. Comput. Stat. Data Anal. **52**(1), 155–173 (2007)
5. Bezdek, J.C.: Pattern Recognition with Fuzzy Objective Function Algorithms. Kluwer Academic Publishers, Norwell, MA (1981)
6. Bezdek, J.C., Ehrlich, R., Full, W.: FCM: the fuzzy c-means clustering algorithm. Comput. Geosci. **10**(2), 191–203 (1984)
7. Boulis, C., Ostendorf, M.: Combining multiple clustering systems. In: Knowledge Discovery in Databases: PKDD, pp. 63–74. Springer, Berlin (2004)
8. Cutler, A., Breiman, L.: Archetypal analysis. Technometrics **36**(4), 338–347 (1994)
9. Dembele, D., Kastner, P.: Fuzzy c-means method for clustering microarray data. Bioinformatics **19**(8), 973–980 (2003)
10. D'Esposito, M., Palumbo, F., Ragozini, G.: Archetypal analysis for interval data in marketing research. Ital. J. Appl. Stat. **18**, 343–358 (2006)
11. Diday, E.: Optimization in non-hierarchical clustering. Pattern Recogn. **6**(1), 17–33 (1974)
12. Eugster, M., Leisch, F.: From spider-man to hero-archetypal analysis in R. J. Stat. Softw. **1**, 1–23 (2009)
13. Fordellone, M., Palumbo, F.: Prototypes definition through consensus analysis between Fuzzy c-Means and Archetypal Analysis. Ital. J. Appl. Stat. **26**(2), 141–162 (2014)
14. Fowlkes, E.B., Mallows, C.L.: A method for comparing two hierarchical clusterings. J. Am. Stat. Assoc. **78**(383), 553–569 (1983)
15. Greenacre, M.J.: Interpreting multiple correspondence analysis. Appl. Stoch. Model Data Anal. **7**(2), 195–210 (1991)
16. Greenacre, M.J.: Correspondence Analysis in Practice, 2nd edn. Chapman and Hall/CRC, Boca Raton (2010)
17. Hubert, L., Arabie, P.: Comparing partitions. J. Classif. **2**(1), 193–218 (1985)
18. Jain, A.K., Murty, M.N., Flynn, P.J.: Data clustering: a review. ACM Comput. Surv. (CSUR) **31**(3), 264–323 (1999)
19. Johnson, C.S.: Hierarchical clustering schemes. Psychometrika **32**(3), 241–254 (1967)
20. Karypis, G., Han, E.H., Kumar, V.: Chameleon: hierarchical clustering using dynamic modeling. Computer **32**(8), 68–75 (1999)
21. Kassin, S.: Psychology. Prentice-Hall, Upper Saddle River, NJ (2003)
22. Loh, W.Y., Shih, Y.S.: Split selection methods for classification trees. Stat. Sin. **74**(4), 815–840 (1997)
23. Rand, W.M.: Objective criteria for the evaluation of clustering methods. J. Am. Stat. Assoc. **66**(336), 846–850 (1971). doi:10.2307/2284239
24. Rosch, E.: Cognitive representations of semantic categories. J. Exp. Psychol. **104**(3), 192 (1975)
25. Rosch, E.: Principles of categorization. In: Concepts: Core Readings, pp. 189–206. The MIT Press, Cambridge, MA (1999)
26. Santos, J.M., Embrechts, M.: On the use of the adjusted Rand index as a metric for evaluating supervised classification. In: Artificial Neural Networks CANN 2009, pp. 175–184. Springer, New York (2009)
27. Scheibler, D., Schneider, W.: Montecarlo tests of the accuracy of cluster analysis algorithms: a comparison of hierarchical and nonhierarchical methods. Multivar. Behav. Res. **20**(3), 283–304 (1985)

28. Steinley, D.: Properties of the Hubert-Arable Adjusted Rand Index. Psychol. Methods **9**(3), 386 (2004)
29. Strehl, A., Ghosh, J.: Cluster ensembles-a knowledge reuse framework for combining multiple partitions. J. Mach. Learn. Res. **3**, 583–617 (2003)
30. Sun, H., Wang, S., Jiang, Q.: FCM-based model selection algorithms for determining the number of clusters. Pattern Recogn. **37**(10), 2027–2037 (1985)
31. Yeung, K.Y., Ruzzo, W.L.: Details of the adjusted Rand index and clustering algorithms, supplement to the paper "An empirical study on principal component analysis for clustering gene expression data". Bioinformatics **17**(9), 763–774 (2001)

Clustering Air Monitoring Stations According to Background and Ambient Pollution Using Hidden Markov Models and Multidimensional Scaling

Álvaro Gómez-Losada

Abstract In order to study the cluster of monitoring sites from an urban air quality monitoring network (AQMN) with respect to the background and ambient pollution of key pollutants, a combined methodology is proposed: firstly, to obtain the ambient and background levels of air pollution from every selected pollutant, time series obtained from the AQMN were modeled with hidden Markov models; secondly, to study the grouping of these monitoring sites according to these levels of pollution, both ambient and background pollution, multidimensional scaling (Smacof MDS) was used and the stability of these solutions obtained with a Jacknife procedure (`smacof` library—R software). Results show that the clustering behaviour of sites is different when studying the ambient from the background pollution. However, sites marked with a distinct pollution contribution could locate them distant from the main cluster of sites as long as they show a marked stability in the MDS solutions.

1 Introduction

Exposure studies rely on detailed characterization of air quality. Pioneering works by Pires et al. [10, 11] to identify city areas with similar air pollution behaviours used principal component analysis and hierarchical cluster analysis. Other studies use models of various complexities to characterize local-scale air quality, but often with poor representation of background concentrations [1]. Background concentration could be defined as the concentration that would be measured if local sources

Electronic supplementary material The online version of this article (doi:10.1007/978-3-319-55723-6_10) contains supplementary material, which is available to authorized users.

Á. Gómez-Losada
European Commission, Joint Research Centre (JRC), Economics of Climate Change, Energy and Transport, Edificio Expo, C/ Inca Garcilaso 3, 41092 Seville, Spain

Department of Statistics and Operational Research, University of Seville, c/ Tarfia s/n, 41012 Seville, Spain
e-mail: alvaro.gomez-losada@ec.europa.eu

F. Palumbo et al. (eds.), *Data Science*, Studies in Classification, Data Analysis, and Knowledge Organization, DOI 10.1007/978-3-319-55723-6_10

of pollution were not present. This fraction of pollution is particularly important since it represents the chronic air pollution to which the population is permanently exposed. The study of the city areas with a similar concentration of background pollution is crucial since it could help connect exposure assessment and health studies to the significant association between pollution, morbidity and mortality.

The aim of this work is to examine the clustering behaviour of monitoring sites with respect to both the levels of ambient and background pollution of key air pollutants over the course of 1 year. To estimate the ambient and background air pollution, the methodology of Gómez-Losada et al. [6] based on hidden Markov models (HMMs) was used, and later, a Smacof-Jacknife Multidimensional scaling (MDS) applied over the previously obtained results.

2 Data and Methods

2.1 Observation Sites and Data

The air quality monitoring network (AQMN) of the metropolitan area of Seville includes nine operating sites. The pollutants CO, NO_2, PM_{10} and SO_2 are mostly monitored at these locations (Table 1). Air quality data (hourly average concentrations of these pollutants) were collected at monitoring sites from the AQMN of Seville for 2013, permitting a wide range of pollution levels, contributions and locations to be considered. Since monitored data are available on an hourly average basis, daily mean concentrations were calculated when at least 80% of the data were attainable (i.e., 19 out of 24 h). The Environmental and Water Agency, which in turn is run by Andalusia's Regional Ministry of Environment and Land Planning, manages this network. The AQMN of Seville is subject to intense maintenance programs, to ensure accurate values, and data are validated in order to their later

Table 1 Analysed pollutants and classification of monitoring sites where data were obtained (locations are given in X, Y ETRS89-UTM coordinates, zone 30)

Site and abbreviation	Classification[a]	Location	Pollutant studied
Alcala de Guadaira (Alc)	U-B	248974 4136631	CO NO_2 PM_{10} SO_2
Aljarafe (Alj)	S-B	230473 4137017	· NO_2 PM_{10} SO_2
Bermejales (Ber)	U-B	236063 4137554	CO NO_2 PM_{10} SO_2
Centro (Cen)	U-B	235156 4142125	CO NO_2 · SO_2
Dos Hermanas (Dos)	U-B	241677 4130413	CO NO_2 · SO_2
Principes (Pri)	U-B	233863 4140741	CO NO_2 PM_{10} SO_2
Ranilla (Ran)	U-T	237965 4141611	CO NO_2 · SO_2
Santa Clara (Sac)	S-B	238720 4143149	CO NO_2 PM_{10} ·
Torneo (Tor)	U-T	234151 4142873	CO NO_2 PM_{10} SO_2

Dots represent unmeasured pollutants (five cases)

[a]Type of area: *S* suburban, *U* urban ; main pollution source: *T* traffic, *B* background, *I* industrial

exploitation. All the data used in this study were obtained from Gómez-Losada et al. [6], from where the length of time series (TS) and their basic statistics and uncertainties can be consulted in the Supplementary Material of this study.

The province of Seville is located in Andalusia, in southern Spain, covering an area of 14,036 km^2. According to the IECA [9], the province had a total population of 1,935,364 the year the study was carried out. The city of Seville is the largest urban agglomeration in Andalusia and has a population of 1,217,811 [14].

2.2 *Application of HMMs to Air Pollutant Concentration Time Series*

HMM represents a flexible method of modeling TS data that exhibits dependence over time, such as daily mean pollutant concentrations collected at AQMNs. The model consists of two parts: firstly, the daily mean pollutant concentrations describing a TS of length T [observations of the TS, $Y_{1:T} = (Y_1, Y_2, \ldots, Y_T)$], and secondly, unobserved state variables [$S_{1:T} = (S_1, S_2, \ldots, S_T)$] responsible for generating the TS observations. The observer only perceives the TS observations; the state variables, on the other hand, are hidden. State variables $S_{1:T}$ are elements from a set $\mathscr{S}$ with K distinct elements of the HMM, K being the number of clusters or states of the model, used to name the HMM (e.g., a K-state HMM).

HMM provides the mean (m_i), standard deviation (sd_i) and weight (π_i) values for the cluster i according to the daily observations comprising the cluster. π_i values are representative of the number of daily observations grouped into cluster i with respect to the TS length (T). Therefore π_i values indicate the incidence of the daily observations from clusters over the annual period. m_i and sd_i values define the different Gaussian distributions assigned to the clusters, and π_i values define the representativeness of every cluster. All three of these statistics provide very useful information as they fully characterize every analysed pollutant TS in a given time period and for a specific monitoring site. Once HMM provides all these values, it is possible to obtain the mean (M) and standard deviation (SD) values of the TS using simple algebraic expressions, as indicated in Appendix, using the mean and standard deviation values of every cluster in a weighted sum, considering as weights the representativeness of every cluster with respect to the overall distribution of the TS data. The mean and standard deviation values of the TS are quantitatively similar to the arithmetic mean ($\bar{x}$) and standard deviation (s) values of the studied data sets, which are obtained without considering temporal correlation. Therefore, M values are synonymous with average ambient pollution concentrations after accounting for temporal correlation in the TS data.

The observations with the lowest daily average concentrations represent the lowest concentration profiles for every TS of pollutants. The mean (m_1) and standard deviation (sd_1) of these observations, which represent the background pollution, are used to characterize this first cluster, but also to define the Gaussian distribution formed by that set of daily concentrations when they are represented as a histogram. The m_1 value is a mean value that may be associated with the magnitude of a kind of chronic exposure concentration experienced by the population, and the sd_1 value with the degree of acuteness of this exposure. Thus, both values offer a thorough characterization of the background air pollution concentration present in cities. The TS model estimation with HMM can be consulted in Gómez-Losada et al. [6], however, is briefly explained in Sect. 2.2.1.

2.2.1 TS Model Estimation with HMM

The computational implementation of Gaussian HMMs was accomplished using the `depmixS4` package [15] from the open-source software R [12]. The appropriate model selection was performed by estimating models for different cluster values or states ($\mathscr{S}$ =1,…,7) and later the optimum solution was selected based on the lowest value of BIC (Bayesian Information Criterion) [13]. The EM algorithm used to obtain the parameters of every HMM was set up with $\epsilon = 10^{-8}$ and a maximum of 2000 iterations. To check the validity of modeling results obtained with the `depmixS4` package, the `HiddenMarkov` [7] and `HMM` [8] packages were also used, and negligible differences were found in the parameter estimates. The statistical accuracy of the HMM parameters was obtained by means of bootstrap through 200 replicates.

2.3 MDS Analyses

Because several researchers [4, 5] have already discussed Smacof-Jacknife MDS analyses in detail, in this paper, we will only describe them briefly. MDS is a set of methods for discovering "hidden" structures in multidimensional data. These are based on a proximity matrix derived from variables measured on objects as input entity [4]. The main output of MDS is a spatial representation (or "map") of distances in typically two or three dimensions (see [2, 3] for an account of the MDS in a general context). In this work, the MDS was applied by using the basic (symmetric) algorithm Smacof and later the stability of the solutions obtained with a Jacknife procedure, using the `smacof` library [4] in R. The spatial representation of the MDS solutions were used to analyse the grouping of sites according to M and m_1 values of air pollutant at the monitoring sites studied.

Being n the number of monitoring sites (Table 1), the Jacknife procedure performs $n + 1$ MDS analyses. The solution to the original problem is $\mathbf{X_0}$, and the solutions to the n additional problems are $\mathbf{X}_i$ ($i = 1, \ldots, n$), where $\mathbf{X}_i$ is the solution to the problem without object i, what is to say, $\mathbf{X}_i$ only have the information on the location of the $n - 1$ points (monitoring sites) [5]. Graphically in Fig. 2, the original MDS solution is given by black dots, and the n remaining MDS analysis are represented by $n - 1$ smaller dots and linked with a dashed line to the original solution. The stability of the solutions given by the Jacknife procedure is represented by a measure as a ratio of between and total variance around the original Smacof solution; thus, the closer the measure is to 1, the more stable the MDS solution is.

Using m_l and M values obtained with HMM from all the air pollutants studied at the different monitoring sites as dissimilarity measures, the MDS was used to detect the clustering of sites from a background and ambient pollution perspective, respectively. The stability of the MDS solutions for ambient and background pollution were later analysed with Jacknife.

2.3.1 MDS Implementation

Two data matrices were designed with the M (ambient pollution) and m_l (background pollution) values, respectively, both of which were obtained from Gómez-Losada et al. [6]—Supplementary Material 5. Rows and columns at each M and m_l matrices represented monitoring sites and air pollutants (a ten-row by four-column matrix, as seen in "Site and abbreviation" and "Pollutant studied" in Table 1). Each of these matrices was missing 5 M or m_l values. Values for M or m_l were scaled due to the different magnitudes of air pollutant concentrations. The `dist` function and the Euclidean distance measure were later used to obtain two distance matrices. These distance matrices were later used as input data for the `smacofSym` function (two dimensions and 1000 the maximum number of iterations). The results were later processed with the `jackknife` function ($\epsilon = 10^{-8}$ and a maximum of 100 iterations). To obtain the information about the grouping of sites with respect to the M and m_l values, a graphical representation of these analyses was created by means of a modified `plot.smacofJK` function.

3 Results and Discussion

3.1 TS Study to Describe the Background PM_{10} Pollution at One Site

Figure 1 illustrates the graphical results of modeling the PM_{10} concentrations at the Torneo (Tor) site during 2013 using HMMs. Three clusters are detected in the Tor TS and the corresponding numeric results are given in Table 2. In Fig. 1a, the daily average PM_{10} concentrations are labeled with the cluster number attributed by the HMMs. The m_l value represents the mean concentration of the background

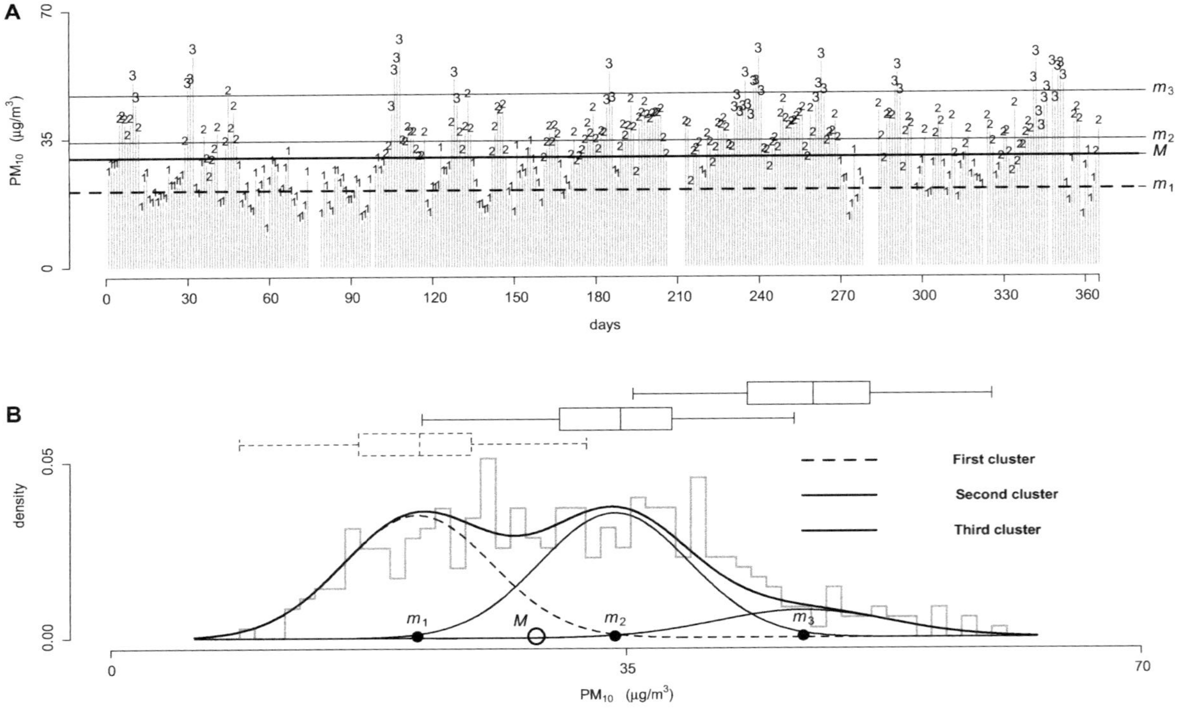

Fig. 1 Grouping of the TS data of daily average PM_{10} concentrations into three clusters at the Tor site (**a**) during 2013 and representation of the TS as a histogram (**b**). In **a** and **b**, a *dashed line* indicates the pollution represented by the first cluster (background pollution). Its average value is denoted as m_1. The mean value of the TS (ambient pollution) is displayed using a continuous, *thicker line* (M). The average values of the second (m_2) and third (m_3) clusters are also given

Table 2 Hidden Markov modeling results for PM_{10} concentrations at Torneo (Tor) site during 2013 (in $\mu g/m^3$)

K=3	Percentage	Size (days)	Range	Cluster mean	sd	M	SD
Cluster 1	41	147	8.7–31.1	20.4±1.6	4.9±0.8		
Cluster 2	47	174	21.2–46.4	33.7±1.9	5.3±0.9	29.8±1.8	10.3±1.0
Cluster 3	12	44	35.5–59.8	46.7±2.5	6.1±0.8		

sd: cluster standard deviations; M: mean (of the TS); SD: standard deviation (of the TS)

pollution at this site. This concentration is indicated with a dashed line, representing the mean value of the cluster with the lowest daily average concentrations. Shown as a histogram, Fig. 1b is equivalent to the TS representations in Fig. 1a. This latter representation reveals the Gaussian curves in which observations belonging to the different clusters are grouped. Therefore, the first cluster is also characterized by the standard deviation value (sd_1), which measures the spread of the data grouped in the first cluster around m_1, or equivalently, the strength of the exposure to PM_{10}.

Table 2 shows that, at Tor site, the incidence value of the first cluster indicates that the daily concentrations belonging to the background pollution were present during 41% of the days of this year. This means that during approximately half of the year, no noticeable influence from external PM_{10} contributions (e.g., Saharan dust intrusions affecting the daily PM_{10} concentrations) was detected.

3.2 *MDS Analyses for the Ambient and Background Pollution*

Following the same approach as Sect. 3.1, the ambient (M) and background (m_1) pollution with respect to CO, NO_2, PM_{10} and SO_2 concentrations were estimated for Seville sites for 2013. The numerical results after modeling every TS with HMMs are given in Gómez-Losada et al. [6] (Supplementary Material 5).

The stability of the MDS solutions for ambient and background pollution were analysed with Jacknife and results are shown in Fig. 2. As this figure indicates, a main cluster of sites is detected with respect to ambient pollution in Fig. 2a. In Fig. 2b, because the sites are more sparsely distributed, this cluster is not detected when the background pollution is analysed. However, the Tor site (an urban traffic site) is separated from the rest of sites in both figures. The cluster formed by Alc-Ber-Pri sites remains approximately constant when analysing the MDS solutions for both, ambient and background pollution. The stability measures for both analyses are 0.90 (**a**) and 0.97 (**b**). It must be noted how Tor site shows a remarkable stability when studying both type of pollutions, a behaviour which is not applicable to the rest of sites. Most likely, there is a distinct contribution from the urban traffic on this site. However, in Fig. 2 the Cen site shows a similar behaviour to Tor, even considering that it is a typical background site without a

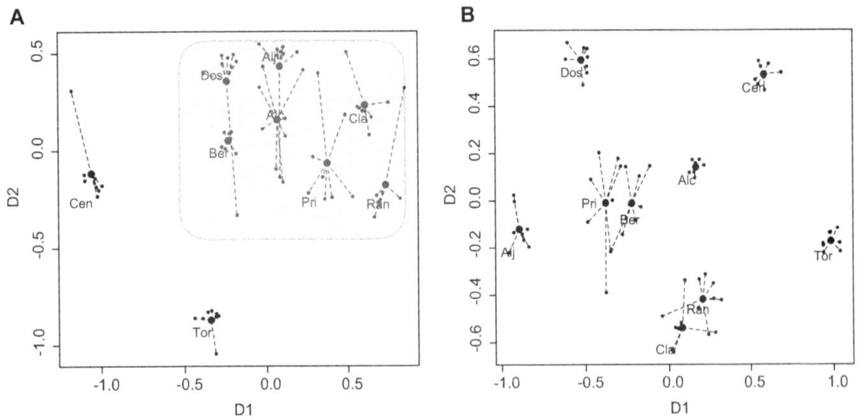

Fig. 2 Stability results for the ambient (**a**) and background (**b**) pollution in the AQMN of Seville. The *gray rectangle* in **a** represents the main cluster of sites, not detected in **b**. *Smallest points* represent the Jacknife results which are linked to the original MDS solutions (*biggest dots*) by a *dashed line*

marked influence from local sources of pollution. The configuration of the stability of the MDS solutions for M and m_1 with respect the rest of sites are considered for further research.

4 Conclusion

With the aim of characterizing the clustering behaviour of monitoring sites with respect to the ambient and lowest concentration profiles of air pollution in the urban area of Seville, TS data of daily average of CO, NO_2, PM_{10} and SO_2 concentrations registered at nine monitoring sites were modeled with HMM following the procedure of Gómez-Losada et al. [6]. MDS solutions with the Smacof algorithm for ambient and background pollution were analysed and the stability of these solutions obtained with the Jacknife results. The research detects that grouping behaviour of sites in MDS analyses are different when studying the ambient and background pollution. These analyses also show that sites with a marked contribution from pollution sources could locate them at a distant position from the main cluster of sites, but also, to have a higher stability in the MDS configurations. However, further discussion and investigation are required considering a broader scope of time.

Acknowledgements The author is grateful to the Regional Ministry of Environment and Land Planning of Andalusia for kindly providing the air quality data.

Disclaimer The views expressed are purely those of the author and may not in any circumstances be regarded as stating an official position of the European Commission.

Appendix

The aim of this appendix is to show how the values TS Mean (M) and TS SD (SD) are calculated for the TS of Torneo site (Table 2):

$$M = \sum_{1}^{3} \pi_i m_i = (0.41 \cdot 20.4) + (0.47 \cdot 33.7) + (0.12 \cdot 46.7) = 29.8$$

$$SD = \left[\sum_{1}^{3} (m_i^2 + sd_i^2)\, \pi_i - M^2 \right]^{1/2}$$

$$= \left[(20.4^2 \cdot 4.9^2) \cdot 0.41 + (33.7^2 \cdot 5.3^2) \cdot 0.47 + (46.7^2 \cdot 6.1^2) \cdot 0.12 - 29.8^2 \right]^{1/2}$$

$$= 10.3$$

References

1. Arunachalam, S., Valencia, A., Akita, Y., Serre, M.L., Omary, M., Garcia, V., Isakov, V.: A method for estimating urban background concentrations in support of hybrid air pollution modeling for environmental health studies. Int. J. Environ. Res. Public. Health **11**(10), 10518–10536 (2014)
2. Borg, I., Groenen, P.J.F., Mair, P.: Applied Multidimensional Scaling. Springer Briefs in Statistics. Springer, Berlin (2013)
3. Cox, T.F., Cox, M.A.A.: Multidimensional Scaling, 2nd edn. Chapman & Hall/CRC, Boca Raton (2001)
4. de Leeuw, J., Mair, P.: Multidimensional scaling using majorization: SMACOF in R. J. Stat. Softw. **31**(3), 1–30 (2009)
5. de Leeuw, J., Meulman, J.: A special jacknife for multidimensional scaling. J. Classif. **3**, 97–112 (1986)
6. Gómez-Losada, Á., Pires, J.C.M., Pino-Mejías, R.: Characterization of background air pollution exposure in urban environments using a metric based on Hidden Markov Models. Atmos. Environ. **127**, 255–261 (2016)
7. Harte, D: HiddenMarkov: hidden Markov models, 2010. R package version 1.8-3 (2015)
8. Himmelmann, L.: HMM: HMM-hidden Markov models. Scientific Software Development, 2010. R package version 1.0. (2010)
9. IECA (Instituto de Estadística y Cartografía de Andalucía). Anuario estadístico de Andalucía. Consejería de Economía, Innovación, Ciencia y Empleo. Junta de Andalucía, Sevilla (2013)
10. Pires, J.C.M., Sousa, S.I.V., Pereira, M.C., Alvim-Ferraz, M.C.M., Martins, F.G.: Management of air quality monitoring using principal component and cluster analysis-Part I: SO_2 and PM_{10}. Atmos. Environ. **42**, 1249–1260 (2008)
11. Pires, J.C.M., Sousa, S.I.V., Pereira, M.C., Alvim-Ferraz, M.C.M., Martins, F.G.: Management of air quality monitoring using principal component and cluster analysis-Part II: CO, NO_2 and O_3. Atmos. Environ. **42**, 1261–1274 (2008)

12. R Core Team. R: A Language and Environment for Statistical Computing. R Foundation for Statistical Computing, Vienna (2015). Available from http://www.R-project.org/. Cited26Dic2015
13. Schwarz, G.: Estimating the dimension of a model. Ann. Stat. **6**, 461–464 (1978)
14. SG (Sevilla Global). Barómetro de economía urbana 2013. Agencia Urbana de Desarrollo Integral. Ayuntamiento de Sevilla, Sevilla (2013)
15. Visser, I., Speekenbrink, M. depmixS4: an R package for Hidden Markov Models. J. Stat. Softw. **36**(7), 1–21 (2010)

Marked Point Processes for Microarray Data Clustering

Khadidja Henni, Olivier Alata, Abdellatif El Idrissi, Brigitte Vannier, Lynda Zaoui, and Ahmed Moussa

Abstract Microarray technologies become a powerful technique for simultaneously monitoring expression patterns of thousands of genes under different conditions. However, it is important to identify gene groups that manifest similar expression profiles and are activated by similar conditions. *ClusterMPP: Clustering by Marked Point Process* is a new microarray data clustering algorithm performed in two steps. The first one detects cluster modes representing regions of high density observations in the raw space. Based on the simulation of a proposed Marked Point Process by the well-known Reversible Jump Markov Chain Monte Carlo algorithm, where we consider several movements like birth and death, this algorithm step identifies prototype observations of each cluster. The second step of *ClusterMPP* is the K nearest neighbors (KNN) assignation that affects the remaining observations to the corresponding clusters. We experiment *ClusterMPP* on several complex and scalable microarray datasets. The results show the efficiency of *ClusterMPP* compared to well-known microarray data clustering methods like K-means, Spectral Clustering, and Mean-Shift.

K. Henni
Department of Computer Science, University of Sciences and Technologies Oran "Mohamed Boudia" USTO-MB, Oran, Algeria
e-mail: khadidja.henni@univ-usto.dz

O. Alata
Hubert Courien Laboratory, UMR 5516, Jean Monnet University, Saint-Étienne, France

A. El Idrissi
LabTIC Laboratory, ENSA-Tangier, Tangier, Morocco

B. Vannier
Receptors, Regulation and Tumor Cells (2RCT), University of Poitiers, Poitiers, France

L. Zaoui
LSSD Laboratory, Department of Computer Science, University of Science and Technology, Oran, Algeria

A. Moussa (✉)
SDET Team, ENSA-Tangier, Tangier, Morocco
e-mail: amoussa@uae.ac.ma

F. Palumbo et al. (eds.), *Data Science*, Studies in Classification, Data Analysis, and Knowledge Organization, DOI 10.1007/978-3-319-55723-6_11

1 Introduction

With the recent advances of biomedical technology, a lot of "OMICS" data from genomic, transcriptomic, and proteomic domain can now be collected quickly and cheaply. Microarray technology analyzes genome-wide gene expression patterns and is used in many areas including biotechnology, pharmacology, medicine, and environment.

Identifying the patterns hidden in gene expression data allows an enhanced understanding of functional genomics. However, the large number of genes and their complexity make difficult comprehending and interpreting the resulting mass of data. Therefore, high-throughput expression profiling exploitation requires advanced analysis tools to extract knowledge from the huge amount of data [8].

When experimental conditions are not known, an unsupervised treatment is recommended. Clustering is a powerful unsupervised technique which needs no a priori information, it helps to understand gene function, gene regulation, cellular processes in cell subtypes [8, 14]. Clustering aims to organize similar data into clusters in which data are similar to each other. This data can be genes, samples, or both genes and samples simultaneously (two-way clustering).

K-means [9], Self Organizing Map (SOM) [8], and Hierarchical clustering [8, 13] are widely used in gene expression analysis field. But they are unable to deal with noise, high dimensionality, complexity, and nonlinear separability associated with microarray gene expression data. K-means [9, 13] is a fast and simple algorithm, but it is sensible to initialization and number of clusters, it converges to local minima. SOM is a very used clustering algorithm with a visual output, but suffers from strong initialization dependency, outputs instability, and is powerless in unbalanced classes' cases. Hierarchical clustering [6, 13] is sensible to data modification, noise, and to outliers; they are also powerless over unbalanced classes and convex shapes. The biclustering algorithms (two-way clustering) [8, 14] are also widely used to cluster simultaneously gene and expression. They are organized in four families: (1) variance minimization methods, (2) two-way clustering methods, (3) motif and pattern recognition methods, and (4) probabilistic and generative approaches.

In gene expression data analysis, the quality and robustness of clustering results is crucial. Thus, the choice of clustering algorithm should be quite selective and depends on several issues such as: the separation of genes or samples whatever the complexity and overlapping level of clusters, the result accuracy, its stability (results do not depend on initialization nor parameters), its scalability, and its ability to handle noise.

Today, a huge number of clustering algorithms are available [8], with an impressive practical performance and desirable theoretical grantees. However, many of them are not able to cover all recent applications needs. Clustering algorithms must be multi-objective and cover all clustering issues, they should consider all existing highlights and minimize theirs disadvantages. Meanwhile, proposing a new clustering algorithm has become a multi-objective optimization problem. The solution is to identify all issues and find a way to take them all into account.

To this aim, we propose a new unsupervised clustering algorithm called "Clustering by Marked Point Process (*ClusterMPP*)," it is a multi-objective algorithm able to deal with cited issues. This algorithm belongs to the density-based algorithms family [10] and it is based on a probabilistic model called Marked Point Process (MPP) [1, 15, 16], used in imaging field to mark and to detect geometrical objects. The main idea of *ClusterMPP* is: first seeking cluster patterns to establish a classification model. To ensure that, it defines cluster as a dense region and simulates a proposed MPP to locate objects (hyper-spheres) on these regions of interest. It gives rise, kills, moves, and resizes objects, under interaction constraints, in attempt to place them in dense regions also called clusters.

At the end of this iterative object manipulating process done by an adapted Reversible Jump Markov Chain Monte Carlo (RJMCMC) [7], objects will be located on clusters. They will delimit cluster fundamental area (cluster mode) by their overlapping (connected component). Mostly, objects do not cover all available data, and covered data are named "Prototypes". The second step of *ClusterMPP* is to assign remaining data (data not covered by objects) to the detected clusters using an improved version of KNN algorithm [4].

In this chapter, Sect. 2 presents a background on clustering problem, introduces some definitions of the MPP theory, and describes a proposed MPP model. Section 3 gives details about "ClusterMPP" implementation. The performance of this clustering algorithm is demonstrated in Sect. 4 using benchmarks of microarray databases.

2 Background and Marked Point Process Model

Let us define the observed data (genes, samples) as multidimensional field Y, with $y = \{y_q\}_{q=1,\dots,Q}$, Q is the number of observations and $y_q = [y_{q,1}, \dots, y_{q,M}]^T \in \chi \subseteq \mathbb{R}^M$. M is the space dimension (number of conditions).

This study aims to organize the Q observations (genes, samples) in k clusters, where a cluster is a set of observations sharing similar M conditions. Each cluster groups genes with similar expression patterns (co-expressed genes) or samples based on the corresponding expression profiles. Clusters have a biological significance. They may further understanding many gene functions, reveal the similar cellular functions or sub-cell types which are hard to identify by traditional morphology-based approaches [8].

This purpose defines cluster by their modes; modes delimit the domains of high local observation concentration, called after prototypes. These cluster modes are assumed to be randomly distributed in the multidimensional raw data space, they reflect carefully the observation patterns inside the raw data space. Thus, the problem comes down to modes seeking by capturing the cumulative density distribution of observations in χ.

2.1 Definitions and Notations

MPP is a random process which models point patterns where the points are mainly positions or centers of geometrical marks in a multidimensional space [1, 15, 16]. If there is no interaction between points and no mark, the process is called Poisson Point Process and it plays a fundamental role in the definition of probability density functions of more advanced point processes [1]. When processes use neighboring relations between points, they belong to the family of Markov Point Processes. This kind of processes was used in statistical physics, under the name of Gibbs processes (see [1–3] for more details).

Let X be an MPP living in a finite simple point process, it is composed of finite random configurations of points; these points arc positions of marks; they are chosen randomly from a compact subspace $\chi \subseteq \mathbb{R}^M$ (χ has a finite measure: $\mu(\chi) < \infty$ with μ the Lebesgue measure). In this paper, a realization of X is a set of hyper-spheres $x = \{x_l\}_{l=1,\ldots,n(x)}$ (also called a configuration). Each hyper-sphere x_l is defined by its center $c_l \in \chi$ (a point) and its radius $r_l \in [r_{\min}, r_{\max}]$ (its mark) which also defines the object neighboring. $n(x)$ is the number of hyper-spheres in the configuration x. A hyper-sphere (or object or marked point) will be denoted $x_l(c_l, r_l)$ and the configuration of points of x, $p_x = \{c_l\}_{l=1,\ldots,n(x)}$. The spherical shape is flexible and easy to adapt in the multidimensional space.

The probability density functions $f(x)$ of the MPP are defined by the reference to the Poisson process. They can be expressed as a Gibbs distribution with an energy $U(x)$ [1]:

$$f(x) \propto e^{U(x)} \tag{1}$$

We assume now that a subset of observations constitutes the positions of a realization of the MPP X; this subset will be the initial set of cluster prototypes. The object of the proposed clustering algorithm is to find this subset by selecting the observations with an associated set of marks that maximize the probability density function of the MPP. This optimal object configuration, $\hat{x}$, is:

$$\hat{x} = argmax_{x, p_x \subseteq \chi} f(x) \tag{2}$$

The density f [see Eq. (1)] is related to a normalizing constant which cannot be calculated. A typical solution for this drawback is the use of Markov Chain Monte Carlo (MCMC) simulator. Several MPP simulators have been proposed in the literature [1, 15], but the one that has been mainly used these twenty last years is the RJMCMC algorithm [7]. This is due to the good convergence speed, its low computational time, and its flexibility. The RJMCMC samplers can employ several movements and use the Green's ratio (*GR*) [7] to accept or decline them. The *GR* depends on the move probability and the process state before and after the move application.

2.2 Proposed Model

The probability density function of the MPP is proportional to the exponential of a Gibbs energy function [see Eq. (1)]. The energy of MPP is classically the sum of two terms [1, 3]:

$$U(x|\theta) = U_{data}(x|\theta) + U_{inter}(x|\theta) \tag{3}$$

where θ is the set of model parameters. $U_{data}(x|\theta)$ is the data driven energy and $U_{inter}(x|\theta)$ is the internal energy.

The Data Driven Energy $U_{data}(x|\theta)$ represents relationships between objects and observed data. For our purpose, objects are placed in high density regions. This energy is the sum of local contributions of each object x_i (detector model):

$$U_{data}(x|\theta) = \sum_{x_i \in x} V(x_i) \tag{4}$$

where $V(x_i)$ is the potential function of hyper-sphere $x_i(c_i, r_i)$.

In order to obtain realizations with objects localized in high concentration areas, this function must favor the acceptation of well-positioned objects. Let us define $no(x_i)$, the number of covered observations by $x_i(c_i, r_i)$ and $do(x_i)$ the density of observations inside $x_i(c_i, r_i)$. We consider that an object x_i is well positioned if it satisfies two criteria: $no(x_i) > n_{\min}$ and $do(x_i) > d_{\min}$, where $n_{\min}$ and $d_{\min}$ belong to the set of model parameters θ.

Therefore, $V(x_i)$ can be expressed as follows:

$$V(x_i) = \begin{cases} no(x_i) & \text{if } no(x_i) > n_{\min} \ \& \ do(x_i) > d_{\min} \\ -no(x_i) - v_{\max} & \text{otherwise.} \end{cases} \tag{5}$$

where $v_{\max}$ is a high value that allows to greatly reduce the probability of a configuration which contains badly positioned objects. In order to strongly penalize objects that are not correctly localized, $v_{\max}$ must be a high value (equal to 1000, for example).

The Internal Energy $U_{inter}(x|\theta)$ interactions between objects are modeled by the use of potential functions which are chosen according to a priori information about searched configurations. Thus, two basic rules are imposed: (1) estimate the exact number of connected components by driving the creation of connected objects to the observed data and their dispersion in space. (2) Prevent the object overlapping phenomenon which leads to exponential growth of objects number. According to those considerations, we propose the following internal energy:

$$U_{inter}(x|\theta) = n(x)\log\beta - |co(x)|\log\gamma + n_v \log\delta \tag{6}$$

where $\beta > 0$ is the point process intensity [15]. The two first terms of this internal energy are those of the connected component process [2] with $co(x)$ the set of connected objects in x, $|co(x)|$ the number of connected components in x, and $\gamma > 0$ the interaction parameter. The connected component process is chosen regarding to the first constraint previously written. In the internal energy, each connected component defined by x, $co_i \in co(x)$, $i = 1, \ldots, |co(x)|$ has a contribution equal to $n(co_i)log\beta - log\gamma$ with $n(co_i)$ the number of hyper-spheres contained in co_i. See [15] for more details.

The third term of the internal energy is inspired from the pairwise point process such as defined in [15], and is used to penalize the hyper-sphere tangle. This term is based on the definition of the following neighboring relation, for $x_i \in x$ and $x_j \in x$, $i \neq j$:

$$x_i \sim x_j \text{ if } d(c_i, c_j) < (r_i + r_j)/5 \tag{7}$$

the denominator value was chosen experimentally. This third term is also based on the following interaction potentials of second order:

$$\phi(x_i, x_j) = \begin{cases} \log\delta \text{ if } x_i \sim x_j \\ 0 \quad\quad \text{otherwise.} \end{cases} \tag{8}$$

The contribution of this term in the internal energy is written as $\sum_{(x_i, x_j), i<j} \phi(x_i, x_j) = n_v \log\delta$ with n_v the total number of neighboring relations in x and $\delta \in [0, 1]$. In the following we will choose δ near to O (such that $\log\delta = -1000$, for example) in order to strongly penalize overlapping objects.

Following the definition of the data driven energy and the internal energy, it is now possible to define the set of parameters associated with the proposed MPP:

$$\theta = \{\beta, \gamma, n_{\min}, d_{\min}, r_{\min}, r_{\max}\}. \tag{9}$$

with $r_{\min}$ and $r_{\max}$, the minimum and maximum radius of the hyper-spheres, respectively.

3 ClusterMPP: A New Clustering Algorithm

ClusterMPP is a new density-based clustering algorithm of microarray datasets able to cover all main clustering requirements. This algorithm is composed of two steps:

- Cluster modes detection: is a simulation of the proposed MPP to capture and delimit the high density regions.
- Classification performing: is a classification process finalization, it classifies the remaining observations to the detected cluster modes.

3.1 Cluster Modes Detection

This section presents the mode detection algorithm (the first step of *ClusterMPP*). It is a sampling algorithm designed to discover regions of high observation concentration by simulating the proposed MPP model (see Sect. 2.2). The proposed mode detection algorithm is a variant of RJMCMC algorithm, it is an iterative and three-step procedure that generates an MPP configuration at each iteration. In what follows, we will write: x^i, $i \geq 0$, the configuration of objects at the ith iteration of *ClusterMPP*. p_{birth}, p_{death}, p_{disp}, and p_{mchg} are, respectively, movement probabilities choose: birth, death, move, or changing radius of an object.

MPP Initialization In this step, *ClusterMPP* generates an initial configuration $x^0 = \{x_l\}_{l=1,\ldots,n(x^0)}$: an objects set which cover all available observations (see Algorithm 2). Parameter estimation() is a function which estimates parameter values as follows:

- β is the intensity parameter of process. The desired number of objects in the proposed MPP realizations depends on observed data, we propose to fix it equal to the cardinality of y ($n(y)$).
- γ is the interaction parameter of a connected component Markov Point Process. γ can be chosen equal to the variance of the observed data. Here we focus on multidimensional datasets, we use the total variance (the trace of the variance–covariance matrix, i.e., the sum of variances).

Algorithm 2 Initialization

Input: Observed data $y = \{y_1, \ldots, y_{n(y)}\}$, $r_{\max}$.
Output: $x^0 = \{x_1, \ldots, x_{n(x^0)}\}$, θ.
$x_1 = (y_1, r_{\max})$
$x^0 = \{x_1\}$
$n(x^0) = 1$
$j = 1$
while ($j \leq n(y)$) **do**
  $l = 1$
  while ($l \leq n(x^0)$) **do**
    if ($d(y_j, c_l) > r_{\max}$) **then**
      $n(x^0) = n(x^0) + 1$
      $x_{n(x^0)} = (y_j, r_{\max})$
      $x^0 = x^0 \cup \{x_{n(x^0)}\}$
    end if
    $l = l + 1$
  end while
  $j = j + 1$
end while
θ =Parameter estimation()

- $n_{\min}$ and $d_{\min}$ are the parameters defining the data driven energy, they are calculated from the initial configuration and equal to the average number of covered points by hyper-spheres and the average density of objects inside hyper-spheres, respectively.
- $r_{\min}$ and $r_{\max}$ describe the object scales: the first one is the minimum value of object radii. We propose to fix it equal to the minimum distance between points in y, in order to have objects that cover at least two observations. $r_{\max}$ is the maximum of the object radii, it is estimated by a learning stage.

MPP Simulation *ClusterMPP* repeats the following steps, until the stabilization of the number of objects and the process energy: at the $(i+1)$th iteration a random draw to select one of the four movements: birth, death, moving, or changing marks. In the *birth* movement, a new object $\omega(c_\omega, r_\omega)$ is created by drawing randomly a center from y and choosing randomly a radius in $[r_{\min}, r_{\max}]$. Next, ω will be added to the configuration $\tilde{x}_b = x^i \cup \{\omega\}$. The *death* is performed if the configuration x^i contains at least one object $(n(x^i) > 0)$. The movement is simulated by selecting an object ω randomly from the current configuration x^i. Then ω is removed from x^i. The proposal configuration in a death case becomes $\tilde{x}_d = x \backslash \{\omega\}$. *Move* and *changing marks* movements are performed if the configuration x^i contains at least one object $(n(x^i) > 0)$. Movements are simulated by selecting an object ω randomly from the current configuration x^i. Choosing randomly a new center $c_{\tilde{\omega}}$ from the observed data field y (*Moving* case) or a new radius $r_{\tilde{\omega}}$ from the interval $[r_{min}, r_{max}]$ (*changing marks* case). Then, ω is replaced by $\tilde{\omega}$, $\tilde{x} = \{x^i \backslash \{\omega\}\} \cup \{\tilde{\omega}\}$.

In order to accept or reject movement, the algorithm computes the *GR* for each move (see Table 1). This iterative algorithm step manipulates the MPP by applying different movements on this process objects. It shifts the MPP from the initial configuration in which objects cover all observations to the configurations where objects move toward the desired regions, the connected components give the searched modes. Details of the MPP simulation step are given in Algorithm 3.

Mode Extraction The final configuration will contain objects located in regions of high concentration of observed data. Thus, the connected components of objects give the searched modes. For each component, *ClusterMPP* extracts all covered observations (prototypes) and assigns them by trivial way to the corresponding clusters. The remaining observations are non-prototypes, they will be classified in the second step of *ClusterMPP*.

Table 1 Green's ratio [7]

Movements	Birth	Death	Move or changing marks
Green's ratio	$GR^{Birth} = \frac{p_{death}}{p_{birth}} \frac{f(\tilde{x}_b)}{f(x^i)} \frac{\mu(\chi)}{n(x)+1}$	$GR^{Death} = \frac{p_{birth}}{p_{death}} \frac{f(\tilde{x}_d)}{f(x)} \frac{n(x)}{\mu(\chi)}$	$GR^{disp/mchg} = \frac{f(\tilde{x})}{f(x)}$

Algorithm 3 MPP simulation algorithm

Input: Observed data $y = \{y_1, \ldots, y_{n(y)}\}$, x^0, θ.
Output: x^f.
$i = 0$
while not converged **do**
  $\tilde{x}$=Movement simulation (y, x^i, θ)
  Compute density $f(\tilde{x})/f(x^i)$ using Eq. (1) and Eq. (3).
  Compute the corresponding Green's ratio GR (Table 1).
  $\alpha = \min(1, GR)$
  Draw $accept \sim U_{(0;1)}$
  if $accept < \alpha$ **then**
    $x^{i+1} = \tilde{x}$
  else
    $x^{i+1} = x^i$
  end if
  $i = i + 1$
end while

3.2 Classification Performing

Prototype observations will be directly assigned to the corresponding clusters. They are mostly well classified in the first stage of "ClusterMPP". So, the problem lies basically in the classification of the remaining observations (non-prototype observations). In order to classify all observations, we propose to use an improved version of the KNN algorithm [4], which assigns non-prototype observations one by one to the nearest cluster, in a specific order, respecting their distances to the prototype observations. *ClusterMPP* detects different prototype observations and assigns non-prototype observations to the corresponding clusters.

4 Experiments

Distance functions are an important factor of clustering procedures. They measure the similarity between two observations. In this chapter, *ClusterMPP* uses the Euclidean distance which measures the geometric relation between two vectors (the generalization of other distances poses no particular problem). *ClusterMPP* is compared with three well-known algorithms *Mean-Shift*, *Spectral Clustering*, and *K-means*. Note that *K-means* [9] and *Spectral Clustering* [12] require the number of clusters K. *Mean-Shift* [17] requires the use of an appropriate value of bandwidth. *ClusterMPP* does not need a priori knowledge and parameters are chosen from observed data. All five algorithms were tested with six benchmarks of microarray datasets [5], used to validate performance of clustering algorithms (see Table 2).

Table 2 Datasets description [5]

Datasets	Rat CNS	Leukemia	Lymphoma	NCI60	Novartis	Yeast cell cycle
Samples	17	100	100	200	1000	72
Attributes	112	38	80	57	103	698
Classes	6	3	3	8	4	5

Assessment methods of clustering algorithms measure how well a computed clustering solution agrees with the gold solution [5] for the given dataset, where the gold solution is a known data partition. We propose to use Receiver Operating Characteristic curve (ROC) [5] and External validation indexes (Balanced Misclassification Index [5] (BMI), Rand index, Jaccard coefficient (JC), Folkes and Mallows index (FM)) [11] to evaluate the performance of *ClusterMPP* algorithm by comparing it with other algorithms.

Receiver Operating Characteristic Curve (ROC) is a graphical technique to compare classifiers and visualize their performance. ROC plane maps the True Positive Rate TPR (sensitivity) versus False Positive Rate FPR (specificity) [5].

External Validation Indexes different external validation indexes have been used: (1) BMI [5] compares the performance of different clustering algorithms by measuring their ability to capture the structure in a dataset. BMI uses the misclassification error rate and the balancing error rate (the average of the errors on each clusters [5]). The BMI index takes values between 0 and 1 and needs to be minimized. (2) RS, JC, and FM [11], these indexes are computed by four terms which indicate if a pair of points in both solutions (gold and resulting solution) share the same cluster. These indexes need to be maximized.

Figure 1 shows the partition in the ROC plane for each considered algorithm and for each dataset. We plot different *ClusterMPP* results obtained by varying the value of r_{max} parameter, and the best result of the other algorithms. Figures 2, 3, and 4 display the comparison of the classification error rates and the external validation indexes (BMI, RS, JC, and MF). *ClusterMPP* has superior performance against *Mean-Shift* and *Spectral Clustering* even in the worst cases. *ClusterMPP* outperforms *K-means* on most datasets and its strength is the estimation of its important parameter (r_{max}) through a learning step. However, *K-means* requires the number of clusters k, which can be computed by several methods like the Bayesian Information Criterion (BIC) [9].

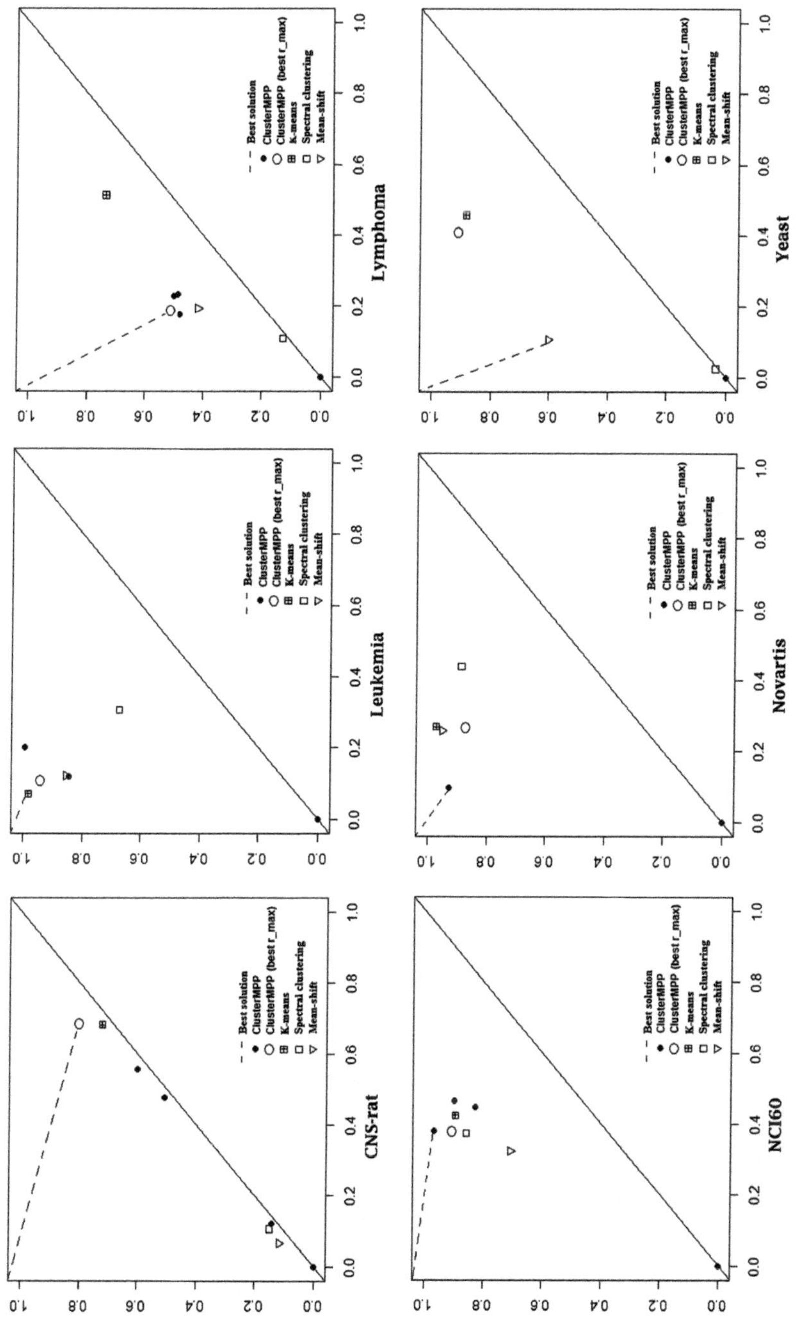

Fig. 1 ROC curves for each dataset. Each sub-figure refers to a dataset and each position in the ROC curves refers to a result solution of each algorithm (*ClusterMPP*, *K-means*, *Mean-Shift*, and *Spectral Clustering*)

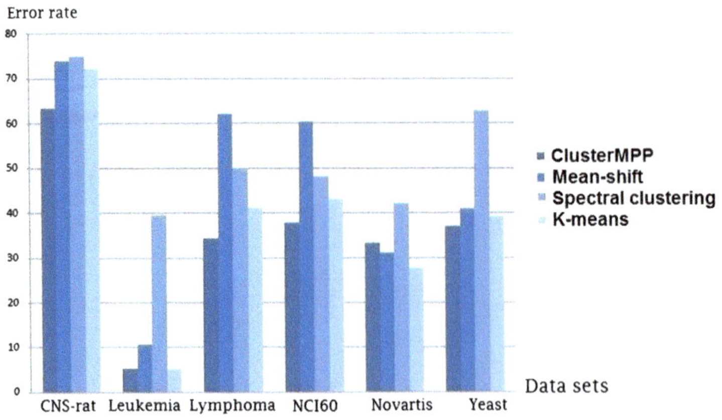

Fig. 2 Comparison of classification results based on error rates

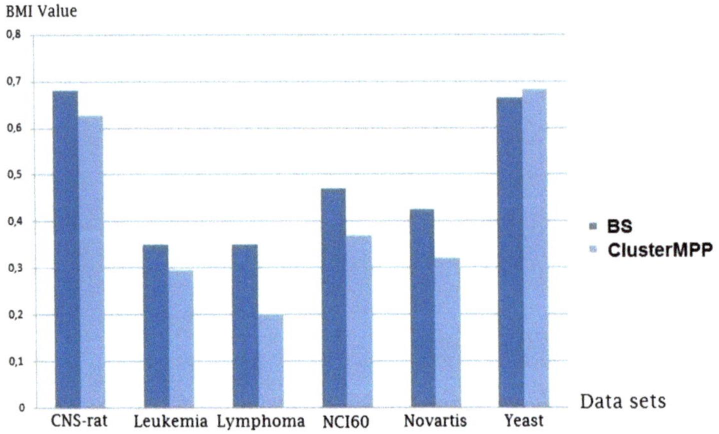

Fig. 3 Comparison of classification results based on BMI values (BS denotes the best solution obtained in [5])

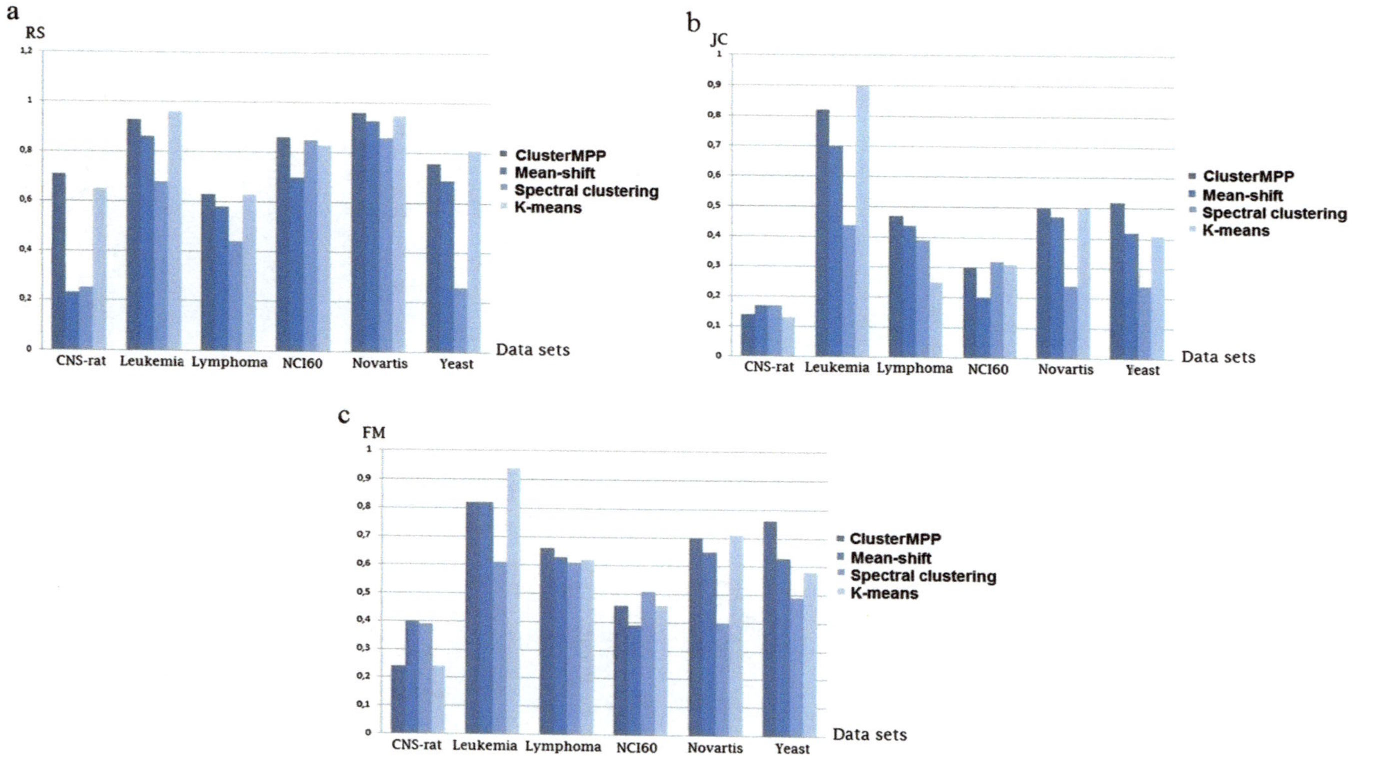

Fig. 4 Comparison of classification results using the external validation indexes: (**a**) Rand static index, (**b**) Jaccard coefficient index, (**c**) Folkes and Mallows index

5 Conclusion

This work was intended to describe a new unsupervised clustering algorithm belonging to density-based family. It also implements a probabilistic technique, which makes it able to solve the clustering problem taking into account different issues. The algorithm seeks cluster modes by the simulation of proposed MPP and use an improved KNN version to finalize the classification process. *ClusterMPP* outperforms the other clustering algorithms. In the future, we will integrate ontological information about genes as an a priori information to improve clustering process of biological data.

Acknowledgements This research was supported in part by the Erasmus Mundus—Al Idrisi II program.

References

1. Alata, O., Burg, S., Dupas, A.: Grouping/degrouping point process, a point process driven by geometrical and topological properties of a partition in regions. Comput. Vis. Image Underst. **115**(9), 1324–1339 (2011)
2. Chin, Y.C., Baddeley, A.J.: Markov interacting component processes. Adv. Appl. Probab. **32**(3), 597–619 (2000)
3. Clifford, P.: Markov random fields in statistics. In: Grimmett, G.R., Welsh, D.J.A. (Eds.) Disorder in Physical Systems, A Volume in Honour of J.M. Hammersley, pp. 19–32. Clarendon Press, Oxford (1990)
4. Ferrandiz, S., Boullé, M.: Bayesian instance selection for the nearest neighbor rule. Mach. Learn. **81**(3), 229–256 (2010)
5. Giancarlo, R., Bosco, L., Pinello, G.L., Utro, F.: A methodology to assess the intrinsic discriminative ability of a distance function and its interplay with clustering algorithms for Microarray data analysis. BMC Bioinformatics **14**(S-1), S6 (2013)
6. Gorunescu, F.: Data Mining: Concepts, Models and Techniques. Intelligent Systems Reference Library, vol. 12, pp. 1–43. Springer, Berlin (2011)
7. Green, P.J.: Reversible jump Markov chain Monte Carlo computation and Bayesian model determination. Biometrika **82**, 711–732 (1995)
8. Harun, P., Burak, E., Andy P., Çetin, Y.: Clustering of high throughput gene expression data. Comput. Oper. Res. **39**(12), 3046–3061 (2012)
9. Kaur, S., Kaur, U.: A survey on various clustering techniques with K-means clustering algorithm in detail. Int. J. Comput. Sci. Mob. Comput. **2**(4), 155–159 (2013)
10. Khaled, S.: TOBAE: a density-based agglomerative clustering algorithm. J. Classif. **32**(2), 241–267 (2015)
11. Liu, Y., Li, Z., Xiong, H., Gao, X., Wu, J.: Understanding of internal clustering validation measures. In: ICDM-10 Proceedings of the 2010 IEEE International Conference on Data Mining, pp. 911–916 (2010)
12. Mouysset, S., et al.: Spectral clustering: interpretation and Gaussian parameter. In: Data Analysis, and Knowledge Organization. Studies in Classification, vol. 4, pp. 153–162 (2013)
13. Reddy, C.K., Vinzamuri, B.: A survey of partitional and hierarchical clustering algorithms. In: Aggarwal, C., Reddy, C.K. (eds.) Data Clustering: Algorithms and Applications, pp. 87–110. CRC (2014)

14. Sepp, H., et al.: FABIA: factor analysis for bicluster acquisition. Bioinformatics. **26**(12), 1520–1527 (2010)
15. Stoica, R.S., Gay, E., Kretzschmar, A.: Cluster pattern detection in spatial data based on Monte Carlo inference. Biom. J. **49**(4), 505–519 (2007)
16. Stoica, R.S., Martinez, V.J., Saar, E.: Filaments in observed and mock galaxy catalogues. Astron. Astrophys. **510**(38), 1–12 (2010)
17. Wu, K.L., Yang, M.S.: Mean shift-based clustering. Pattern Recogn. **40**(11) 3035–3052 (2007)

Social Differentiation of Cultural Taste and Practice in Contemporary Japan: Nonhierarchical Asymmetric Cluster Analysis

Miki Nakai

Abstract This paper is an application of a new procedure of cluster analysis which deals with asymmetric similarities to sociological data. The aim of this study is to explore how cultural practices in contemporary Japan are classified, and to assess how social class and gender are related to the difference in the constellation of cultural taste. Nonhierarchical asymmetric cluster analysis is used to gain better understanding the underlying structure of cultural taste of some social subgroups in contemporary Japan. Drawing on the national survey which provides detailed information on a broad range of cultural participation of the 2915 respondents from a nationally representative sample in Japan collected in 2005, the paper demonstrates that there are some notable dissimilarity in cultural participation practices between genders, as well as among social classes. The result shows that middle-class women has distinct cultural taste patterns from other social class groups. This finding suggests that they define boundaries and form cultural segmentation. The results obtained from the application of the clustering approach to the data are compared to those obtained from the asymmetric multidimensional scaling. Results are basically consistent.

1 Introduction: Cultural Consumption and Social Class

In sociological debate on culture and social class, cultural consumption patterns and lifestyles can be seen to generate social boundaries so as to separate people into groups. In Bourdieu's well-known theory of social and cultural reproduction, he differentiated "highbrow", or elite culture, and popular culture [3]. People in higher social class position tend to patronize the fine arts and avoid contact with popular entertainment. On the other hand, people in lower position tend to enjoy "brutish" popular entertainment.

M. Nakai
Department of Social Sciences, College of Social Sciences, Ritsumeikan University, 56-1 Tojiin Kitamachi, Kyoto 603-8577, Japan
e-mail: mnakai@ss.ritsumei.ac.jp

F. Palumbo et al. (eds.), *Data Science*, Studies in Classification, Data Analysis, and Knowledge Organization, DOI 10.1007/978-3-319-55723-6_12

Other sociologists have emphasized that the cultural omnivorousness, or consumption of wider range of cultural activities, is a necessary attribute of the elite in contemporary post-industrial diverse societies. Numerous studies have attested the shift from elitist snobbishness to omnivorous cultural taste among higher status people (e.g. [4, 15–17]).

2 Structure of Cultural Taste: Theory and Hypothesis

While the relationship between one's class position and the cultural tastes is supported by a lot of empirical evidence, there are persistent and strong skeptical commentaries on the relationship between social class and cultural consumption in Japan. Some scholars have questioned the idea of class-based patterns of lifestyle and consumption in post-modern societies.

The aim of this paper is to explore the structure of cultural taste based on survey data on cultural consumption. Nonhierarchical asymmetric cluster analysis is used to gain better understanding the way in which cultural consumption profiles are associated with each other and how it varies by social groups of the respondents [12].

In this study, we hypothesize that the differences in cultural consumption patterns are interpretable not only in terms of social class but also in terms of gender. A lot of studies of cultural participation have consistently found that women are more likely to participate in traditional highbrow cultural activities than men (e.g. [5, 7, 8]). What is appropriate and what is improper is highly segregated by gender regarding not only job, occupation and working life, but also cultural consumption. Such may be the case in many countries, but the situation surrounding gendered division of labour remains much more striking in Japan. Given the fact that the ideology of gendered division of labour is strong in Japanese society, clusters might emerge differently according to gender. We expect that cultural participation varies according to the gender of the respondents.

3 Data and Method

3.1 Data

The data is from a nationally representative survey of the Social Stratification and Social Mobility (SSM) conducted in 2005 in Japan. Of approximately 13,000 men and women sampled, aged 20–69, 5742 (2660 men and 3082 women) were successfully interviewed. We analysed 2915 (1317 men and 1598 women) respondents to portray the structure of cultural participation.[1]

This data provides a wide range of socio-demographic characteristics and information about the frequency of participation in cultural activities and consumption of each respondent. Respondents are asked to indicate how often they had participated in each cultural activity during the past several years. Answers have originally scored on a five-point scale. We grouped into two categories: engaged in and not engaged in.

The following nine variables on the activities of cultural consumption have been selected for investigation: (1) classical music performances and concerts **(Classical)**, (2) museums and art exhibitions **(Museum)**, (3) taking hobby and cultural lessons **(Lessons)**, (4) travel abroad **(Abroad)**, (5) volunteer activities **(Volunteer)**, (6) public library use **(Library)**, (7) reading novels, or books about history **(Novels)**, (8) karaoke **(Karaoke)**, (9) reading tabloid papers or downmarket women's weeklies **(Tabloid)**.

Among these cultural items, the first two items (1) and (2) can often be regarded as highbrow activities, from (3) to (7) as middlebrow, and (8) and (9) as lowbrow activities in terms of the so-called cultural prestige. The words in bold in parentheses are used to represent each cultural practice in the following tables.

Because it is of interest here to look at class and gender related differences of cultural engagement pattern, respondents are subgrouped into seven groups based on their educational level and gender. We define people's social class in terms of their educational credentials, not in terms of their occupational status, since quite a few female respondents are not in the work force at the time of the survey. The groups are: (a) male primary education graduates, (b) male secondary education graduates, (c) male tertiary education graduates, (d) female primary education graduates, (e) female secondary education graduates, (f) female lower tertiary education (2-year college) graduates, and (g) female upper tertiary education (4-year college or university) graduates.

The number of respondents in each group is 178, 635, 425, 231, 927, 191, and 148, respectively.

The input data consists of seven 9×9 contingency tables shown in Table 1 (three out of seven tables are shown). Each of seven proximity matrices is one-mode two-way similarity for the mutually exclusive groups of the respondents based on social class and gender, and thus input data set is two-mode three-way data. The (i, j)-th element in the asymmetric proximity matrices indicates the percentage

[1]We analysed 2915 respondents, approximately half of the respondents collected, who were asked and responded to all the questions concerning a wide range of cultural consumption and practices as well as their socioeconomic characteristics. The SSM survey in 2005 implemented a module design. The questionnaire has two parts: (1) the core component, consisting of fixed questions such as respondents' socioeconomic background and educational and occupational history, which are asked repeatedly across survey years, and (2) optional modules, which are additional questions on specific themes. Cultural consumption was chosen for the theme for the module, and therefore half of the people sampled were asked for information about their cultural activities.

Table 1 Part of seven 9 × 9 contingency tables of cultural consumption

	(1)	(2)	(3)	(4)	(5)	(6)	(7)	(8)	(9)
(a) Male primary education graduates (n = 178)									
(1) Classical	100.0	52.9	23.5	35.3	41.2	20.0	58.8	82.4	88.2
(2) Museum	25.7	100.0	29.4	47.1	52.9	44.1	68.4	70.6	82.4
(3) Lessons	16.7	40.0	100.0	25.0	45.8	34.8	48.0	52.0	62.5
(4) Abroad	21.4	57.1	21.4	100.0	42.9	21.4	67.9	57.1	64.3
(5) Volunteer	15.2	39.1	25.0	27.3	100.0	22.2	54.3	60.9	80.0
(6) Library	10.3	51.7	26.7	20.0	33.3	100.0	65.5	75.9	82.1
(7) Novels	15.6	37.5	18.8	29.7	39.1	30.6	100.0	61.9	84.4
(8) Karaoke	15.2	25.8	14.1	17.6	30.4	24.4	41.9	100.0	75.8
(9) Tabloid	12.1	22.4	12.2	14.6	28.8	19.2	43.2	56.1	100.0
(b) Male secondary education graduates (n = 635)									
(1) Classical	100.0	75.0	48.8	46.9	4.8	61.3	88.8	76.3	90.1
(2) Museum	26.8	100.0	34.2	43.4	45.6	55.9	83.1	76.3	85.8
(3) Lessons	23.8	46.1	100.0	36.9	49.4	44.3	68.9	75.7	80.5
(4) Abroad	21.1	54.1	34.3	100.0	42.3	48.6	70.2	76.9	82.9
(5) Volunteer	19.7	51.5	41.2	37.7	100.0	44.3	74.3	73.7	83.3
(6) Library	23.1	58.8	34.7	41.3	42.3	100.0	83.3	75.1	88.8
(7) Novels	19.0	50.0	30.7	33.9	39.9	48.0	100.0	75.5	88.0
(8) Karaoke	13.6	38.1	28.4	31.3	33.6	35.8	63.0	100.0	84.0
(9) Tabloid	14.7	39.1	27.5	30.3	34.2	38.5	66.6	76.2	100.0
(c) Male tertiary education graduates (n = 425)									
(1) Classical	100.0	85.6	40.4	55.8	39.4	76.7	94.2	84.6	80.8
(2) Museum	36.6	100.0	35.0	47.1	37.3	77.7	95.1	79.4	84.4
(3) Lessons	33.9	68.5	100.0	59.2	43.2	77.2	87.9	82.3	78.4
(4) Abroad	32.0	63.2	40.7	100.0	40.7	63.5	89.1	81.9	83.6
(5) Volunteer	30.1	66.4	38.6	52.9	100.0	69.3	92.1	81.0	85.6
(6) Library	31.1	74.0	37.3	44.9	37.1	100.0	94.9	80.8	86.3
(7) Novels	27.1	64.1	30.0	44.5	35.4	67.5	100.0	76.0	83.5
(8) Karaoke	27.8	61.1	32.2	47.0	35.0	65.2	86.8	100.0	84.2
(9) Tabloid	24.8	60.6	28.8	44.7	34.9	65.1	88.9	78.5	100.0

of respondents who participate in cultural activity j among the respondents who participate in cultural activity i. Therefore diagonal elements are 100.

3.2 Method

We analysed data by using nonhierarchical asymmetric cluster analysis [12]. This is an extension of k-means cluster analysis which deals with asymmetric similarities. Each cluster consists of a dominant (central) object and other less dominant objects for the predetermined number of clusters. The purpose of this procedure is to find K clusters which maximize the goodness of fit with taking the concept of dominant object into account as well [12–14].

The dataset here appears to have asymmetric properties, which means that, for example, a person who engages in classical music concert tends to engage in reading novels, and a person who engages in reading novels tends to engage in Karaoke; but opposite direction might not be the case. When data are intrinsically asymmetric, which frequently happens in case of preferences, exchanges, migration data, and confusion data, we should not dealt with the data by ignoring the asymmetry (e.g. [20]).

Moreover, we compare the results obtained by this technique with those obtained by the Asymmetric Multidimensional Scaling (MDS). A number of researchers have recommended the application of both MDS and cluster analysis separately to the same proximity data to uncover complementary aspects of the underlying structure in the data (e.g. [1, 2, 6, 10]). In the present research, we first employ a clustering technique and then we compare the findings with those we already got in earlier study at which we applied asymmetric MDS to the same dataset [9] since it has been suggested that the relationship between MDS and cluster analysis may be complementary.

4 Results

4.1 Determining the Number of Clusters

The data were analysed by the cluster analysis above-mentioned for the number of clusters $K = 1$ through 5. To determine the number of clusters, we first examine the goodness of fit (GOF) statistics of proposed models. Obtained GOFs for $K = 1$ through 5 are shown in Tables 2 and 3. According to the GOF statistics, the models with one or two clusters might fit the data for all the male groups (a, b, and c) and the female groups who graduate from primary and secondary education (d and e), whereas the models with two or three clusters might fit the data for the female tertiary-level graduates (f and g), because the preferred models are those that maximize the values of GOF.

Table 2 Goodness of fit: male groups, $K = 1$ through 5

	(a) Primary	(b) Secondary	(c) Tertiary
$K = 1$	233,313,737	251,472,611	248,728,089
$K = 2$	226,561,899	249,911,211	247,861,402
$K = 3$	212,443,769	238,965,436	244,769,994
$K = 4$	197,066,940	206,201,260	224,969,630
$K = 5$	169,429,629	172,917,014	200,613,431

Table 3 Goodness of fit: female groups, $K = 1$ through 5

	(d) Primary	(e) Secondary	(f) Lower tertiary	(g) Upper tertiary
$K=1$	193,100,372	214,132,961	170,474,708	141,330,134
$K=2$	185,035,624	209,612,046	187,132,730	147,383,659
$K=3$	174,499,692	198,490,554	177,700,154	146,400,289
$K=4$	153,132,648	177,341,133	166,293,053	141,918,558
$K=5$	126,970,651	151,598,194	150,728,686	128,379,691

With careful examination of both interpretation of the content of the clusters and the values of GOF, the one cluster results were adopted as the solution for all the male groups as well as the female groups who graduate from primary and secondary level of education. For the two female groups of tertiary-level graduates, the two cluster results were adopted as the solution.

4.2 *Asymmetric Clustering of Cultural Consumption*

The overall results for each of the social groups are shown in Table 4. Without ignoring the asymmetry, and taking a dominant object and other less dominant objects into account, we determine that the number of cluster is one in regard to all the social groups except the groups of female tertiary-level graduates. Moreover, we find that the dominant cultural activity is "Tabloid" for all the one cluster results. In this data, whether or not particular activity is chosen as dominant object seems to reflect the differences in popularity of the cultural activity in particular cluster. Therefore, the result suggests that reading tabloid papers or down-market women's weeklies is likely the most mass appeal leisure activity for most social groups except some groups of highly educated women. Having said that, no matter what a material that is an object of reading may be selected, engagement in reading activity is prevalent cultural activity for all the groups of respondents.

In contemporary Japan, mass culture, several examples of which are manga (comics), karaoke, and pachinko, has achieved widespread popularity. Furthermore, the common assertion about newspapers in Japan is that there is no distinction between quality papers and popular papers as in European countries. At the same time, Japan has many sensationalist, scandal-hungry, and exposé-oriented tabloid newspapers and weekly magazines whose approaches contrast sharply with those of the established and sanitized broadsheet newspapers [18]. These papers and magazines target commuting white-collar workers as well as general readers and sell in large quantities. The result here can be understood in the light of that argument.

On the other hand, for the female lower tertiary (2-year college) graduates, the dominant cultural activities are "Novels" in cluster 1, and "Tabloid" in cluster 2, both of which, again, appeal to a large percentage of consumers. This means that among women, especially highly educated, connection among cultural genres is

Table 4 Cluster solution by social groups

(a) Male primary education graduates		
Number of cluster = 1		
Cluster	Dominant object	Other objects
1	Tabloid	Classical, Museum, Lessons, Abroad, Volunteer, Library, Novels, Karaoke
(b) Male secondary education graduates		
Number of cluster = 1		
Cluster	Dominant object	Other objects
1	Tabloid	Classical, Museum, Lessons, Abroad, Volunteer, Library, Novels, Karaoke
(c) Male tertiary education graduates		
Number of cluster = 1		
Cluster	Dominant object	Other objects
1	Tabloid	Classical, Museum, Lessons, Abroad, Volunteer, Library, Novels, Karaoke
(d) Female primary education graduates		
Number of cluster = 1		
Cluster	Dominant object	Other objects
1	Tabloid	Classical, Museum, Lessons, Abroad, Volunteer, Library, Novels, Karaoke
(e) Female secondary education graduates		
Number of cluster = 1		
Cluster	Dominant object	Other objects
1	Tabloid	Classical, Museum, Lessons, Abroad, Volunteer, Library, Novels, Karaoke
(f) Female lower tertiary education graduates		
Number of cluster = 2		
Cluster	Dominant object	Other objects
1	Novels	Classical, Museum, Volunteer
2	Tabloid	Lessons, Abroad, Library, Karaoke
(g) Female upper tertiary education graduates		
Number of cluster = 2		
Cluster	Dominant object	Other objects
1	Library	Volunteer
2	Novels	Classical, Museum, Lessons, Abroad, Karaoke, Tabloid

different from other social groups. For the female upper tertiary (4-year college or university) graduates, the dominant objects are "Library" in cluster 1, and "Novels" in cluster 2. This seems to indicate that female groups of tertiary graduates have a quite distinct cultural consumption structure from other groups: the cultural preference or social activities is differentiated.

As for the lower tertiary graduates, Novels, Classical, Museum, and Volunteer belong together in the same cluster, and other activities in the other cluster.

The former styles of cultural consumption (the first cluster) may be identified as inclusive elitist highbrow cultural consumption, meaning that appreciation of both highbrow and middlebrow activities. While the latter activities range from popular lowbrow to middlebrow.

On top of that, as for the women with the highest educational credentials, Library and Volunteer are classified into the same cluster, and other cultural consumption practices are in the other cluster. The former seems to mean preference for a limited range of relatively middlebrow cultural taste, whereas the latter range from highbrow to lowbrow. Taking another look at the cluster results from a different perspective, the former consists of non-profit-making or non-commercially motivated leisure activities, whereas the latter consists of items that are consumed as cultural commodities with a great variety of cultural activity from highbrow to popular mass culture. Not only the forms of popular entertainment such as karaoke or tabloid newspapers, but also highbrow culture such as visiting classical music concerts or exhibition at art galleries and museums are products to be consumed in most industrialized societies. Cluster 2 corresponds to that taste. For this social group, however, popular entertainment such as reading tabloid papers or down-market women's weeklies is not the dominant object. This means that they tend not to consume lowbrow popular cultural activity as much as other social groups consume. Women are expected to devote themselves to the production of symbolic status through the so-called vicarious consumption [19]. The reason why highly educated middle-class women perform differently from other social groups is that the wives of middle-class men tend to behave in order to become the ideal "good wife, wise mother". Many Japanese highly educated women still withdraw from the labour force upon the birth of children and stay outside the labour market when children are young. They often cannot resume their regular employment pattern. Many of them presumably engage in certain activities run by volunteers or non-profit-making organization to achieve their role and to differentiate themselves.

4.3 Examining the Results by Clustering and MDS

The results by the nonhierarchical asymmetric cluster analysis procedure are then compared with the results obtained in our earlier study using alternative methods of clustering and scaling, or asymmetric MDS, to the same data sets [9, 11]. We chose the two-dimensional result as the best solution in regard to asymmetric MDS based on interpretability as well as the value of the badness-of-fit measure. The two-dimensional common object configuration is illustrated in Fig. 1. Each cultural activity is represented by a point and a circle in a two-dimensional configuration. The distance between the points represents the symmetric proximity relationship, and the distance between the radii of the circles represents the asymmetry. Larger radius means that people who consume a certain activity have a larger tendency of consuming other cultural activities, and that people who consume other activities have a lesser tendency of consuming that activity. In contrast, small radius means

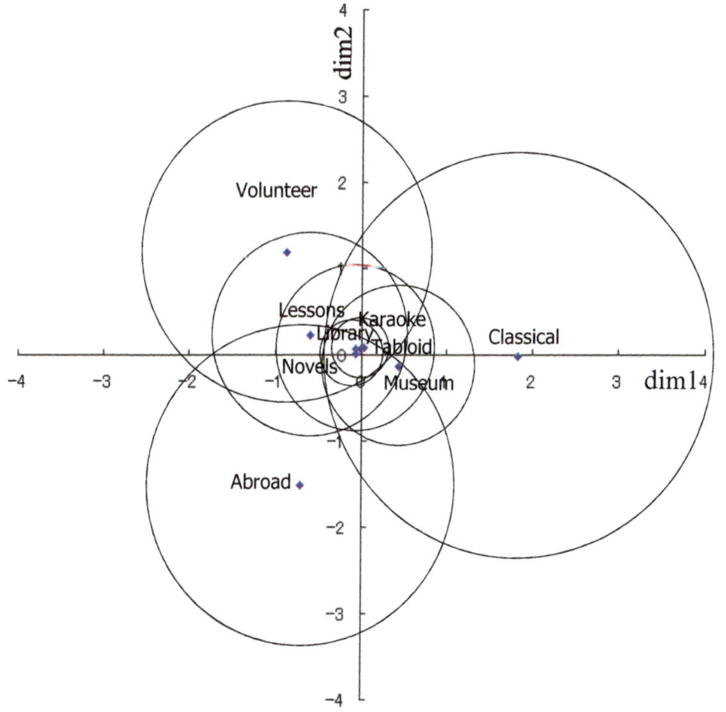

Fig. 1 Two-dimensional plots of asymmetric MDS analysis of cultural practices

that people who consume other activities than that cultural activity tend to consume that activity, but those who consume that activity have lower probability to consume other activities simultaneously. Moreover, the items near the origin of the configuration in the results of MDS are considered to be more popularized cultural activities rather than having limited consumer. Figure 1 shows that the activity which has the smallest radius is "Tabloid" and it means that reading tabloid or women's weeklies is the most popular activity in general.

The cluster results obtained here contain the same information as the results obtained based on MDS. Clustering also reveals that the dominant object is Tabloid for the most social groups as mentioned before. MDS also revealed that Tabloid had the smallest radius and located close to the origin in the two-dimensional configuration. Therefore, the results of clustering technique are consistent with those of MDS.

However, cluster results disclose that among highly educated women, there are two clusters, one of which contains of non-commercially motivated activities and the other cluster contains broader cultural activities, and that their dominant objects are different from other social groups.

MDS does not detect the clusters of the cultural activities which reflect people's practices as well as boundaries among social groups. Therefore, the findings based on nonhierarchical asymmetric cluster analysis add to our knowledge about the differentiated cultural taste in terms of class and gender.

5 Conclusion

In this paper, we have classified taste and cultural lifestyle and shown the difference in connecting pattern among cultural activities in terms of social groups in Japan based on nonhierarchical asymmetric cluster analysis. Social class related and gender related differences in the constellation of cultural taste revealed here seem to be explained partly by deep-rooted gender role expectation in Japanese society. Japan is known to be one of the least gender-equal societies among advanced industrial societies, where people believe in conservative gender roles and are much less supportive of gender egalitarianism than other societies. Moreover, early socialization process in the family, as well as in the school, is highly gendered in Japan. With the strong ideology of gendered division of labour in Japanese society, women's cultural activities and characteristics tend to conform to gendered cultural norms. Men might be inclined to more popular genres for gaining career building advantages irrespective of their social status. On the other hand, women are expected to play an important role to reproduce their social status, and are socialized to conform to certain gender roles. As a result, there exists certain differentiation in cultural taste.

The exact mechanism of the gender differences in cultural tastes remains to be solved in future research. Further investigation must be undertaken in these areas utilizing the data of a follow-up study.

Acknowledgements This work is supported by Grant-in-Aid for Scientific Research (C; No. 26380658) from the Japan Society for the Promotion of Science. The author thanks the Social Stratification and Social Mobility (SSM) 2005 Committee for the use of data.

References

1. Arabie, P., Hubert, L.: Cluster analysis in marketing research. In: Bagozzi, R.P. (ed.) Advanced Methods in Marketing Research, pp. 160–89. Blackwell, Oxford (1994)
2. Arabie, P., Carroll, J.D., DeSarbo, W.S.: Three-Way Scaling and Clustering. Sage Publications, Newbury Park, CA (1987)
3. Bourdieu, P.: La Distinction: Critique Sociale du Jugement. Minuit, Paris (1979)
4. Bryson, B.: Anything but heavy metal: symbolic exclusion and musical dislikes. Am. Sociol. Rev. **61**, 884–899 (1996)
5. Collins, R.: Women and the production of status cultures. In: Lamont, M., Fournier, M. (eds.) Cultivating Differences, pp. 213–231. University of Chicago Press, Chicago (1992)
6. Kruskal, J.: The relationship between multidimensional scaling and clustering. In: Van Ryzin, J. (ed.) Classification and Clustering, pp. 17–44. Academic, New York (1977)

7. Nakai, M.: Social stratification and gendered cultural consumption and lifestyles. In: Sugano, T. (ed.) Social Class and Disparities in Quality of Life, The 2005 SSM Research Series, vol. 10, pp. 1–28 (2008)
8. Nakai, M.: Social stratification and consumption patterns: cultural practices and lifestyles in Japan. In: Ingrassia, S., Rocci, R., Vichi, M. (eds.) New Perspectives in Statistical Modeling and Data Analysis, pp. 211–218. Springer, Berlin (2011)
9. Nakai, M.: Class and gender differences in cultural participation: asymmetric multidimensional scaling of cultural consumption. Book and abstracts. In: The 8th International Meeting of the Classification and Data Analysis Group of the Italian Statistical Society, p. 40 (2011)
10. Okada, A.: A review of cluster analysis and multidimensional scaling research in sociology. Sociol. Theory Methods **17**(2), 167–181 (2002)
11. Okada, A., Imaizumi, T.: Asymmetric multidimensional scaling of two-mode three-way proximities. J. Classif. **14**, 195–224 (1997)
12. Okada, A., Yokoyama, S.: Nonhierarchical asymmetric cluster analysis procedure. In: Minerva, T., Morlini, I., Palumbo, F. (eds.) Books of Abstract, CLADAG 2013, pp. 353–356 (2013)
13. Olszewski, D.: Asymmetric k-means algorithm. In: Dobnikar, A., Lotric, U., Ster, B. (eds.) Adaptive and Natural Computing Algorithms, pp. 1–10. Springer, Heidelberg (2011)
14. Olszewski, D.: k-means clustering of asymmetric data. In: Corchado, E., Snasel, V., Abraham, A., Wozniak, M., Grana, M., Cho, S.-B. (eds.) Hybrid Artificial Intelligent Systems, pp. 243–254. Springer, Heidelberg (2012)
15. Peterson, R.A.: Understanding audience segmentation: from elite and mass to omnivore and univore. Poetics **21**, 243–258 (1992)
16. Peterson, R.A., Kern, R.: Changing highbrow taste: from snob to omnivore. Am. Sociol. Rev. **61**, 900–907 (1996)
17. Peterson, R.A., Simkus, A.: How musical tastes mark occupational status groups. In: Lamont, M., Fournier, M. (eds.) Cultivating Differences: Symbolic Boundaries and the Making of Inequality, pp. 152–186. University of Chicago Press, Chicago (1992)
18. Sugimoto, Y.: An Introduce to Japanese Society, 3rd edn. Cambridge University Press, Cambridge (2010)
19. Veblen, T.B.: The Theory of the Leisure Class: An Economic Study in the Evolution of Institutions. Modern Library, New York (1889)
20. Vicari, D.: Classification of asymmetric proximity data. J. Classif. **13**, 386–420 (2014)

The Classification and Visualization of Twitter Trending Topics Considering Time Series Variation

Atsuho Nakayama

Abstract This study attempted to detect trending topics and temporal variation in web communication topics regarding new products among consumers using social media. This was done by classifying words into clusters based on their co-occurrence. We collected Twitter entries about new products based on their specific expressions of sentiment or interest. Because of the desire to identify market trends, the analysis of consumer tweet data has received much attention. To construct appropriate words, we used a complementary similarity measure, a classification method that is widely applied in character recognition. We classified the words extracted from Twitter data using non-negative matrix factorization as a dimensionality reduction model. To help interpret the results, we proposed a visualization method for text classification using a multidimensional scaling model.

1 Introduction

The aim of this study was to detect trending topics in web communications among consumers using social media, with a focus on topics related to new products. This was done by classifying words into clusters based on their co-occurrence. We collected Twitter entries about new products based on their specific expressions of sentiment or interest. Twitter is an online social networking and microblog service that enables users to post and read tweets, which are text-based messages of up to 140 characters in length. Twitter has been spreading recently in Japan. To help identify market trends, analysis of consumer tweet data has received much attention. In this study, we examined temporal variation in topics regarding new products by classifying words into clusters based on the co-occurrence of words in Twitter entries. Twitter is an online social networking and microblog service that enables users to post and read text-based messages, known as tweets. Although a single Twitter entry is limited to 140 characters, this is sufficient to express ideas and even to write a short story in the Japanese language. In Japanese, just a few

A. Nakayama
Tokyo Metropolitan University, 1-1 Minami-Ohsawa, Hachioji-shi 192-0397, Japan
e-mail: atsuho@tmu.ac.jp

F. Palumbo et al. (eds.), *Data Science*, Studies in Classification, Data Analysis, and Knowledge Organization, DOI 10.1007/978-3-319-55723-6_13

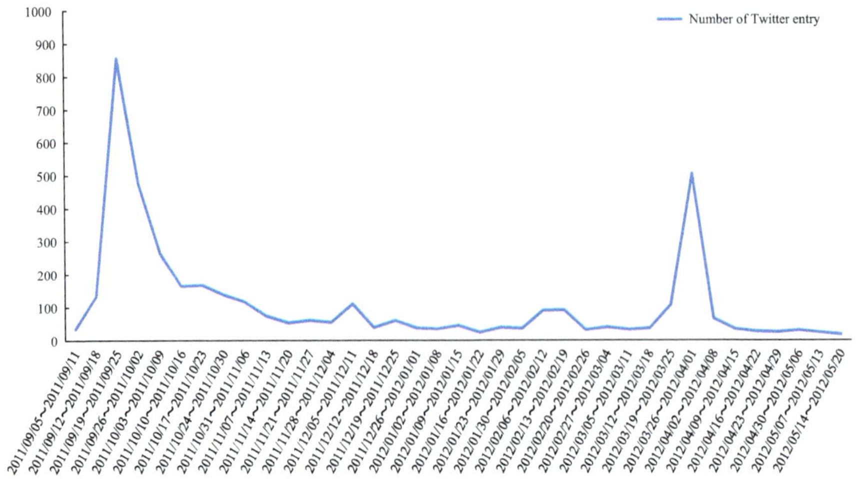

Fig. 1 The weekly number of Twitter entries regarding a new type of inexpensive, beer-like beverage

characters can convey a considerable amount of information. The Japanese writing system uses characters derived mainly from Chinese characters. For example, "経済 (keizai)," which means "economics" in Japanese, consists of two characters, whereas the equivalent English word contains nine. Thus, the limitation on the number of characters per tweet is less restrictive in Japanese, stimulating tweet posting in Japan. Tweeters can easily post short entries at any time and from any location using their smartphones or other mobile devices.

It is important to consider the temporal variation in trending topics when detecting such trending topics by classifying words into clusters based on co-occurrence of words in Twitter entries. Personal concerns are influenced by product strategies, such as marketing communication strategies, and thus change over time. For example, Fig. 1 shows the weekly number of Twitter entries regarding a new product, an inexpensive, beer-like beverage. On September 22, 2011, a new product was launched, and TV commercials for the product began running that week. A peak in Twitter entries was reached a few weeks after the product release. The number of entries per week slowly decreased after this peak. The gross rating point declined during the 2 months following the release, and the number of entries also decreased. Small peaks, however, were triggered by the release of new TV commercials. These data show how the weekly number of Twitter entries exhibited temporal change. Thus, to understand topic characteristics, it is important to consider temporal variation in trending topics and to establish criteria to select appropriate words that are representative of such temporal variation. We chose keywords representing various topics from Twitter entries and tracked the weekly variation in these topics. We then classified the words extracted from Twitter data using non-negative matrix factorization as a dimensionality reduction model. Finally, we used a visualization method for text classification to interpret the results, employing a multidimensional scaling model.

Table 1 An example entry × word matrix

	heat	hot	is	next	persists	the	today	very	week
Entry 1	0	1	1	0	0	0	1	1	0
Entry 2	1	0	0	1	1	1	0	0	1

2 The Data

The text data of Twitter entries regarding certain product names were searched and collected at 5-min intervals. We created a system for data cloning that was programmed in Ruby (https://www.ruby-lang.org/en/). Due to changes in specifications of Twitter API, our system is not currently operational. However, as of recently, we are able to easily collect Twitter entries by using the R package "twitterR," and libraries are updated and released in many programming languages. We searched for Twitter entries regarding a new brand of inexpensive, beer-like beverage named "金のオフ" (Kin no Off) produced by Sapporo Breweries, Ltd., and we collected 4622 tweets from September 2, 2011 through May 18, 2012. Our reasons for focusing on Twitter entries regarding new beverage products were twofold: it is a useful means of diminishing the effect of past product strategies such as merchandising and advertising, and we have found it particularly easy to evaluate time series variation in personal concerns for newly released beverage products.

In this study, we looked for topics associated with new products by classifying words into clusters based on the entry × word matrix of Twitter entries. For example, the entry × word matrix shown in Table 1 consists of the following nine words: heat, hot, is, next, persists, the, today, very, and week. Entry 1 is "Today is very hot," and Entry 2 is "The heat will persist next week." To detect topics more easily, we tokenized each tweet message that was written in sentences or sets of words. However, one of the most difficult natural language-processing problems in Japanese is tokenization. This is referred to as the "wakachigaki" problem. In most Western languages, words are delimited by spaces and punctuation. In Japanese, words are not separated by spaces. Consider the following sentence, "今日はとても暑い" (kyouhatotemoatsui). The English translation of this sentence is, "Today is very hot." In contrast, there are no spaces or separation symbols between the Japanese words. We used morphological analyses such as tokenization, stemming, and part-of-speech tagging to separate the words as follows.

今日	は	とても	暑い
kyou	ha	totemo	atsui
noun	Japanese particle	adverb	adjective

In our study, we used the Japanese morphological analyzer ChaSen to separate words in passages and to distinguish all nouns, verbs, and adjectives. ChaSen (http://chasen.naist.jp/) is a fast, customizable Japanese morphological analyzer that takes

Table 2 Example of a dataset to demonstrate the method used to calculate CSM

	Week *i*	Week *j*
Frequency of the word X	*a*	*b*
Frequency of words other than X	*c*	*d*

the form of a hierarchical structure. It is designed for generic use, and can be applied to a variety of language-processing tasks. A detailed discussion of ChaSen can be found in [6].

Next, we selected keywords representative of our chosen topics. To better understand topic characteristics, it was important to establish criteria to choose appropriate words representing temporal variation. We performed a statistical analysis based on the complementary similarity measure (CSM; [9]). CSM has been widely applied in the area of character recognition, and was originally developed for the recognition of degraded machine-printed characters. To construct appropriate word-set topics each week, we estimated the associations within word pairs. CSM is able to measure the inclusion relation between weeks i and j to recognize characters and identify word trends on a weekly basis. Given the following table of data, CSM is defined as follows:

$$CSM(Week\ i, Week\ j) = (ad - bc) \Big/ \sqrt{(a + d)(b + c)}. \tag{1}$$

CSM is an asymmetric measure (Table 2). Chi-square values have often been used to estimate the relation between two words, and are defined as follows:

$$\chi^2 = N(ad - bc)^2 \Big/ \sqrt{(a + b)(c + d)(a + c)(b + d)}, \tag{2}$$

where $N = a + b + c + d$. The formulas for CSM and chi-square are quite similar. However, the chi-square analysis is more likely to select words occurring with low frequency compared to the CSM method when analyzing data that contain a large spread in the occurrence frequency of words. Certain words occurred only rarely, whereas others occurred quite frequently in the text data of Twitter entries used in this study. Thus, the frequency of occurrence of some words was hundreds of times larger than that of others. For this reason, we decided to use CSM in this study. We collected the words receiving the top 10 CSM scores each week, and retained words with a total selection frequency of eight or more. The CSM score depends on word frequency. Thus, it was possible for words with low total frequency of occurrence to be selected as distinct words during a particular week, provided that the words occurred frequently that week. We extracted 359 words and removed all entries that did not include any of these words. The dataset comprised 4232 entries $\times$ 351 words. The data showed co-occurrences among 351 words in the selected entries.

3 The Analysis

The entry × word matrix obtained from the Twitter entries was sparse and of high dimensionality, so it was necessary to perform a dimensionality reduction analysis. We employed some excellent computing resources to help analyze the highly dimensional and sparse matrices. In addition, these matrices often contained noise, making it difficult to uncover the underlying semantic structure. Because of these difficulties, we found it necessary to implement dimensionality reduction. To reduce dimensionality, procedures such as Latent Semantic Analysis (LSA) or Latent Semantic Indexing (LSI) [2] and Probabilistic Latent Semantic Analysis (PLSA) or Probabilistic Latent Semantic Indexing (PLSI) (e.g., [5]) are often applied. LSA reduces the dimensionality of the entry × word matrix by applying a singular value decomposition (SVD), and it then expresses the result in an intuitive and comprehensible form. However, it can take a long time to perform LSA on a large matrix. In PLSA, a probabilistic framework is combined with LSA. This method uses mixture decomposition (the convex combination of aspects), which has a well-defined probability distribution. The factors have clear probabilistic interpretations in terms of the distribution of mixture components. We analyzed the entry × word matrix using Non-negative Matrix Factorization (NMF) to reduce the dimensionality [8]. Similar to principal component analysis (PCA), NMF consists of positive coefficients in linear combination. The computation of NMF is based on a simple iterative algorithm, which is particularly useful for applications involving large, sparse matrices. Ding et al. [3] have shown that both NMF and PLSI (PLSA) optimize the same objective function, ensuring that the use of NMF and PLSI is equivalent.

NMF is used for dimensionality reduction as follows:

$$V_n \approx W_n \times H_r, r < nm/(n+m) \tag{3}$$

The matrix V consists of non-negative data, such as that in an entry × word matrix. The matrix W contains non-negative basis vectors and shows the strength of associations between words and topics. The matrix H contains non-negative coefficients and shows the strength of associations between entries and topics. We can detect topics involving new products using the basis vector coefficients. The results are conceptually similar to those of PCA, but the basis vectors are non-negative. Here, the original data are represented purely through additive combinations of the basis vectors. This characteristic of NMF, i.e., data representation based on additive combinations, is effective because it suggests the intuitive notion of combining parts to form a whole. NMF computation is based on this simple iterative algorithm, and it is very efficient for applications involving large matrices.

Personal concerns are influenced by new product strategies, such as marketing communication strategies, and they change over time. It is important to consider the temporal variation in trending topics when detecting trending topics by classifying words into clusters based on co-occurrence of words. To assist us in the interpretation of the effects of temporal variation, we visualized the results of text classification using multidimensional preference scaling (MDPREF; [1]), which provided SVD of the scalar products of the preference ratings data. MDPREF provided a "point-vector" representation, such that the columns (i.e., stimuli) were represented as points, and the rows (i.e., subjects) as unit vectors. We revealed temporal variation by analyzing the coefficients in the matrix H, regarding these associations among entries and topics as preference data.

Lattin et al. [7] have formalized the MDPREF model as follows. Let s_{ij} denote the preference expressed by an individual i for a stimulus j. According to the vector model, the subjective utility can be represented by

$$S_{ij} = y_i' x_j \tag{4}$$

where y_i' is a row vector representing the relative preference of individual i, and x_j represents the location of objects j in multidimensional space. We can write the model in matrix form as follows. According to the vector model, the subjective utility can be represented by

$$S = YX' \tag{5}$$

where Y is a matrix with m rows (one for each individual), X' is a matrix with n columns (one for each object), and S is the $m \times n$ matrix of subjective utilities. The rows of Y can be normalized to unit length so that they correspond to unit vectors. If we assume that the subjective utilities expressed by the individual have metric properties, then we can solve for Y and X' by factoring S using matrix decomposition. According to the vector model, the subjective utility can be represented by

$$S = U\beta V' \tag{6}$$

We then set $Y=U\beta$ and $X=V$, defining X and Y that yield the best ordinary least squares r-dimensional vector model representation of the S matrix. Carroll [1] points out that this procedure, outlined by Eckart and Young [4], produces matrices X and Y such that $\hat{S}=YX'$ is indeed the best ordinary least squares r-dimensional vector model approximation of S.

3.1 Analysis of Topic Classification

Note that throughout this section, Japanese words will be followed by their English translations in parentheses. We classified the words extracted from the tweet data regarding a new brand of inexpensive, beer-like beverage named "金のオフ" (Kin no Off) produced by Sapporo Breweries, Ltd. We implemented NMF to reduce dimensionality using an R package "NMF" based on Lee's model [8]. "Kin no Off" contains 50% less purine and 70% less carbohydrates than other inexpensive, beer-like beverages. It is thus classified as a third-category beer, containing ingredients such as corn, soybeans, and peas rather than malt for the purpose of price reduction. For Japanese taxation purposes, brewed malt beverages in Japan fall into one of three categories: beer, Happoshu, or third-category beer. Alcoholic beverages made from malt are classified as beer if their malt content exceeds 67%. If a beverage contains less than 67% malt content, it falls under the tax category of Happoshu. Japanese breweries have produced even lower-taxed and non-malt brews made from soybeans and other ingredients, which do not fall under either of these classifications. These lower-taxed, non-malt brews, referred to by the mass media as third-category beers, were developed to compete with Happoshu.

Lee's model is an algorithm based on Euclidean distance that uses simple multiplicative updates. We determined that the maximum number of topics was 10, and the minimum as 4. In this analysis, eight topics are discussed for interpretation purposes. Table 3 lists the eight topics and shows the top 10 heavily weighted words in the basis vector W. Spellings using the Roman alphabet as well as English translations of the Japanese words are also shown in Table 3. From results such as these, we are able to identify the one or two words that are most heavily weighted. As a result of Twitter's 140-character limit, each topic consists of a small, core set of words. We can detect the prevalence of certain topics based on observations of which words are most heavily weighted.

We were able to divide the eight topics into three groups. One was the review topics, which consisted of Topics 1, 3, 7, and 8. Topic 1 was the review containing a link to an external website and product images, Topic 3 was the review of purchasing behavior and information about the new product, Topic 7 was the review of the brewery's release of the new product, and Topic 8 was the review of experiences actually drinking the product. The second group was the topics associated with advertising, which consisted of Topics 2, 5, and 6. Topic 2 was about advertisements on the train, Topic 5 was about TV commercials, and Topic 6 was concerned with performers in TV commercials. The third group consisted only of Topic 4. Topic 4 was not associated with inexpensive beer-like beverages, and the product name used as a keyword to extract Twitter entries that occurred in a different context.

Topics 1, 3, 7, and 8 are all based on reviews, though in various ways. In Topic 1, the words associated with the review containing a link to an external website and product images are heavily weighted. The most heavily weighted word was "http" (http), so it is the core word of Topic 1. Other words with comparatively large weights, ranking within the top 10, were often found along with the core word in tweets. Some Twitter entries were posted containing links to the external website and in-line product images, as well as phrases such as "the new product 'Kin no Off' was released, and I updated my blog about it"; "the new TV commercial for the new product 'Kin no Off' was broadcast"; or "'Kin no Off' is a new release and it tastes good." Some users posted links to external websites, such as their own blogs or the manufacturer's homepage. Others added in-line product images to their tweets. We therefore believe that it would be possible to infer the topic of these tweets, namely reviews containing a link to an external website as well as product

Table 3 The eight topic results and the top 10 weighted Japanese words in the basis vector W

Topic 1				*Topic 2*			
Japanese	*Roman alphabet*	*English translation*	*Weight*	*Japanese*	*Roman alphabet*	*English translation*	*Weight*
ｈｔｔｐ	ｈｔｔｐ	http	0.48	可愛い	kawaii	cute	0.28
オフ	ofu	off	0.02	広告	koukoku	advertisement	0.21
更新	koushin	update	0.02	見る	miru	see	0.08
ブログ	blog	blog	0.02	良い	yoi	good	0.05
良い	yoi	good	0.02	電車	densha	train	0.05
発売	hatubai	release	0.01	永作	Nagasaku	Nagasaku	0.04
なう	nau	now	0.01	電車内	denshanai	on the train	0.03
ひる	hiru	daytime	0.01	人	hito	people	0.03
新ＣＭ	shinCM	new TV commercial	0.01	ー	ー	ー	0.02
新発売	shinhatubai	new release	0.01	ω	ω	ω	0.02
Topic 3				*Topic 4*			
Japanese	*Roman alphabet*	*English translation*	*Weight*	*Japanese*	*Roman alphabet*	*English translation*	*Weight*
ビール	biiru	beer	0.13	ＲＴ	RT	RT	0.21
買う	kau	purchase	0.08	なう	nau	now	0.04
オフ	ofu	off	0.07	w	w	w	0.03
美味しい	oisii	delicious	0.06	金	kin	Friday	0.02
プリン体	purintai	purine	0.04	オフ会	ofukai	alcoholic party	0.02
味	azi	taste	0.04	予定	yotei	schedule	0.02
50%	50%	50%	0.04	いる	iru	stay	0.02
発泡酒	happoushu	low-malt beer	0.04	下さる	kudasaru	do	0.02
上手い	umai	tasty	0.03	お願い	onegai	please	0.02
糖質70	toushitu70	carbohydrate 70	0.02	宜しい	yoroshii	kind regards	0.02
Topic 5				*Topic 6*			
Japanese	*Roman alphabet*	*English translation*	*Weight*	*Japanese*	*Roman alphabet*	*English translation*	*Weight*
ＣＭ	CM	TV commercial	0.51	永作	Nagasaku	Nagasaku	0.56
見る	miru	see	0.06	可愛いすぎる	kawaiisugiru	way too cute	0.12
ー	ー	ー	0.03	見える	mieru	appear	0.03
可愛い	kawaii	cute	0.03	ポスター	Posuta	poster	0.02
出る	deru	perform	0.03	大島優子	Oshima Yuuko	Yuuko Oshima	0.02
似る	niru	resemble	0.03	似る	niru	resemble	0.02
やる	yaru	do	0.02	好き	suki	like	0.01
好き	suki	like	0.02	車内広告	shanaikoukoku	advertisement on the train	0.01
パフ	Pafu	Puff	0.01	男装	dansou	dressing as a man	0.01
曲	kyoku	music	0.01	ひる	hiru	daytime	0.01

(continued)

Table 3 (continued)

Topic 7				*Topic 8*			
Japanese	*Roman alphabet*	*English translation*	*Weight*	*Japanese*	*Roman alphabet*	*English translation*	*Weight*
サッポロ	Sapporo	Sapporo	0.48	飲む	nomu	drink	0.43
上手い	umai	tasty	0.02	寝る	neru	sleep	0.02
味	azi	taste	0.01	見る	miru	see	0.02
発泡酒	happoushu	low-malt beer	0.01	味	azi	taste	0.02
出る	deru	release	0.01	ー	ー	ー	0.02
金麦	Kinmugi	Kinmugi	0.01	美味しい	oishii	delicious	0.02
発売	hatubai	release	0.01	好き	suki	like	0.01
ひる	hiru	daytime	0.01	笑	wara	laugh	0.01
良い	yoi	good	0.01	なう	nau	now	0.01
こだわる	kodawaru	pursue	0.01	いる	iru	stay	0.01

images, solely from the most heavily weighted words of Topic 1. In Topic 3, the words associated with the review of purchasing behavior and information about the new product were heavily weighted. The most heavily weighted word was "ビール" (beer), followed by "買う" (buy). These words are the core words of Topic 3. Other words with comparatively large weights, ranking within the top 10, were often found along with the core words in tweets. "Kin no Off" contains 50% less purine and 70% less carbohydrates than other third-category beers, and is thought to be a healthier product. We believe that these features of the new product can be inferred from the list of heavily weighted words of Topic 3. Actual Twitter entries corresponding to this topic include "I bought the third-category beer named 'Kin no Off,' and it features 50% reduced purine and 70% reduced carbohydrate"; "the catch-phrase of the third-category beer named 'Kin no Off' is that it is delicious, though the purine and carbohydrate are reduced, so we should purchase it if its taste is as delicious as that of low-malt beer or especially normal beer"; and "the features of the third-category beer named 'Kin no Off' include are 50% reduced purine and 70% reduced carbohydrate, and it is as tasty as normal beer." To repeat, Topic 3 is reflected in tweets concerning purchasing behavior and information about the new product. In Topic 7, the words associated with the review of the release of the new product from Sapporo Breweries, Ltd., are heavily weighted. The most heavily weighted word is the Brewery's name, "サッポロ" (Sapporo), and it is the core word of Topic 7. Other words with comparatively large weights, ranking within the top 10, were often found along with the core word in tweets. "金麦" (Kinmugi) is a rival third-category beer. Further Twitter entries include "Sapporo 'Kin no Off' is tasty, and the taste is better than other low-malt beers, so I think the materials to make it were selected carefully"; and "I made a trial purchase of Sapporo 'Kin no Off' that had been newly released, and its taste is good." To repeat, Topic 7 is associated with the release of the new product from Sapporo Breweries, Ltd., and reviews of its taste. In Topic 8, the words associated with reviews of drinking the product have the heaviest weight. The most heavily weighted word is "飲む"

(drink), and it is the core word of Topic 8. Other words with comparatively large weights, ranking within the top 10, were often found along with the core word in tweets. Some examples of Twitter entries associated with Topic 8 include "I like to drink 'Kin no Off'"; "personally, it is my very favorite taste"; "I will sleep well after drinking 'Kin no Off' because I am tired today"; and "I drank 'Kin no Off,' and it was more delicious than other third-category beers." To repeat, Topic 8 is associated with reviews of product consumption.

Topics 2, 5, and 6 are associated with advertising. Topic 2 regards advertisements on the train, Topic 5 is associated with TV commercials, and Topic 6 concerns a TV commercial performer. In Topic 2, the words associated with advertisements on the train have the heaviest weight. The most heavily weighted word is "可愛い" (cute), followed by "広告" (advertisement). These words are the core words of Topic 2. Other words with comparatively large weights, ranking within the top 10, were often found along with the core words in tweets. Hiromi Nagasaku (永作 博美), a popular Japanese actress, appeared in the advertisements on the train. We believe that it would be possible to infer this by observing the top words of Topic 2. In Topic 5, the words associated with advertising have the heaviest weight. The most heavily weighted word is "CM" (TV commercial), and it is the core word of Topic 5. Other words with comparatively large weights, ranking within the top 10, were often found along with the core word in tweets. The Twitter entries generally contained positive feedback in regard to the performer in the TV commercial. The song "Puff, the Magic Dragon" played during the TV commercial, and Twitter entries addressing the music were also posted. We believe that the top words of Topic 5 reflect tweeters' impressions of the performer and music in the commercial. In Topic 6, the words associated with the performer in the advertisement are most heavily weighted. The most heavily weighted word is "永作" (Nagasaku), the name of the performer, followed by "可愛すぎる" (way too cute). These words are the core words of Topic 6. Other words with comparatively large weights, ranking within the top 10, were often found along with the core words in tweets. In the advertisement, Nagasaku is dressed as a man. Twitter entries regarding this topic have generally been positive, and have included phrases such as "I like the TV commercial performer 'Nagasaku'"; "the TV commercial performer 'Nagasaku' dressed as a man in the advertisement is cute"; or "the TV commercial performer 'Nagasaku' resembles 'Yuuko Oshima' (a popular Japanese actress and singer)." Therefore, we believe that tweeters' general impressions of Nagasaku in the commercial can be inferred by observing the list of top words for Topic 6.

3.2 *Analysis of Topic Predictions*

We identified weekly variation in topics by analyzing the averaged data using MDPREF. The weight coefficients indicating entries' contributions to topics were averaged on a weekly basis. This analysis used the maximum dimensionality of categories three through seven. The largest variance accounted for (VAF) in assessing

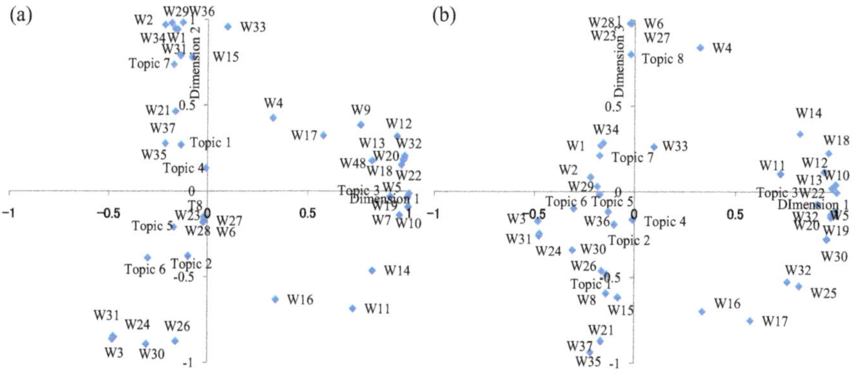

Fig. 2 Two-dimensional view of the three-dimensional configuration. The left-hand figure (**a**) defines a plane using a two-dimensional configuration for dimensions 1 and 2 derived from the results for three dimensions. The right-hand figure (**b**) defines a two-dimensional configuration for dimensions 1 and 3 derived from the results for three dimensions

the goodness of fit in each dimensionality space was selected as the maximum VAF for that space. The VAF represents the ratio of the sum of eigenvalues to the total amount of on-diagonal elements in SS'. For example, the VAF in two dimensions represents the ratio of the sum of the two largest eigenvalues to the total amount of on-diagonal elements in SS'. The VAF ratio can be represented by a configuration of variances in the preference data s_{ij}, and it measures the goodness of fit in each dimensional space. The resulting maximum VAFs from five one-dimensional spaces were 0.889, 0.766, 0.639, 0.492, and 0.329. Examination of these five VAFs initially encouraged us to adopt four- or higher-dimensional configurations as solutions. However, we did not believe that increasing the dimensionality of space beyond three dimensions would help to understand the weekly variation in topics. Three-dimensional space is sufficient to understand the weekly variation, so we chose to use three-dimensional results for the solutions. The three-dimensional configuration of the results is represented in two parts: configurations for dimensions 1 and 2 and for dimensions 1 and 3. Figure 2a shows a two-dimensional plot for dimensions 1 and 2 of the three-dimensional results. Figure 2b shows a two-dimensional plot for dimensions 1 and 3 of the three-dimensional results. In these figures, "W" represents weeks, numbered from 1 to 37, where W1 is the first week of data collection, and W37 is the last. This result provides a "point-vector" representation such that the columns (stimuli, or topics) are represented as points, and the rows (subjects, or weeks) are represented as unit vectors. The vector of topics indicates the direction of increasing preference, and the weeks are characterized by these topics.

Topic 3, which focuses on the review of purchasing behavior and information about the new product, is found in the right half of Fig. 2a. Topic 7, which is associated with a review of the brewery's release of the new product, is found in the upper half of Fig. 2a. Topics 3 and 7 are similar in terms of their characteristics, but they were the subjects of tweets during different weeks. Note the presence of corresponding vector weeks near Topic 3, during the period 1–3 months after the product release. During this period, consumers accepted the new product, and it became more widely recognized. Accordingly, the review of purchasing behavior and information about the new product were the subject of tweets during these weeks. On the other hand, certain vector weeks experienced gross rating point increases near Topic 7 only a few weeks after the product release. Of course, topics associated with the review of the brewery's release of the new product experienced more activity during the first few weeks following the product release because of the advertisements and marketing associated with the product launch. Topics 2, 5, and 6, which were associated with advertising reviews, are shown in the lower left quadrant of Fig. 2a. During the broadcast of new TV commercials, certain vectors of weeks experienced increased gross rating points for tweets associated with topics 2, 5, and 6. These topics associated with advertising reviews are sensitive to the amount of advertising present. Topic 1, shown in the lower half of Fig. 2b, is associated with tweets containing links. Topic 8, which relates to the review of experiences of actually drinking the product, is located in the upper half of Fig. 2b. The vector weeks occurring a few weeks to several months following the product release are located near Topic 1. This shows that topics associated with sharing links in tweets were active during the weeks following the release. Presumably, the various weeks closest to Topic 8 and other topics associated with the review of experiences of actually drinking the product correspond to the times when consumers were trying the new product.

4 Conclusion

We detected trending topics related to a new product by classifying words into clusters based on the co-occurrence of words in Twitter entries. Each topic consisted of a small set of core words. The topics of Twitter entries were divided into two categories: those associated with reviews, and those associated with advertising. These topics were further classified by the characteristics of their core words. We then detected weekly trends in topics related to new products by classifying words into clusters based on the co-occurrence of words in Twitter entries. We found that the personal concerns and tweet contents of Twitter users were influenced by new product strategies, such as marketing communication strategies, and they changed over time.

Acknowledgements We express our gratitude to the anonymous referees for their valuable reviews. This work was supported by a Grant-in-Aid for Scientific Research (C) (No. 16K00052) from the Japan Society for the Promotion of Science. We are grateful for financial support from the 45th Yoshida Hideo Memorial Foundation. We wish to thank Video Research, Ltd., for allowing us to make use of the GRP data. We are also greatly indebted to Hiroyuki Tsurumi of Yokohama National University and Jyunya Masuda of INTAGE, Inc., for their great support and advice in analyzing data.

References

1. Carroll, J.D.: Individual differences and multidimensional scaling. In: Shepard, R.N., et al. (eds.) Multidimensional Scaling, Vol. I Theory, pp. 105–155. Seminar Press, New York (1972)
2. Deerwester, S., Dumais, S., Furnas, G.W., Landauer, T.K., Harshman, R.: Indexing by latent semantic analysis. J. Am. Soc. Inf. Sci. **41**(6), 391–407 (1990)
3. Ding, C., Li, T., Peng, W.: Nonnegative matrix factorization and probabilistic latent semantic indexing: equivalence, chi-square statistic, and a hybrid method. In: Proceedings of the 21st National Conference on Artificial Intelligence and the 18th Innovative Applications of Artificial Intelligence Conference (AAAI'06), pp. 342–347 (2006)
4. Eckart, C., Young, G.: The approximation of one matrix by another of lower rank. Psychometrika **1**, 211–218 (1936)
5. Hofmann, T.: Probabilistic latent semantic analysis. In: Proceedings of the 15th Conference on Uncertainty in Artificial Intelligence, pp. 289–296 (1999)
6. Kudo, T., Yamamoto, K., Matsumoto, Y.: Applying conditional random fields to Japanese morphological analysis. In: Proceedings of the 2004 Conference on Empirical Methods in Natural Language Processing (EMNLP-2004), pp. 230–237 (2004)
7. Lattin, J.M., Carroll, J.D., Green, P.E., Green, P.E.: Analyzing Multivariate Data. Thomson Brooks/Cole, Pacific Grove, CA (2003)
8. Lee, D.D., Seung, H.S.: Algorithms for non-negative matrix factorization. In: Leen, T.K., Dietterich, T.G., Tresp, V. (eds.) Advances in Neural Information Processing Systems, vol. 13, pp. 556–562. MIT Press, Cambridge (2000)
9. Sawaki, M., Hagita, N.: Recognition of degraded machine-printed characters using a complementary similarity measure and error-correction learning. IEICE Trans. Inf. Syst. **E79-D**(5), 491–497 (1996)

Handling Missing Data in Observational Clinical Studies Concerning Cardiovascular Risk: An Insight into Critical Aspects

Nadia Solaro, Daniela Lucini, and Massimo Pagani

Abstract In observational clinical studies, subjects' health status is empirically assessed according to research protocols that prescribe aspects to investigate and methods for investigation. Commonly to many fields of research, these studies are frequently affected by incompleteness of information, a problem that, if not duly handled, may seriously invalidate conclusions drawn from investigations. Regarding cardiovascular risk assessment, coronary risk factors (e.g. high blood pressure) and proxies of neurovegetative domain (e.g. heart rate variability) are individually evaluated through direct measurements taken in laboratory. A major cause of missingness can be ascribed to the fact that overall sets of collected data typically derive from aggregation of a multitude of sub-studies, undertaken at different times and under slightly different protocols that might not involve the same variables. Data on certain variables can thus be missing if such variables were not included in all protocols. This issue is addressed by referring to a clinical case study concerning the role of Autonomic Nervous System in the evaluation of subjects' health status.

1 Introduction

Cardiovascular Diseases (CVD) are the major cause of morbidity, disability and mortality worldwide. Over the last years, the trend has been appearing decreasing at least in European countries, although CVD remains the leading cause of death. CVD Standardized Mortality Rate of EU-28 member states, computed with the 2012 European Standard Population (ESP), was 46.3 deaths per 10,000 inhabitants in 2007, 41.8 in 2010 and 39.4 in 2012 [10]—updated SMRs with the 2013 ESP are given for each EU country in [19]. Plausibly, prevention has played a fundamental role against controllable CVD risk factors, such as obesity,

N. Solaro (✉)
Department of Statistics and Quantitative Methods, University of Milano-Bicocca, Milano, Italy
e-mail: nadia.solaro@unimib.it

D. Lucini • M. Pagani
BIOMETRA Department, University of Milan, Milano, Italy
e-mail: daniela.lucini@unimi.it; massimo.paganiz@gmail.com

F. Palumbo et al. (eds.), *Data Science*, Studies in Classification, Data Analysis, and Knowledge Organization, DOI 10.1007/978-3-319-55723-6_14

hypertension and lifestyles (e.g. smoking, wrong diet habits, sedentary life), and a multitude of clinical studies have been undertaken for assessing CVD prevention through lifestyle changes. These studies are typically observational, as opposed to clinical trials. In observational studies, subjects' health status is assessed per se according to research protocols that prescribe aspects to investigate and methods for investigation, while they do not provide for recruiting participants in order to receive specific treatments.

Commonly to many fields of research, observational studies are frequently affected by incompleteness of information, a problem that, if not duly handled, may seriously invalidate conclusions drawn from investigations. Regarding CVD risk assessment, coronary risk factors (e.g. high blood pressure) and proxies of neurovegetative domain (e.g. heart rate variability) are individually evaluated through direct measurements taken in laboratory. Apart from subjects refusing to undergo tests, a major cause of missingness can be ascribed to the fact that overall sets of collected data typically derive from aggregation of a multitude of studies, which may be undertaken at different times over a long period and under protocols that might not involve the same variables. All of this clearly causes missing data.

As a practice commonly adopted, clinical researchers carry out statistical analyses by applying, often uncritically, standard imputation methods settled as default in the main statistical packages. Statistical research should then suggest the most suitable approaches for handling missingness within specific fields of application, such as the CVD risk assessment. Referring to a specific case study, this issue is addressed by considering a few critical aspects in statistical handling of missing data and involving a multitude of imputation methods of different types (non-parametric and parametric methods [12, 14], data fusion methods [4]). Then, two questions relevant to the context will be addressed: First, how do the considered imputation methods perform? This involves the assessment of the imputation performance under specific schemas of missingness through a Monte Carlo (MC) simulation study. Second, are the imputation results consistent with the considered real situation? This is a more crucial aspect based on the assessment of the quality of imputation, for which we propose a strategy of analysis strictly related to the CVD context.

2 Case Study: Autonomic Nervous System and Health Status

Critical aspects concerning missing data imputation in CVD context and subsequent assessment of quality of imputation are considered with reference to a clinical case study. This is part of an ongoing project on the role of autonomic assessment in the context of CVD prevention through lifestyle changes. The main focus is on the connection between Autonomic Nervous System (ANS)—a branch of the nervous system that works without conscious control—and hypertensive states, because altered autonomic profiles are usually observed in hypertensive subjects [13]. To study more in-depth how ANS profiles link to subjects' health condition, several clinical groups other than hypertensive individuals were involved in the project.

The overall set of data considered here consists of a collection of 88 observational clinical studies undertaken over the period 1999–2014 and involving a total number of $n = 1314$ individuals. Each of these studies aimed specifically at investigating ANS profiles on one among competitive athletes, non-smoking healthy individuals, smokers without pathologies, and stressed, obese and hypertensive individuals, respectively. In each study, subjects' health status was assessed empirically according to a specific Research Protocol prescribing aspects to investigate, measures to collect and methods for investigation. Protocols can be regarded as aggregations of different clinical studies, because a same protocol was applied to, and then shared by, more than one study. We will then refer to the aggregation in protocols, rather than the single studies, thus overcoming the problem of having clinical studies of low size, as occurred, e.g. for the athletes of particular disciplines.

Evaluation of subjects' health condition was mostly carried out at the Exercise Medicine Unit, C.T.N.V.—Research Centre for the Neurovegetative Therapy and Medicine of Exercise—, "Luigi Sacco" Hospital, University of Milano (in the past), and at the Exercise Medicine Unit, Humanitas University Hospital, BIOMETRA Department, University of Milano (currently). Participants were recruited into the studies in several different ways. A large part of subjects was referred to the clinic for evaluation from their family physicians. Other subjects decided to participate in a specific clinical study on voluntary basis. In other cases, a few Italian companies interested in applying wellness programs to their workers complied with a Web-based health promotion initiative, so that workers could fulfill a self-administered Web questionnaire by logging on to a health portal at the work-site. In still other cases, recruitment occurred through agreement to a stop-smoking program.

Measurements considered here are ANS proxies, i.e. Heart Rate Variability (HRV) and baroreflex gain indices, blood pressure measures, Body Mass Index (BMI) and personal data (age and gender). HRV (or RR variability) measures are the outcomes of the spectral analysis carried out on the electrocardiogram traces, while baroreflex gain refers to the mechanism that helps blood pressure remain stable at nearly constant levels. Variables are 17 in all and are listed in Table 1.

Subjects in the overall set of data are classified into the nine clinical groups described in Table 2. These groups were formed posteriorly to the studies according to their clinical value. Special attention was paid to age (mean age 37.65 ± 13.59 years) and gender (F: 51.3%, M: 48.7%), since they are fundamental biological

Table 1 Variables considered in the case study

Variables	Labels
Neurovegetative (or Autonomic) domain variables—Neuro vars	
HRV at rest	HR, RRMean, RRTP, RRLFnu, RRHFnu, RRLFHF, RRLFHz, RRHFHz
RR rest/stand	ΔRRLFnu
Pressure domain variables—Press vars	
Baroreflex gain	AlphaM
Blood pressure (P)	SAP (Systolic Arterial P), DAP (Diastolic AP), SAPMean, SAPLFa
Anthropometrics	BMI (Body Mass Index) = weight in kg/height in m^2
Personal data	age, gender ($0 =$ Female, $1 =$ Male)

Table 2 Description of the clinical groups involved in the case study

Clinical groups	Frequency	Percentage (%)	Description
Athlete	109	8.30	CONI[a] competitive sport players, basket players, football players, badminton players
Normal	517	39.35	Non-smoking subjects without pathologies
Smoker	128	9.74	Smoking normal subjects
Obese	98	7.46	Subjects with BMI $\geq$ 30 (kg/m^2)
Stress	168	12.79	Psychological dimension of stress. Subjects reporting stress symptoms according to stress and control perception scales
Hypertensive (Ht)	187	14.23	Subjects with SAP $\geq$ 140 mmHg, or DAP $\geq$ 90 mmHg, or both
Ht-Athlete	15	1.14	Athletes with high blood pressure
Ht-Obese	54	4.11	Obeses with high blood pressure
Ht-Stress	38	2.89	Stressed subjects with high blood pressure
Total	1314	100.00	

[a]Comitato Olimpico Nazionale Italiano—Italian National Olympic Committee

parameters greatly affecting the autonomic, pressure and anthropometric measures in Table 1. Clearly, age and gender are not uniformly distributed over the clinical groups. For instance, 98.2% of Athletes (F: 97.7% and M: 98.5%) are under 34 years, while 46.0% of Hypertensives (F: 54.3% and M: 41.0%) are over 50 years. As a consequence, analyses on these data required us to introduce proper adjustments in order to account for the effects of age, gender and their interaction.

In Table 3, the five protocols aggregating the 88 clinical studies are described. The complete case is denoted with Protocol 0, under which all the 17 variables in Table 1 were completely observed on 836 subjects (63.6% of the total set). On the other hand, under the other four protocols, subsets of autonomic proxies, or pressure variables, or both, were not systematically collected, because at the time of the study they were not admitted as official aspects to investigate.

Incompleteness of Clinical Information The overall dataset with $p = 17$ variables and $n = 1314$ subjects contains 6.73% missing values in all. With the exception of a small percentage of MCAR (Missing Completely At Random) values, equal to 1.29% (Code 99, Table 3), incompleteness of information is mainly caused by the types of protocols under which the studies were undertaken, because protocols, having changed over the considered period of time, did not include the same set of variables. The missing pattern can be nonetheless assimilated to a MAR (Missing At Random) mechanism. Missingness does not depend on unobserved values of the variables not included in the various protocols. It is indeed conditional on the adopted protocol and then on the clinical study in which a subject was recruited. Since each clinical study was addressed to examining subjects in a similar clinical condition, missingness is then ultimately linked to the subjects'

Table 3 Description of the five protocols aggregating the clinical studies

Protocols	Description	Frequency	Percentage (%)
Protocol 0	All the 17 variables (vars) measured	836	63.62
Protocol 1	The 2 neuro vars: RRLFHz, RRHFHz, not included	126	9.59
Protocol 2	The 3 press vars: AlphaM, SAPMean, SAPLFa, not included	234	17.81
Protocol 12	The 2 neuro and 3 press vars not included: Prot.12 = Prot.1 ∪ Prot.2	83	6.32
Protocol 22	All the 5 press vars not included: Prot.22 = (SAP, DAP) ∪ Prot.2	18	1.37
Code 99	MCAR data concerning all the vars—age, gender and HR excepted	17	1.29
Total		1314	100.00

individual membership to a clinical group. In other terms, subjects who belong to certain clinical groups are more likely to have missing values on certain variables. For instance, from the within-group percentage distribution of protocols (table omitted), it can be seen that athletes and hypertensives do not have values on the three pressure variables excluded by Protocol 2 (Table 3) more frequently than on average—22.0% and 41.7% of their group size, respectively, against 17.8% on average—and similarly smokers on the two neurovegetative variables excluded by Protocol 1 (Table 3)—36.7% of their group size against 9.6% on average. On the other hand, almost all stressed subjects in "Stress" and "Ht-Stress" groups are under Protocol 0—95.8% and 94.1% of their respective group sizes against 63.6% on average—and only small percentages of MCAR values occurred for them overall.

3 Imputation Methods

Imputation performance was tested through an MC simulation study carried out on the complete part of the dataset, formed by $n_c = 836$ subjects and $p = 17$ variables, and with reference to a multitude of imputation methods selected from the vast literature on missing data [12, 14]. To have a wide overview, we dealt with methods of very different types, i.e. non-parametric methods for single imputation, parametric methods for multiple imputation and data fusion methods. The methods tested up to now, and implemented in the R environment [15] through dedicated packages, are summed up as follows:

- *Non-parametric methods*. Two different approaches were considered: (1) The imputation methods developed in [11] and based on factorial techniques, which are available in the R package *missMDA* [9], in particular: IPCA (Iterative Principal Component Analysis), applied here in both the non-parametric (Regularized method) and the parametric versions (EM algorithm), and FAMD (Factorial Analysis for Mixed Data), which carries out imputation on mixed-type data;

(2) Forward Imputation (ForImp) [18], a sequential distance-based approach based on the nearest-neighbour imputation method and available in the R package *GenForImp* [17], with its two variants for quantitative data: ForImp with PCA (FIP) and ForImp with the Mahalanobis distance (FIM). In order to involve the clinical grouping variable into the imputation process, we have introduced a third variant: Within-Group FIP (WG.FIP), that is, FIP run within each clinical group, rather than on the whole dataset.

- *Parametric methods*. The EM algorithm [2] is a very popular imputation method in medical research [14], although we are of the opinion that sometimes it is misused. Basic multinormality assumption is not always inspected, or even does not hold, in the data, and the presence of at least a MAR mechanism generating missing data is not often ascertained. One of the main attractive features is the multiple imputation process, by means of which uncertainty of imputation can be assessed through generation of a number m of imputed datasets. Clinical researchers handling statistical data, however, need to work with a single imputed matrix, so that a synthesis (usually, the mean) of the m imputed datasets is computed. Accordingly, we considered parametric methods in a single imputation perspective, in particular the EM method implemented in the R package *Amelia II*, which combines the classical EM algorithm with a bootstrap approach [6, 7]. In addition, we dealt with the multiple imputation version of IPCA (MIPCA) [9, 11], in the two non-parametric and parametric versions, both based on bootstrap.
- *Data fusion approach*. Data fusion is part of the statistical matching problem [4, 16] and, strictly speaking, is not a real imputation methodology. The main objective is to integrate two or more data sources referred to different sets of units having a set of variables in common and a set of variables not jointly observed. In [4], the various approaches to statistical matching and their key assumptions are extensively treated. For our purposes, we relied on the non-parametric approach based on hot-deck methods, such as random hot-deck and distance hot-deck techniques. These are implemented in the R package *StatMatch* [3].

Given the multitude of experimental conditions and schemas of missingness considered in the simulation study, as well as the vast collection of the achieved results, we cannot present and discuss the main findings in this context. We limit to say that the imputation methods more fitting to our situation turned out to be the considered non-parametric methods, since overall they produced smaller imputation errors.

4 Quality of Imputation

One crucial aspect in our study was the assessment of the "Quality of Imputation" (QoI). We consider QoI as a concept strictly related to the field of application, so that the "best" analysis strategies for QoI assessment should be suggested by a prior knowledge of the context, such as subjects' membership to clinical groups and the relationship among ANS proxies within groups. Strategy for QoI assessment adopted in this study then consisted of the following stages:

1. *Set-up of within-group adjusted median profiles on the complete part of the data:* In order to highlight the main differences, we characterized the clinical groups in Table 2 by means of their median profiles. Age and gender are biological parameters greatly influential in this clinical context. To account for their effects, a 2-way full ANOVA model including age, gender and their interaction was fitted to each variable in Table 1. Within-group medians were then computed on the ANOVA residuals after standardization. We denote ANOVA residuals as adjusted variables and the profiles formed by their medians after standardization as adjusted median profiles. Given the presence of missing values, this analysis was necessarily performed on the complete part of the data, i.e. on the 836 subjects recruited under Protocol 0. Besides comparing the clinical groups, this analysis aimed at disclosing potential within-group relationships between the following two sets of variables: (A) the variables collected completely over all the protocols (i.e. common complete variables), and: (B) the variables having missing values in any of the other protocols. This inspection was performed through the canonical correlation analysis (CCA) [5] applied to adjusted standardized variables within each clinical group. Variables in the set A that linked more strongly with variables in the set B to the first two canonical variates were treated as *distinguishing partial pattern* of a specific clinical group, and subsequently used to assess QoI results. Given our prior knowledge of the clinical groups, we expected partial patterns to be greatly different across the groups, especially Athletes in comparison with Hypertensives.
2. *Comparisons, within each clinical group and for each protocol, between adjusted median profiles of complete and incomplete subjects concerning common complete variables:* Analyses in this stage were based on all the 1314 subjects, complete and incomplete, and concerned variables in the set A. We re-fitted the 2-way full ANOVA model with age and gender to each complete variable, standardized the ANOVA residuals, and set up the within-group adjusted median profiles as described at Stage 1. Subsequently, by means of the non-parametric Fisher–Pitman (FP) permutation, Wilcoxon–Mann–Whitney (WMW) and Brown–Mood (BM) median tests [8], we made pairwise comparisons, within each clinical group, between the adjusted median profiles of:

   (a) Protocol 0 built on all the 1314 subjects ("Prot.0-all") and Protocol 0 built on the subset of the 836 complete subjects ("Prot.0-compl");
   (b) Prot.0-all and each of any other incomplete protocol—i.e. Prot.1, 2, 12 or 22—that could occur in a specific clinical group.

   These comparisons aimed at:

   (a) appraising whether Prot.0-compl adjusted median profiles were representative of Prot.0-all. In such a case, the non-parametric tests would have led to acceptance of the null hypotheses and supported the partial patterns detected at Stage 1 as reasonable references for the incomplete part of the data;

(b) using Prot.0-all adjusted median profiles in comparison with the other protocols for disclosing potential specificities of incomplete subjects concerning common complete variables. In particular, if Prot.0-compl were representative in the sense given above, and Prot.0-all were basically similar to Prot.0-compl, then the partial patterns detected at Stage 1 could be used reasonably to "clinically infer" the trend of the median profiles of the variables in the set B.

3. *Consistency analysis concerning imputed variables:* This stage is at the core of QoI analysis. The main aim was to ascertain if imputed values of incomplete variables in the set B were consistent with the conclusions drawn in the previous analyses about their within-group clinical trend. A major aspect of concern was the admissibility of imputed values, that is, whether they proved to lie within appropriate clinical ranges. Analyses pertinent to this stage were performed through:

   (a) inspection of box plots, for each incomplete variable, of within-group distributions of the values imputed with the various methods. These distributions were also compared with their corresponding complete counterpart under Protocol 0. In addition, admissibility of imputed values was carefully examined;
   (b) closer comparisons, for each incomplete variable, between the within-group distributions of known values of complete subjects and of imputed values of incomplete subjects by using adjusted imputed variables. This was accomplished by computing ANOVA residuals of the imputed variables over all the $n = 1314$ subjects, estimating their densities non-parametrically through smoothing techniques [1], and then applying Bowman–Azzalini's permutation procedure [1] for testing the equality of estimated within-group densities between complete and incomplete subjects;
   (c) appraisal, for each incomplete variable, of the correspondence between imputed values and inferred clinical trend by using adjusted variables and Jonckheere–Terpstra (JT) permutation test for ordered alternatives [20], in which the direction of the inequality is suggested by the inferred clinical trend;
   (d) for each incomplete protocol and imputation method, evaluation of the consistency between within-group adjusted median profiles built on the complete variables in the set A and those built on the imputed variables in the set B.

QoI Results As an example, we show QoI results achieved for the Athlete group using the strategy described above. Concerning Stage 1, Fig. 1 displays within-group adjusted median profiles set up on the complete part of the data. It can be noticed that Athlete and Hypertensive groups tend to present opposite clinical traits, their profiles being nearly specular. Here we focus specifically on index AlphaM (Table 1), which is known to assume higher levels in healthy subjects and lower

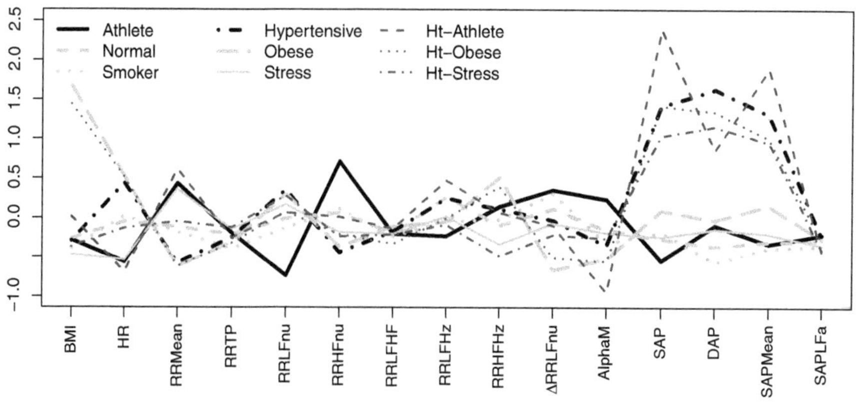

Fig. 1 Within-group adjusted median profiles on the complete part of the data ($n_c = 836$ subjects)

in unhealthy subjects. Consistently, in Fig. 1, Athletes show the highest adjusted AlphaM median values, while the four Ht-groups the lowest median values.

Common complete variables (set A) are given by the first seven measures, from BMI to RRLFHF, in Fig. 1. The rest of the variables, from RRLFHz to SAPLFa, are the incomplete ones (set B). Focusing specifically on the Athletes, CCA is applied to the adjusted standardized variables within this group in order to disclose relationships between the two sets A and B. Relevant results are displayed in Fig. 2 in the form of correlation circle. Variables in the set A (bold labels in Fig. 2) that are outside the inner correlation circle with radius equal to 0.5 form the distinguishing partial pattern of the Athlete group. In particular, since AlphaM is correlated much more highly ($> |0.5|$) with the second canonical variate, we use this dimension to connote the clinical trend involving AlphaM within this group. As can be seen, AlphaM is positively associated with RRMean, RRTP and RRHFnu, and negatively with HR, RRLFnu and RRLFHF.

Concerning Stage 2, Fig. 3 reports the adjusted median profiles for the Athletes pertaining to the variables in the set A, i.e. common complete variables, and distinguished by the protocol, namely, Prot.0-compl, Prot.0-all and Prot.2+22, this latter referred to the subjects examined under the protocols 2 or 22 in which AlphaM is missing. Two remarks are worth making. First, the three profiles have the same trend. Second, the Prot.2+22 profile has the highest median values of BMI, RRMean and RRHFnu, and the lowest median values of HR and RRLFnu. We know that the athletes studied under Prot.2+22 had “more vigorous” traits, as they were involved in competitive sports requiring longer and greater physical efforts. They are thus expected to have imputed values of AlphaM basically higher than athletes with known AlphaM values. Non-parametric FP, WMW and BM tests prove that Prot.0-compl is representative of Prot.0-all (all the tests lead to the acceptance of the null hypothesis), while Prot.2+22 turns out to be significantly different to Prot.0-all, thus confirming that Prot.2+22 involves a set of athletes with their own peculiarities.

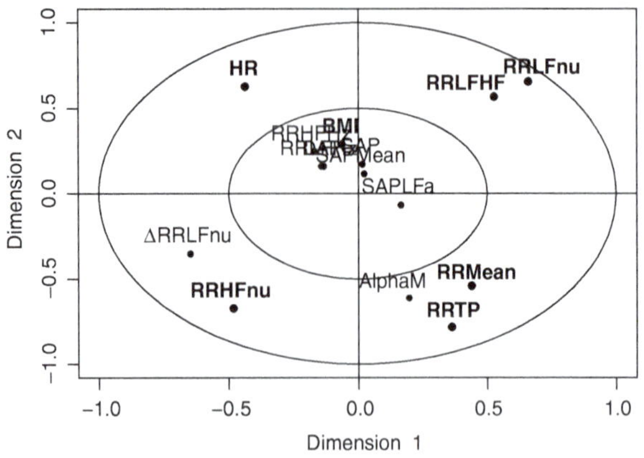

Fig. 2 Athlete group: CCA for the detection of the distinguishing partial pattern, given by the relationship between common complete variables in the set A (*black bold labels*) and incomplete variables in the set B (*gray labels*)

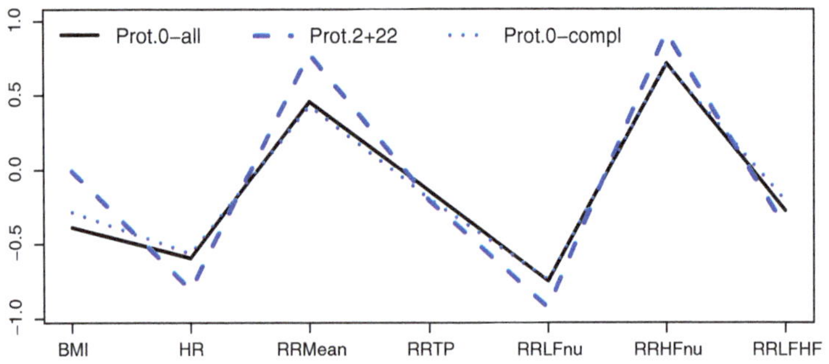

Fig. 3 Athlete group: Comparison between adjusted median profiles of complete and incomplete subjects on common observed variables and with reference to Prot.0-compl, Prot.0-all and Prot.2+22

Regarding Stage 3, step a, Fig. 4 displays box plots of AlphaM values imputed for the athletes under the Protocols 2 or 22 with the various methods. Results therein shown concern the methods with their best performing options as revealed by the simulation study. The first white box plot regards the distribution of the known AlphaM values of the athletes examined under Protocol 0. The reference dashed line drawn at zero indicates that values imputed below it are unacceptable, since AlphaM must be positive. All the imputation methods appear to fulfill this requirement. The crucial point is, however, to identify the method that produces more reliable imputations, consistently with the previous conclusions. As before argued, higher values of AlphaM are expected for this subset of athletes than in the complete part. At a first insight, the WG.FIP method appears to meet this requirement slightly better, although other methods also seem to be "good candidates", i.e. IPCA, FAMD and MIPCA, since they have among the highest medians of imputed values.

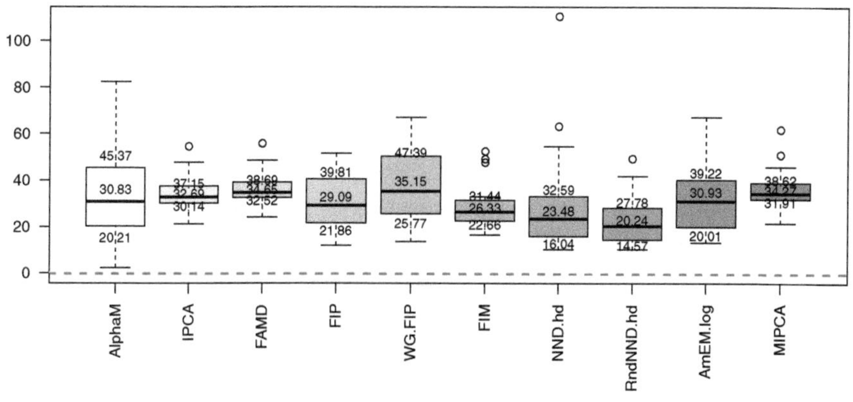

Fig. 4 Athlete group: box plots of distributions of AlphaM values imputed with various methods

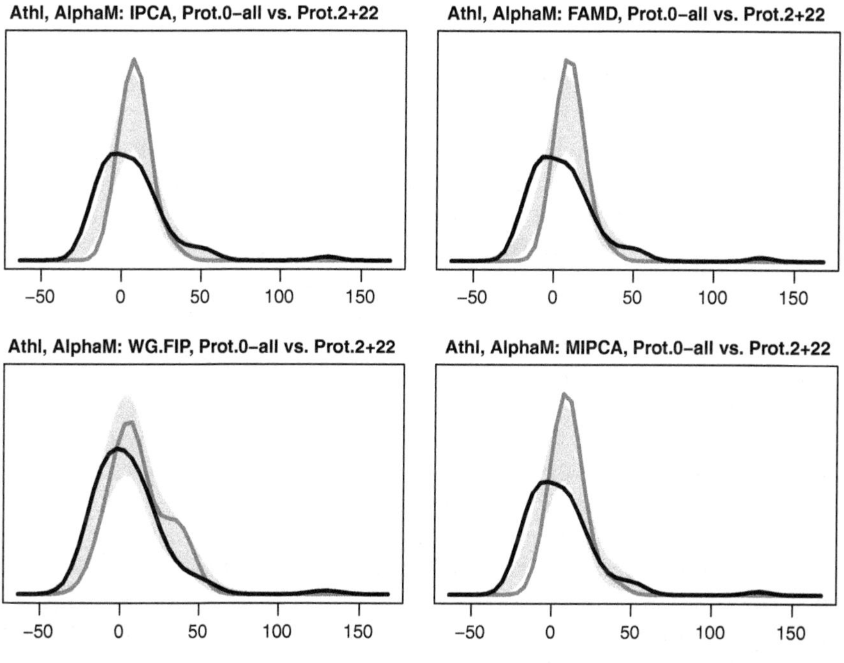

Fig. 5 Athlete group: Non-parametric density estimates of adjusted AlphaM in the comparison between known values (*black curve*, "Prot.0-all") and imputed values (*gray curve*, "Prot.2+22"), respectively. *Light gray region* is the reference band for BA equality test (IPCA: $P < 0.001$, FAMD: $P < 0.001$, WG.FIP: $P = 0.099$, MIPCA: $P < 0.001$)

A more fitting analysis must, however, take into account age and gender effects, as already pointed out in step b, Stage 3. Figure 5 reports the results of this further appraisal, here confined to the four methods resulting in significant BA or JT test results (tables omitted), i.e. IPCA, FAMD, WG.FIP and MIPCA. According to the inferred clinical trend, the null hypothesis tested by JT test on adjusted AlphaM values (step c) is $H_0 : F_{\text{Prot.2+22}}(x) = F_{\text{Prot.0-all}}(x)$ against $H_1 : F_{\text{Prot.2+22}}(x) \geq F_{\text{Prot.0-all}}(x)$, with a strict inequality holding for at least one x. In this case, H_0 is accepted for IPCA ($P = 0.0644$), and is rejected for FAMD ($P = 0.0146$), WG.FIP ($P = 0.0368$) and MIPCA ($P = 0.0232$). Consequently, we have empirical support in the case of Athlete group for regarding FAMD, WG.FIP and MIPCA as reliable imputation methods for AlphaM. The outcome obtained with WG.FIP is nonetheless more attractive from a clinical point of view. The obtained distribution of imputed values is positively skewed and has a longer right tail than the other methods (step d, analyses omitted).

Finally, it is worth pointing out that in the case of the Hypertensive group imputation has resulted in a problem of admissibility, since IPCA, FAMD and MIPCA have imputed several negative values for AlphaM, which indeed must be always positive, as stressed before (results omitted).

5 Discussion and Conclusions

Evaluation of the quality of imputation (QoI) performed in this work was based on a strategy of analysis that is strictly related to the context of CVD risk assessment. Prior knowledge of the field of application was fundamental for QoI analysis. What we proposed depended basically on knowledge of the clinical groups into which individuals are classified, along with the main characteristics of the subjects observed under the various protocols, and the availability of a set of common complete variables good enough to detect the distinguishing partial patterns, by means of which the clinical trends could be inferred for incomplete variables. Imputation was then regarded as overall reliable in the cases where the clinical trends were reflected, as best as possible, in the imputed values. Involvement of clinical groups as auxiliary information was instrumental in performing QoI analysis in a realistic and pertinent way. First of all, the clinical groups were highly representative of the health condition of their target subjects. Second, they were available in a substantial number, and this allowed the variation of ANS proxies to be represented in wide ranges.

While our strategy for QoI would require further refinements, a few still open questions would need additional investigations. One of the main problems was that a same imputation method could produce satisfactory results for a clinical group and a protocol, but not for another group or protocol. For example, normal individuals recruited under protocols Prot.2+22 had an adjusted median profiles much closer to hypertensives than the complete normals under Prot.0-all (Stage 2, Sect. 4). We thus expected AlphaM values for these subjects to be imputed at lower levels than their complete counterpart. The imputation method that best met this requirement was the non-parametric hot-deck method with the Mahalanobis distance (NND.hd), instead of WG.FIP chosen for the athletes. Is it then better to switch from one imputation method to another depending on their performance on specific clinical groups and protocols, or to fix a short list of general criteria that an imputation method should satisfy in order to be regarded as the one with the overall satisfactory QoI?

A further problem regards the evaluation of the imputation performance. This appraisal needs necessarily that the complete part of the data is representative of the incomplete part. In our case, this occurs when the complete part contains not only all the clinical groups, but also groups that are internally heterogeneous enough to be fully indicative of their target subjects. A strictly related point is the admissibility of imputed values. An imputation method could perform better than another when tested on the complete part of the data, but it could then impute inadmissible values for any variable, estimating several or more values outside clinical ranges.

Finally, it would be important, especially for users, to have a set of interactive diagnostic tools capable to ease the QoI assessment, particularly for disclosing the link between partial profiles of complete subjects and profiles of incomplete subjects.

References

1. Bowman, A.W., Azzalini, A.: R package 'sm': nonparametric smoothing methods (2014). Version 2.2-5.4 http://www.stats.gla.ac.uk/~adrian/sm
2. Dempster, A.P., Laird, N.M., Rubin, D.B.: Maximum likelihood estimation from incomplete data via the EM algorithm. J. Roy. Stat. Soc. B **39**, 1–38 (1977)
3. D'Orazio, M.: StatMatch: statistical matching (2015). R package version 1.2.3 http://CRAN.R-project.org/package=StatMatch
4. D'Orazio, M., Di Zio, M., Scanu, M.: Statistical Matching Theory and Practice. Wiley, New York (2006)
5. González, I., Déjean, S.: CCA: canonical correlation analysis (2012). R package version 1.2. http://CRAN.R-project.org/package=CCA
6. Honaker, J., King, G.: What to do about missing values in time-series cross-section data. Am. J. Polit. Sci. **54**, 561–581 (2010)
7. Honaker, J., King, G., Blackwell, M.: Amelia II: A program for missing data. J. Stat. Softw. **45**, 1–47 (2011). http://www.jstatsoft.org/v45/i07/
8. Hothorn, T., Hornik, K., van de Wiel, M.A., Zeileis, A.: Implementing a class of permutation tests: the coin package. J. Stat. Softw. **28**, 1–23 (2008). http://www.jstatsoft.org/v28/i08/
9. Husson, F., Josse, J.: missMDA: handling missing values with/in multivariate data analysis (principal component methods) (2015). R package version 1.8.2. http://CRAN.R-project.org/package=missMDA
10. Istat.it: Noi Italia – 100 statistiche per capire il Paese in cui viviamo. 2016 edition: http://noi-italia.istat.it/. 2015 edition: http://noi-italia2015.istat.it/
11. Josse, J., Pagès, J., Husson, F.: Multiple imputation in principal component analysis. Adv. Data Anal. Classif. **5**, 231–246 (2011)
12. Little, R.J.A., Rubin, D.B.: Statistical Analysis with Missing Data, 2nd edn. Wiley, New York (2002)
13. Lucini, D., Solaro, N., Pagani, M.: May autonomic indices from cardiovascular variability help identify hypertension? J. Hypertens. **32**, 363–373 (2014)
14. Molenberghs, G., Kenward, M.G.: Missing Data in Clinical Studies. Wiley, Chichester (2007)
15. R Core Team: R: A language and environment for statistical computing. R Foundation for Statistical Computing, Vienna (2016). http://www.R-project.org
16. Saporta, G.: Data fusion and data grafting. Comput. Stat. Data An. **38**, 465–473 (2002)

17. Solaro, N., Barbiero, A., Manzi, G., Ferrari, P.A.: GenForImp: a sequential distance-based approach for imputing missing data (2015). R package version 1.0.0. http://CRAN.R-project.org/package=GenForImp
18. Solaro, N., Barbiero, A., Manzi, G., Ferrari, P.A.: A sequential distance-based approach for imputing missing data: forward imputation. Adv. Data Anal. Classif. **1–20** (2016) doi:10.1007/s11634-016-0243-0
19. Townsend, N., Nichols, M., Scarborough, P., Rayner, M.: Cardiovascular disease in Europe – epidemiological update 2015. Eur. Heart J. **36**, 2696–2705 (2015)
20. Venkatraman, E.S.: clinfun: Clinical trial design and data analysis functions (2015). R package version 1.0.10. http://CRAN.R-project.org/package=clinfun

Part III
Multivariate Methods and Applications

Prediction Error in Distance-Based Generalized Linear Models

Eva Boj, Teresa Costa, and Josep Fortiana

Abstract In generalized linear models, the mean squared prediction error can be approximated by the sum of two components: variability in the data (process variance) and variability due to estimation (estimation variance). The estimation variance can be calculated by using the corresponding formula or, alternatively, it can be approximated by using bootstrap methodology. When we use bootstrap methodology we are able to obtain, in addition, the predictive distribution of estimations. We apply these concepts to the actuarial problem of claim reserving, where data are collected in a run-off triangle, and it is of interest the use of generalized linear models and the calculus of prediction error. We illustrate computations with a well-known data set. Distance-based generalized linear model is fitted using the `dbglm` function of the `dbstats` package for `R`.

1 Introduction

We have two objectives. The first one is to find a mechanism of estimating prediction error (PE) for new observations in the distance-based generalized linear model (DB-GLM). The second one is to propose the use of DB-GLM to solve the claim reserving problem.

The DB-GLM, defined in [6], extends the ordinary GLM allowing information on predictors to be entered as interdistances between observation pairs instead of as individual coordinates. The estimation process of a DB-GLM is schematically as follows: a Euclidean configuration, X_w, is obtained by a metric multidimensional scaling-like procedure, then the linear predictor of the underlying GLM is a linear combination of the resulting Euclidean coordinates, latent variables in the model.

E. Boj (✉) • T. Costa
Facultat d'Economia i Empresa, Universitat de Barcelona, Avinguda Diagonal 690, 08034 Barcelona, Spain
e-mail: evaboj@ub.edu; tcosta@ub.edu

J. Fortiana
Facultat de Matemàtiques, Universitat de Barcelona, Gran Via de les Corts Catalanes 585, 08007 Barcelona, Spain
e-mail: fortiana@ub.edu

F. Palumbo et al. (eds.), *Data Science*, Studies in Classification, Data Analysis, and Knowledge Organization, DOI 10.1007/978-3-319-55723-6_15

Let $\Omega = (\Omega_1, \ldots, \Omega_n)$ be a population of n individuals; let $Y : (Y_1, \ldots, Y_n)^T$ be the random response variable, and $(y_1, \ldots, y_n)^T$ be the observed response variable of size $n \times 1$; let $(w_1, \ldots, w_n)^T$ be a priori weights of individuals of size $n \times 1$ with $w_i \in (0, 1)$; let $(F_1, \ldots, F_p)$ be the set of p observed mixed predictors; and let Δ be an $n \times n$ matrix, whose entries are the squared distances $\delta^2(\Omega_i, \Omega_j)$. For the sake of expediency, henceforth we will refer to Δ as the "distance matrix."

The distance matrix Δ is calculated from the observed predictors by means of a distance function with the euclidean property. It contains predictor's information and it is the only information entered in the model in the predictor's space $\wp$. Therefore, DB prediction can be applied to mixed (qualitative and quantitative) explanatory variables or when the regressor is of functional type.

In DB-GLM we assume that the response distribution is in an exponential dispersion family, as in any GLM. Besides, the relation between the linear predictor, η, and the expected response, μ, is given by a link function: $\mu = g(\eta)$. We calculate the inner products matrix $G_w = -\frac{1}{2} J_w \cdot \Delta \cdot J_w$ where $J_w = I - 1 \cdot w^T$ is the w-centering matrix. A DB-GLM consists of random variables $(Y_1, \ldots, Y_n)^T$ whose expectation, $(\mu_1, \ldots, \mu_n)^T$, transformed by the link function and w-centered, is a vector in the column space of G_w. This space coincides with the column space of any *Euclidean configuration* X_w of Δ, by definition any matrix such that $G_w = X_w \cdot X_w^T$.

DB-GLM contains GLM as a particular case: if we start from a w-centered matrix X_w of continuous predictors and we define Δ as the matrix of squared Euclidean l^2 distances between rows of X_w, then X_w is trivially a Euclidean configuration of X_w, hence hat matrix, response, and predictions with DB-GLM and ordinary GLM are the same. Distance-based generalized linear models can be fitted using the `dbglm` function of the `dbstats` package for `R` (see [3]).

The paper is structured as follows: In Sect. 2 we define the PE for the DB-GLM when we have new observations in the cases of using analytic and bootstrap estimation. In Sect. 3 we obtain the expressions of the PE for the origin year provisions and for the total provision in the problem of claim reserving. In this section, we propose the DB-GLM using pairs bootstrap as a generalized Chain-Ladder stochastic method of claim reserving. We make an application with a well-known run-off triangle. In addition, we calculate the predictive distribution of provisions in the framework of Solvency II (Directive 2009/138/EC of the European Parliament and of the Council on the taking-up and pursuit of the business of Insurance and Reinsurance). Finally, in Sect. 4 we conclude.

2 Prediction Error

Assume a DB-GLM and the parametric family for the variance function, the power family:

$$V(\mu_i) = \mu_i^{\xi}, \tag{1}$$

for $i = 1, \dots, n$. Mean and variance of the random response variable are:

$$\mu_i = E[y_i],$$
$$Var[y_i] = (\varphi/w_i)\, V(\mu_i) = (\varphi/w_i)\, \mu_i^{\xi},$$

where φ is the dispersion parameter and w_i are data a priori weights of the data. Particular cases in this family are: Poisson distribution when $\xi = 1$; Gamma distribution when $\xi = 2$; and Inverse Gaussian distribution when $\xi = 3$.

Throughout this section we follow formulations from [9]. We define the mean squared error (MSE) of prediction as:

$$E\left[(y_i - \hat{\mu}_i)^2\right]. \tag{2}$$

To estimate the MSE for the original data, $i = 1, \dots, n$, we only need to compare the original response values, y_i, with the corresponding fitted values, $\hat{\mu}_i$, for $i = 1, \dots, n$. But we are interested in calculating the PE for a new observation: $i = n+1$. PE, by definition, is the squared root of the MSE. For new observations, when observed predictor values do not coincide with those of the original sample we need an approximated estimated MSE. In the following, we construct an expression for the MSE for new observations.

First we expand expression (2) as follows:

$$E\left[(y_i - \hat{\mu}_i)^2\right] = E\left[((y_i - E[y_i]) - (\hat{\mu}_i - E[y_i]))^2\right].$$

Plugging in $\hat{\mu}_i$ instead of y_i in the final expectation and expanding once again gives:

$$E\left[(y_i - \hat{\mu}_i)^2\right] \cong E\left[(y_i - E[y_i])^2\right] - 2E\left[(y_i - E[y_i])(\hat{\mu}_i - E[\hat{\mu}_i])\right] + E\left[(\hat{\mu}_i - E[\hat{\mu}_i])^2\right].$$

Assuming that new observations are independent of original observations gives that the MSE (2) is the sum of the *process variance* and the *estimation variance*:

$$E\left[(y_i - \hat{\mu}_i)^2\right] \cong E\left[(y_i - E[y_i])^2\right] + E\left[(\hat{\mu}_i - E[\hat{\mu}_i])^2\right].$$

Then, the MSE of predictions is given by the sum:

$$E\left[(y_i - \hat{\mu}_i)^2\right] \cong Var[y_i] + Var[\hat{\mu}_i]. \tag{3}$$

Using the delta method can be derived that:

$$Var[\hat{\mu}_i] \cong \left|\frac{\partial \mu_i}{\partial \eta_i}\right|^2 Var[\eta_i],$$

and then, the MSE (2) can be approximated as:

$$E\left[(y_i-\hat{\mu}_i)^2\right] \cong (\varphi/w_i)\,\mu_i^{\xi} + \left|\frac{\partial\mu_i}{\partial\eta_i}\right|^2 Var\left[\eta_i\right]. \quad (4)$$

Finally, the expression of the PE is given by the square root of the MSE (4):

$$\sqrt{E\left[(y_i-\hat{\mu}_i)^2\right]} \cong \sqrt{(\varphi/w_i)\,\mu_i^{\xi} + \left|\frac{\partial\mu_i}{\partial\eta_i}\right|^2 Var\left[\eta_i\right]}. \quad (5)$$

Consider the power family of link functions as in formula (2.10) of McCullagh and Nelder [11]:

$$\eta_i = g(\mu_i) = \begin{cases} \mu_i^{\lambda},\ \lambda \neq 0, \\ \log(\mu_i),\ \lambda = 0. \end{cases} \quad (6)$$

The particular expressions of PE (5) are:

- If $\mu_i^{\lambda} = \eta_i$: the predicted values μ_i are estimated from $\mu_i = \eta_i^{-\lambda}$. Then

  $$\frac{\partial\mu_i}{\partial\eta_i} = -\lambda\eta_i^{-\lambda-1} = -\lambda\eta_i^{-\lambda}\eta_i^{-1} = -\lambda\mu_i\left(\mu_i^{-\lambda}\right)^{-1} = -\lambda\mu_i\mu_i^{\lambda} = -\lambda\mu_i^{(\lambda+1)},$$

  and the PE can be calculated as:

  $$\sqrt{(\varphi/w_i)\,\mu_i^{\xi} + \lambda^2\mu_i^{(2\lambda+2)} Var\left[\eta_i\right]}. \quad (7)$$

- If $\log\mu_i = \eta_i$: the predicted values μ_i are estimated from $\mu_i = \exp(\eta_i)$. Then

  $$\frac{\partial\mu_i}{\partial\eta_i} = \exp(\eta_i) = \mu_i,$$

  and the PE can be calculated as:

  $$\sqrt{(\varphi/w_i)\,\mu_i^{\xi} + \mu_i^2 Var\left[\eta_i\right]}. \quad (8)$$

To estimate the scale parameter $\hat{\varphi}$ from the sample data we can use the mean of the Pearson residuals:

$$\hat{\varphi}^P = \frac{1}{n - eff.rank - 1}\sum_{i=1}^{n}\left(r_i^P\right)^2 = \frac{1}{n - eff.rank - 1}\sum_{i=1}^{n} w_i\frac{(y_i-\hat{\mu}_i)^2}{\hat{\mu}_i^{\xi}},$$

where the squared Pearson residuals are defined as:

$$r_i^P = \frac{y_i - \hat{\mu}_i}{\left(V(\hat{\mu}_i)/_{w_i} \right)^{1/2}} = \frac{y_i - \hat{\mu}_i}{\left(\hat{\mu}_i^{\xi} / w_i \right)^{1/2}},$$

and the *eff.rank* is the effective rank of the euclidean latent configuration, i.e., the dimension of the predictor space in the DB-GLM, a quantity analogous to the number of parameters in the linear predictor of an ordinary GLM without taking into account the intercept term.

An alternative way to obtain the MSE (3) and in consequence the PE (5) is to estimate $Var[\hat{\mu}_i]$ by bootstrap. We propose to use the resampling technique of pairs bootstrap or resampling cases in which each bootstrap sample consists of n response-predictor pairs from the original data (see, e.g., [7]). This technique is adequate for distance-based models as is shown in [2] where an F-test was defined for the DB-LM, and in [5] where a Wald test was defined for influence coefficients in DB-GLM.

We can estimate the predictive distribution of $\hat{\mu}_i$, called $\hat{\mu}_i^{boot}$. Then the estimation variance, the second component of formula (3), can be approximated by the variance of the predictive distribution: $Var\left[\hat{\mu}_i^{boot}\right]$. And the bootstrap estimation of PE (5) is:

$$\sqrt{\left(\hat{\varphi}^P / w_i\right) \hat{\mu}_i^{\xi} + Var\left[\hat{\mu}_i^{boot}\right]}. \tag{9}$$

3 Actuarial Application

Consider a portfolio of risks and assume that each claim is settled either in the accident year or in the following k development years. Consider a family of random variables $\{c_{ij}\}_{i,j \in \{0,1,\ldots k\}}$, where c_{ij} is the amount of claim losses of accident year i which is paid with a delay of j years and hence in development year j and in calendar year $i+j$. Assume that the incremental losses c_{ij} are observable for calendar years $i + j \leq k$ and that they are collected in a run-off triangle, where the rows correspond to the accident years, the columns to the development years, and the against-diagonals to the calendar years.

The provisions for the different accident years $i = 1, \ldots, k$ are obtained by adding the predicted future incremental losses in the corresponding row of the square:

$$P_i = \sum_{j=k-i+1}^{k} \hat{c}_{ij}. \tag{10}$$

The total provision is calculated by adding all the predicted future incremental losses in the bottom-right part of the run-off triangle:

$$P = \sum_{i=1}^{k} \sum_{j=k-i+1}^{k} \hat{c}_{ij}. \quad (11)$$

Several often used methods to complete a run-off triangle can be described by a GLM: the Chain-Ladder method, the arithmetic and geometric separation methods, and the de Vylder's least squares method. In particular, the classical Chain-Ladder deterministic method (see [13]) can be derived from a GLM assuming over-dispersed Poisson errors and logarithmic link function (see, e.g., [1, 4, 8, 9] and [10]). The linear predictor includes the dummies of the origin and of the development years, and the prior weights are equal to one, $w_i = 1$. With the aim to reproduce the Chain-Ladder method we assume a DB-GLM with an over-dispersed Poisson distribution, the logarithmic link function, and the Euclidean metric, l^2. We can consider the DB-GLM as a stochastic version of the Chain-Ladder deterministic method of claim reserving.

Applying (4), and assuming the logarithmic link as in (8), the MSE for the accident year provisions for $i = 1, \ldots, k$ are:

$$\begin{aligned} E\left[\left(P_i - \hat{P}_i\right)^2\right] &\cong \sum_{\substack{j=1,\ldots,k \\ i+j>k}} \varphi\mu_{ij}^{\xi} + \mu_i^T Var[\eta_i]\,\mu_i \\ &= \sum_{\substack{j=1,\ldots,k \\ i+j>k}} \varphi\mu_{ij}^{\xi} + \sum_{\substack{j=1,\ldots,k \\ i+j>k}} \mu_{ij}^2 Var\left[\eta_{ij}\right] + 2 \sum_{\substack{j_1,j_2=1,\ldots,k \\ j_2>j_1 \\ i+j_1>k,i+j_2>k}} \mu_{ij_1}\mu_{ij_2} Cov\left[\eta_{ij_1}, \eta_{ij_2}\right] \end{aligned} \quad (12)$$

And the MSE for the total provision is:

$$\begin{aligned} E\left[\left(P - \hat{P}\right)^2\right] &\cong \sum_{\substack{i,j=1,\ldots,k \\ i+j>k}} \varphi\mu_{ij}^{\xi} + \mu^T Var[\eta]\,\mu \\ &= \sum_{\substack{i,j=1,\ldots,k \\ i+j>k}} \varphi\mu_{ij}^{\xi} + \sum_{\substack{i,j=1,\ldots,k \\ i+j>k}} \mu_{ij}^2\left[\eta_{ij}\right] + 2 \sum_{\substack{i_1,j_1,i_2,j_2=1,\ldots,k \\ i_1+j_1>k,i_2+j_2>k \\ i_1j_1\neq i_2j_2}} \mu_{i_1j_1}\mu_{i_2j_2} Cov\left[\eta_{i_1j_1}, \eta_{i_2j_2}\right] \end{aligned} \quad (13)$$

The bootstrap estimations (9) of the PE for the accident year provisions (10), $PE^{boot}(P_i)$ for $i = 1, \ldots, k$, are:

$$PE^{boot}(P_i) \cong \sqrt{\sum_{\substack{j=1,\ldots,k \\ i+j>k}} \hat{\varphi}^P \hat{c}_{ij}^{\xi} + Var\left[\hat{P}_i^{boot}\right]}. \quad (14)$$

And the bootstrap estimation of the PE for the total provision (11), $PE^{boot}(P)$, is:

$$PE^{boot}(P) \cong \sqrt{\sum_{\substack{i,j=1,\ldots,k \\ i+j>k}} \hat{\varphi}^P \hat{c}_{ij}^{\xi} + Var\left[\hat{P}^{boot}\right]}. \tag{15}$$

In Table 1 there is the often used in the actuarial literature run-off triangle of Taylor and Ashe [12] with 55 incremental losses. In Table 2 there are the Chain-Ladder predictions of the future incremental losses, which coincide with the predictions of the assumed DB-GLM. In Appendix we include the instruction to fit the DB-GLM using the `dbglm` function of the `dbstats` package for `R` and we also include the summary of the model named `CL.dbglm`. We include too the corresponding Chain-Ladder GLM, `CL.glm`, fitted with the `glm` function of the `stats` package for `R`. In the instructions of Appendix, the data `cij` are the incremental losses of Table 1 introduced by rows. The two models reproduce the Chain-Ladder estimation of reserves.

In the first column of Table 3 there are the Chain-Ladder estimations of the origin year provisions and of the total provision which coincide with the estimations of provisions of the GLM (`CL.glm`) and of the DB-GLM (`CL.dbglm`) assuming the over-dispersed Poisson, the logarithmic link, and the l^2 metric. In the second column there are the PE of provisions calculated using the analytic formula (8). Finally, in the third column there are the associated coefficients of variation, i.e., the quotient between the PE over the provisions, in percentage.

In Table 4 there are the results of using pairs bootstrap to estimate the predictive distributions of provisions. In the first column there are the mean of the estimated distributions for the origin year provisions and the mean of the total provision. In the second column there are the PE calculated using formula (9), where the estimation variance is obtained with the empirical variance of the predictive distributions. Finally, in the third column there are the associated coefficients of variation. For comparison, in Table 5 we include the same quantities than those of Table 4 but calculated for a standard GLM, `CL.glm`, and using residual bootstrap, following the indications of, e.g., [8, 9] and [10].

It is well known that one deficiency of the pairs bootstrap is that, compared with the residual bootstrap (when it is valid) and with the wild bootstrap, generally it does not yield very accurate results. But the pairs bootstrap is less sensible to the hypotheses of the model than the residual bootstrap. And the estimated standard error via the pairs bootstrap offers reasonable results when some hypotheses of the model are not satisfied. In the problem of claim reserving we have a small data set and it is probably that it does not follow the hypotheses of the assumed GLM. Then, it seems to be adequate to use pairs bootstrap for the estimation of the *real* distribution of predictions.

To illustrate graphically one of the predictive distributions, we show in Fig. 1 the predictive distribution of the total provision estimated with the DB-GLM `CL.dbglm` and using pairs bootstrap. In Appendix we include some descriptive statistics of the distribution as are the mean, the standard deviation, the coefficient

Table 1 Run-off triangle of Taylor and Ashe [12] with 55 incremental losses

	0	1	2	3	4	5	6	7	8	9
0	357848	766940	610542	482940	527326	574398	146342	139950	227229	67948
1	352118	884021	933894	1183289	445745	320996	527804	266172	425046	
2	290507	1001799	926219	1016654	750816	146923	495992	280405		
3	310608	1108250	776189	1562400	272482	352053	206286			
4	443160	693190	991983	769488	504851	470639				
5	396132	937085	847498	805037	705960					
6	440832	847361	1131398	1063269						
7	359480	1061648	1443370							
8	376686	986608								
9	344014									

Table 2 Chain-Ladder estimation of the future incremental losses of the run-off triangle of Taylor and Ashe [12]

	1	2	3	4	5	6	7	8	9
1									94633.8
2								375833.5	93677.8
3							247190	370179.3	92268.5
4						334148.1	226674.1	339455.9	84610.6
5					383286.6	351547.5	238477.3	357131.7	89016.3
6				605548.1	414501	389349.1	264120.5	395533.7	98588.2
7			1310258.2	725788.5	508791.9	466660	316565.5	474072.7	118164.3
8		1018834.1	1089616	603568.6	423113.4	388076.4	263257.2	394240.8	98265.9
9	856803.5	897410.1	959756.3	531635.7	372687	348125.7	231182.4	347255.4	86554.6

Table 3 Origin year provisions and total provision, prediction errors, and coefficients of variation for the GLM and DB-GLM assuming an over-dispersed Poisson, the logarithmic link, and the l^2 metric, using analytic formula of the run-off triangle of Taylor and Ashe [12]

Origin year	Provision	Prediction error	Coefficient of variation (%)
1	94633.8	110099.6	116.34
2	469511.2	216042.8	46.01
3	709637.8	260871.3	36.76
4	984888.6	303549.1	30.82
5	1419459.4	375012.8	26.42
6	2177640.6	495376.8	22.75
7	3920301.0	789959.7	20.15
8	4278972.2	1046512.6	24.46
9	4625810.6	1980100.7	42.81
Total	18680855.6	2945659.0	15.77

Table 4 Origin year mean provisions and total mean provision, prediction errors, and coefficients of variation for the DB-GLM assuming an over-dispersed Poisson, the logarithmic link, and the l^2 metric, using pairs bootstrap with size 1000 of the run-off triangle of Taylor and Ashe [12]

Origin year	Mean provision	Prediction error	Coefficient of variation (%)
1	197097.3	155179.5	163.97
2	567831.6	229653.5	48.91
3	802433.7	292340.3	41.19
4	1101788.1	317124.6	32.19
5	1563680.6	391937.7	27.61
6	2311566.8	489300.1	22.46
7	3926323.4	835373.9	21.30
8	4339191.5	660743.7	15.44
9	4826806.1	677215.6	14.63
Total	19554135.3	2231053.6	11.94

of variation, and some quantiles. If we compare the coefficients of variation of the DB-GLM and of the GLM in Tables 4 and 5 when using bootstrap approximation, we observe that we obtain less variation for the DB-GLM, 11.94%, than for the GLM, 15.43%.

It is of relevance the quantile 99.5% of the predictive distribution of the total provision, 26301996 (see Appendix). It represents the *value at risk* (VaR) of technical provisions in the actual context of Solvency II.

Table 5 Origin year mean provisions and total mean provision, prediction errors, and coefficients of variation for the GLM assuming an over-dispersed Poisson and the logarithmic link, using residual bootstrap with size 1000 of the run-off triangle of Taylor and Ashe [12]

Origin year	Mean provision	Prediction error	Coefficient of variation (%)
1	100415.9	108422.0	114.57
2	477357.4	213628.5	45.50
3	727897.6	257700.5	36.31
4	978122.3	301692.6	30.63
5	1438384.2	369127.7	26.00
6	2194055.3	491173.6	22.55
7	3934897.4	787571.2	20.08
8	4236250.6	1032951.4	24.14
9	4711135.7	2081503.2	44.99
Total	18757855.9	2882412.8	15.43

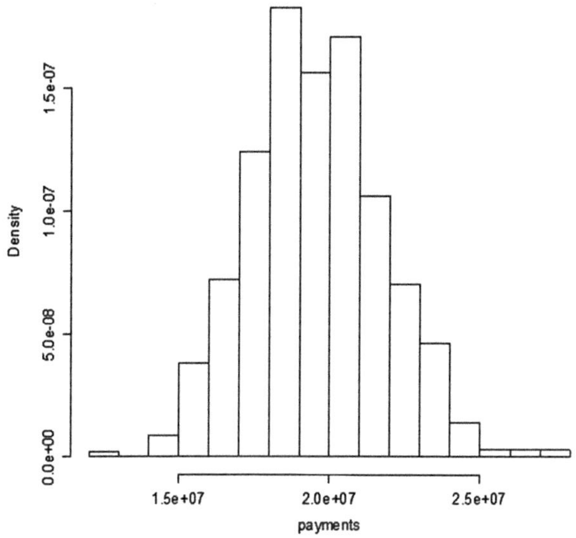

Fig. 1 Predictive distribution of the total provision of the run-off triangle of Taylor and Ashe [12]

4 Concluding Remarks

We show the formulation of the MSE (3) for the DB-GLM when we have new observations. We assume the parametric family of distributions (1) and the parametric family of link functions (6). We obtain the general expression (4).

We calculate PE with the square root of the MSE. We obtain expressions (7) and (8) for the PE when we use analytic formulation, and expression (9) when we use bootstrap estimation. We propose to employ the technique of pairs bootstrap with the DB-GLM, in the same way that has been employed in [2] where an F-test was defined for the DB-LM, and in [5] where a Wald test was defined for influence coefficients in the DB-GLM.

We obtain the formulas of the PE for the origin year provisions (10) in the case of using analytic formula, (12), and in the case of using bootstrap approximation, (14). And we obtain the formulas for the total provision (11) in both cases too, when using analytic formula, (13), and when using bootstrap approximation, (15).

We apply the formulas to solve the claim reserving problem in the context of Solvency II. We propose a claim reserving method by using the DB-GLM with a pairs bootstrap methodology. The method has the Chain-Ladder classical method as a particular case, when an over-dispersed Poisson distribution, the logarithmic link, and the l^2 Euclidean distance between factors are assumed. As it occurs with GLM, DB-GLM contains other classical methods of reserving as particular instances: least squares of De Vylder and the geometric and arithmetic separation methods. But as an additional remark, with DB-GLM it is possible to use another distance function between factors (origin and development years) of the run-off triangle with the aim to obtain better provisions estimation.

Acknowledgements Authors have partially been supported by the Spanish Ministerio de Educación y Ciencia under grant MTM2014-56535-R and by the Research Group of Multivariate Analysis and Classification (AMyC), http://amyc.seio.es/, of the Spanish Society of Statistics and Operations Research (SEIO) with José Fernando Vera Vera as principal investigator.

Appendix

```
R> # Fitted Chain-Ladder GLM and DB-GLM
R> n <- length(cij)
R> k <- trunc(sqrt(2*n))
R> i <- rep(1:k, k:1); i <- as.factor(i)
R> j <- sequence(k:1); j <- as.factor(j)
R> CL.glm <- glm(cij ~ i + j, family = quasipoisson)
R> CL.glm

  Call:  glm(cij ~ i + j, family = quasipoisson)

  Coefficients:
  (Intercept)          i2          i3          i4          i5
    12.506405    0.331272    0.321119    0.305960    0.219316
           i6          i7          i8          i9         i10
     0.270077    0.372208    0.553333    0.368934    0.242033
           j2          j3          j4          j5          j6
     0.912526    0.958831    1.025997    0.435276    0.080057
           j7          j8          j9         j10
    -0.006381   -0.394452    0.009378   -1.379907

  Degrees of Freedom: 54 Total (i.e., Null);  36 Residual
  Null Deviance:      10700000
```

```
  Residual Deviance: 1903000      AIC: NA

R> CL.dbglm <- dbglm(cij ~ i + j, family = quasipoisson,
 metric = "euclidean", method = "rel.gvar", rel.gvar = 1)
R> CL.dbglm

  Call:  dbglm(cij ~ i + j, family = quasipoisson,
  metric = "euclidean", method = "rel.gvar", rel.gvar = 1)

  family: quasipoisson
  metric: euclidean

  Degrees of Freedom: 54 Total (i.e., Null);  36 Residual
  Null Deviance:      10700000
  Residual Deviance: 1903000

  AIC:          NA
  BIC:          NA
  GCV:          76454.18

R> # Predictive distribution of the total provision
R> quantile(payments, c(0.5, 0.75, 0.90, 0.95, 0.995))
       50%      75%      90%      95%    99.5%
   19409902 20987943 22518643 23302403 26301996
R> mean(payments)
 [1] 19554135
R> sd(payments)
 [1] 2241767
R> cv <- (sd(payments) / mean(payments)) * 100; cv
 [1] 11.46441
```

References

1. Boj, E., Costa, T.: Provisions for claims outstanding, incurred but not reported, with generalized linear models: prediction error formulation by calendar years. Cuadernos de Gestión (2015, to appear). doi:10.5295/cdg.150526eb
2. Boj, E., Claramunt, M.M., Fortiana, J.: Selection of predictors in distance-based regression. Commun. Stat. A Theory Methods **36**, 87–98 (2007)
3. Boj, E., Caballé, A., Delicado, P., Fortiana, J.: dbstats: distance-based statistics (dbstats). R package version 1.4 (2014). http://CRAN.R-project.org/package=dbstats
4. Boj, E., Costa, T., Espejo, J.: Provisiones técnicas por años de calendario mediante modelo lineal generalizado. Una aplicación con RExcel. Anales del Instituto de Actuarios Españoles **20**, 83–116 (2014)
5. Boj, E., Costa, T., Fortiana, J., Esteve, A.: Assessing the importance of risk factors in distance-based generalized linear models. Methodol. Comput. Appl. Probab. **17**, 951–962 (2015)
6. Boj, E., Delicado, P., Fortiana, J., Esteve A., Caballé, A.: Global and local distance-based generalized linear models. TEST (2015). doi:10.1007/s11749-015-0447-1
7. Efron, B., Tibshirani, J.: An Introduction to the Bootstrap. Chapman and Hall, New York (1998)
8. England, P.D.: Addendum to 'Analytic and bootstrap estimates of prediction errors in claim reserving'. Insur. Math. Econ. **31**, 461–466 (2002)

9. England, P.D., Verrall, R.J.: Analytic and bootstrap estimates of prediction errors in claims reserving. Insur. Math. Econ. **25**, 281–293 (1999)
10. England, P.D., Verrall, R.J.: Predictive distributions of outstanding liabilities in general insurance. Ann. Actuar. Sci. **1**(II), 221–270 (2006)
11. McCullagh, P., Nelder, J.A.: Generalized Linear Models, 2nd edn. Chapman and Hall, London (1989)
12. Taylor, G., Ashe, F.R.: Second moments of estimates of outstanding claims. J. Econ. **23**, 37–61 (1983)
13. van Eeghen, J., Greup, E.K., Nijssen, J.A.: Loss reserving methods. Surveys of Actuarial Studies 1, National Nederlanden (1981)

An Inflated Model to Account for Large Heterogeneity in Ordinal Data

Stefania Capecchi, Rosaria Simone, and Domenico Piccolo

Abstract In sample surveys where people are asked to express their opinions, a high level of indecision among respondents may generate sub-optimal statistical analyses caused by large heterogeneity in the responses. We discuss a model belonging to the class of generalized CUB models that is suitable for this kind of surveys. Then, we examine a real case study where the observed heterogeneity as well as respondents' indecision can be analyzed within the theoretical framework of the proposed model leading to convincing interpretations. A comparison with current literature and some concluding remarks end the paper.

1 Introduction

Ordinal data are common in many fields involving Social Sciences and Humanities, Medicine and Marketing, among others. As a matter of fact, this typology of responses is collected in sample surveys concerning personal beliefs, habits, opinions, preferences, tastes, political orientation, well-being, work related issues, job satisfaction, etc. In such cases, although quite often the answer modalities are designed to be expressed by natural numbers to simplify coding and synthesis, ordinal data convey categorical information; thus, adequate statistical analysis should be performed to avoid oversimplified interpretations [1, 27].

The *response style* is a prominent issue in psychological and marketing studies when people respond to questionnaires according to a subjective disposition that could hide the "true" score. The issue is widely investigated, in different disciplines and fields. Generally, it has been argued [2, 22] that the response styles are tendencies to respond systematically to questionnaire items on the basis other than what the items were specifically designed to measure; that is, very briefly, a tendency to provide responses to the questionnaire items regardless of their specific content. Thus, a significant proportion of respondents tends to use only a smaller number of the rating scale options [9].

S. Capecchi (✉) • R. Simone • D. Piccolo
Department of Political Sciences, University of Naples Federico II, Via L. Rodinò 22, I-80138 Naples, Italy
e-mail: stefania.capecchi@unina.it; rosaria.simone@unina.it; domenico.piccolo@unina.it

F. Palumbo et al. (eds.), *Data Science*, Studies in Classification, Data Analysis, and Knowledge Organization, DOI 10.1007/978-3-319-55723-6_16

More specifically, we discuss a distinctive typology of responses characterized by a sharp preference towards the item (expressed by a proportion of the respondents) and by diffuse indecision manifested by a large group of the other interviewees. For a number of different motivations, a quota of interviewees selects a specific category which can be considered a sort of "refuge" (*shelter option*). According to different circumstances, this behavior may be caused by indecision, desire of privacy, or a real approval/disapproval of the item (for extreme categories) and it can be referred to one of the most common response styles.

If the item concerns evaluation/judgement which is of interest for a limited portion of the population, responses are concentrated in just one or very few categories. As an instance, consider a questionnaire designed to analyze relational goods and leisure habits. In this case, it has been asked "How important for you is to walk the dog" with an ordinal rating scale from 1 to 10; then, it is common to observe a very high frequency of respondents selecting 1 (since they don't have a dog) and low frequencies of responses spread over the other categories. A sample of this evidence from real sample surveys for different topics is presented in Fig. 1.

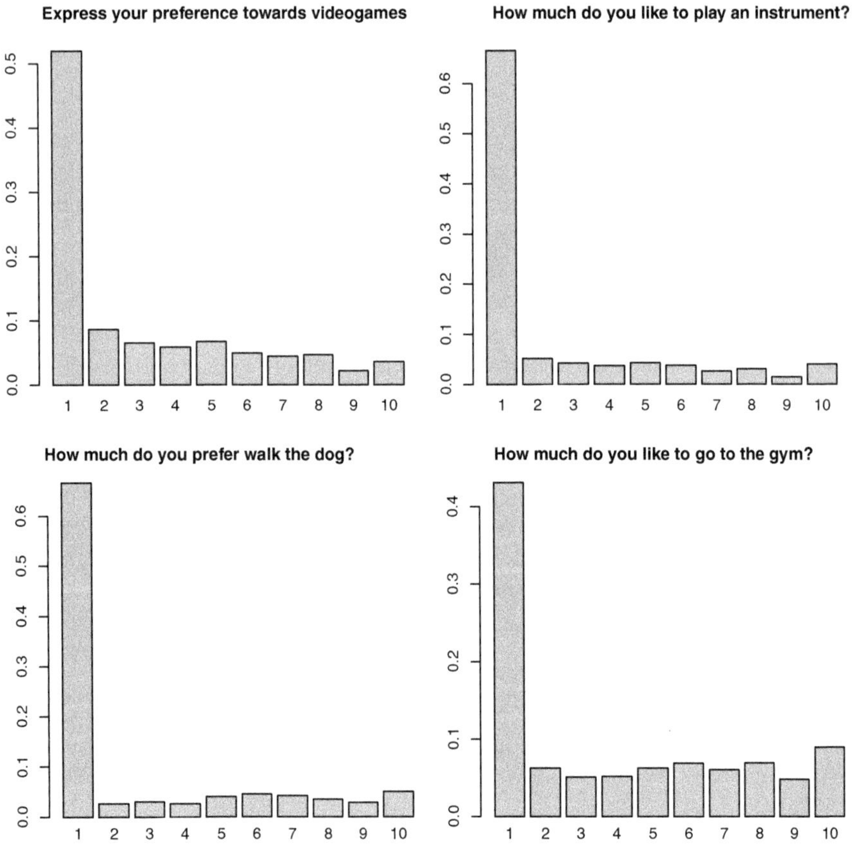

Fig. 1 Observed distributions with large heterogeneity

Similarly, in working conditions surveys, if some of the items concern the possibility of workers to select their own partners for a specific job/duty, most of the people would select the category "Never" (a large proportion of respondents indeed does not have the chance to choose their job partners or co-workers) whereas all the others would pick different categories. It is evident that without a specific approach, the analysis of these responses may cause bias in the interpretation.

Thus, it seems useful to take these information into specific account and search for adequate statistical modelling of this biased behavior and to relate it to subjects' characteristics when significant. In this regard a focal point is the relationship between individual indecision and observed heterogeneity of the responses. In fact, if the reaction of the interviewees is substantially homogeneous with respect to the item and their responses are mostly concentrated in few categories, then we observe a very low heterogeneity; on the contrary, interviewees express quite different responses when their indecision is high and thus we notice high heterogeneity in the observed distribution.

The paper is organized as follows: in the next section, we motivate the introduction of a parsimonious parametrization in modelling ordinal data in presence of large heterogeneity; then, in Sect. 3 we discuss the inferential issues related to estimates and test of the parameters that characterize the model. In Sect. 4 a real case study is presented and the role of significant covariates is discussed. Some concluding remarks end the work.

2 A Model for Large Heterogeneity

Motivated by psychological aspects of the decisional process, an alternative framework for modelling preferences and evaluations expressed as ordinal data has been proposed [5, 23, 24]. The main features of the approach are: parsimony in the number of parameters, high flexibility of shapes in the derived distributions, sharp visualization of the estimated models (an open source software is available as R package: [14]).

More specifically, the generating process leading to ordinal data is interpreted as a mixture where the indecision of the choice and the attractiveness/repulsion towards the item are explicitly modelled according to discrete distributions [13]. This class of models includes several generalizations and extensions to cope with real situations [10, 12, 19, 20, 25].

A model may be specified as the mixture of two distributions allowing for both *shelter effect* and extreme uncertainty, respectively, in case data originating from sample surveys suggest a large indecision among respondents (except for the ones who do not select a response category to be interpreted as a *shelter* one: [11]). This model is a Combination of a discrete *U*niform distribution with a *SH*elter effect and will be called CUSH model.

If R is the ordinal random variable defined on the support $\{1, 2, \ldots, m\}$, for a given m and $c \in \{1, 2, \ldots, m\}$ is the known location of the *shelter effect*, a CUSH model is defined by:

$$Pr(R = r \mid \boldsymbol{x}_i) = \delta_i\, D_r^{(c)} + (1-\delta_i)\,\frac{1}{m}\,, \qquad r = 1, 2, \ldots, m; \quad logit(\delta_i) = \boldsymbol{x}_i \boldsymbol{\omega}\,;$$
$$i = 1, \ldots, n\,. \tag{1}$$

where $D_j^{(c)} = I(r = c)$ and $I(A)$ is the indicator function such that is 1 when A is true and is 0 elsewhere. The row vector $\boldsymbol{x}_i = (x_{i0}, x_{i1}, \ldots, x_{is})$ includes the explanatory covariates of the *shelter effect* for the ith subject, with the convention: $x_{i0} = 1$, $i = 1, \ldots, n$.

If the previous model is conditioned on a specific pattern of the subjects' covariates (so that $\delta_i = \delta$), a CUSH model without covariates is:

$$Pr\,(R = r \mid \delta) = \delta\, D_r^{(c)} + (1-\delta)\,\frac{1}{m}\,, \qquad \delta \in [0, 1]\,. \tag{2}$$

As a matter of fact, δ measures the differential effect of a preferred category with respect to all the others. A CUSH model may be considered as a *c-inflated* model with respect to the discrete Uniform distribution in the same line of reasoning leading to the well-known zero-inflated models for Poisson [16] and Negative Binomial distributions [8], for instance. In Fig. 2 CUSH models of different shapes are obtained by varying c and δ.

For given m and c, a CUSH model is fully characterized by the parameter δ; thus, expectation and variance are given by:

$$\mathbb{E}\,(R) = \delta\, c + (1-\delta)\,\frac{m+1}{2}; \qquad Var(R) = (1-\delta)\left[\delta\left(c - \frac{m+1}{2}\right)^2 + \frac{m^2-1}{12}\right].$$

The mean value is a convex combination of the mean of the discrete Uniform distribution and a degenerate random variable at the shelter category, whereas the variance is a parabolic function whose maximum depends on m and c. The variance is 0 when $\delta = 1$ since the CUSH model collapses to a degenerate distribution at $R = c$.

More noticeably, the (normalized) heterogeneity [7] index turns out to be:

$$G = \frac{m}{m-1}\left(1 - \sum_{r=1}^{m} [Pr(R = r)]^2\right) = 1 - \delta^2\,;$$

thus, it is independent of the location c of the shelter and also of the number m of categories. As a consequence, δ may be interpreted as an inverse measure of heterogeneity: for increasing δ the probability mass distribution becomes more and more concentrated on the shelter category.

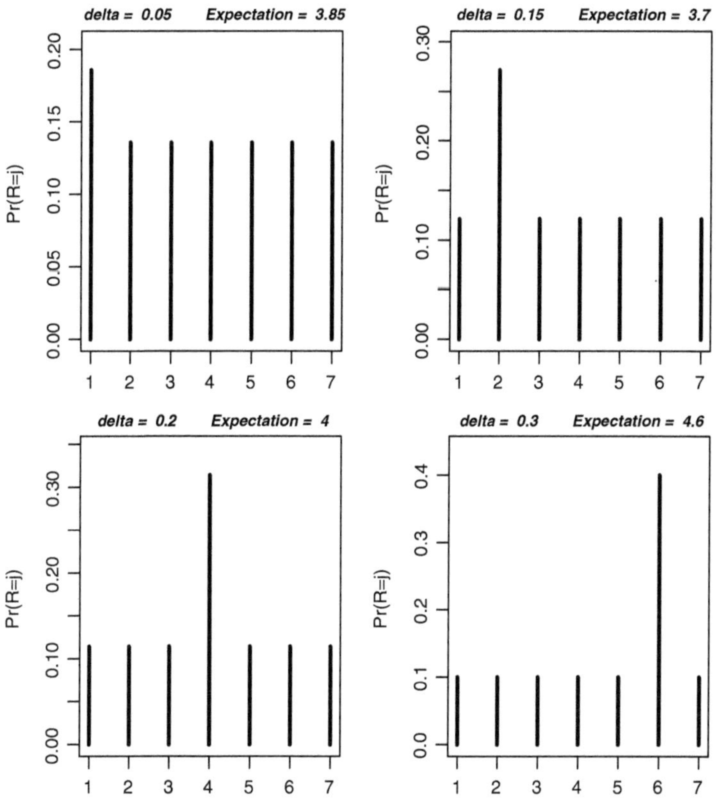

Fig. 2 A sample of typologies for CUSH distributions

However, Gini index is not so selective (it is invariably high in most of the circumstances) and this invariance with respect to m may be a limitation. Thus, the normalized [15] index may be preferred. In general, and for CUSH models, this index turns out to be:

$$\mathcal{H} = \frac{1}{m-1}\left[\left(\sum_{i=1}^{m}[Pr(R=r)]^2\right)^{-1} - 1\right] = \frac{1-\delta^2}{1+(m-1)\,\delta^2}\,.$$

This measure is monotonically related to the Gini index, since $\mathcal{H} = G/[m-G(m-1)]$, but it is more selective and depends on m.

It is possible to derive preliminary estimators of δ from the sample Gini index, and this measure may be useful also for the selection of covariates in the logit link for the model (1) as fully discussed in [3].

3 Inferential Issues for CUSH Models

If we denote by $(f_1, \ldots, f_m)'$ the vector of the relative frequencies obtained from the sample of ordinal data $(r_1, \ldots, r_n)'$, the Maximum Likelihood (ML) estimator of δ is obtained by maximizing the log-likelihood function:

$$\ell(\delta) = \sum_{i=1}^{n} \log\left(Pr\left(R = r_i \mid \delta\right)\right) = \sum_{r=1}^{m} n_r \log\left(Pr\left(R = r \mid \delta\right)\right).$$

It is simple to prove that ML estimator exists, it is unique and defined by:

$$\hat{\delta} = \begin{cases} \frac{m f_c - 1}{m-1} = \frac{f_c - 1/m}{1-1/m}, & \text{if } f_c \geq 1/m; \\ 0, & \text{otherwise.} \end{cases}$$

Since for common values of n the probability to observe $f_c < 1/m$ is virtually 0, in the following we consider the first expression as the ML estimator. Notice that $\hat{\delta}$ has a simple interpretation: it compares the relative frequency f_c at the *shelter category* c with the discrete Uniform hypothesis ($f_c = 1/m$), and then normalizes this difference.

It is possible to prove that the ML estimator $\hat{\delta}$ is an unbiased estimator of δ with a variance given by:

$$Var(\hat{\delta}) = \frac{1}{n} \frac{1-\delta}{m-1} \left[1 + (m-1)\delta\right].$$

The standard error of $\hat{\delta}$ is evaluated by plugging the ML estimate into the last expression:

$$es(\hat{\delta}) = \sqrt{\frac{1}{n} \frac{1-\hat{\delta}}{m-1} \left[1 + (m-1)\hat{\delta}\right]} = \frac{f_c}{\sqrt{n}} \sqrt{\frac{m}{m-1}}.$$

In order to test the presence of a *shelter effect*, both Wald and Likelihood Ratio test may be exploited, and they are asymptotically equivalent. Thus, the Wald test for checking $H_0 : \delta = 0$ is:

$$W = \frac{\hat{\delta}}{es(\hat{\delta})} = \sqrt{n}\, \frac{f_c\, m - 1}{f_c\, \sqrt{m(m-1)}}. \tag{3}$$

After some algebra, the maximized log-likelihood function turns out to be:

$$\ell(\hat{\delta}) = n \left[f_c \log(f_c) + (1 - f_c) \log\left(\frac{1 - f_c}{m-1}\right) \right].$$

Since the log-likelihood function under H_0 is $\ell(0) = -n\log(m)$, the Likelihood Ratio test (*LRT*) is:

$$LRT = 2\left(\ell(\hat{\delta}) - \ell(0)\right) = 2n\left[f_c \log(f_c) + (1-f_c)\log\left(\frac{1-f_c}{m-1}\right) + \log(m)\right]. \quad (4)$$

As a function of f_c, for a given c, *LRT* is U-shaped with a unique minimum at $f_c = 1/m$. So, for a given α-level, the critical region $LRT > c_{\alpha;n}$ is strictly equivalent to $|\, f_c - 1/m \,| > d_{\alpha;n}$, for a convenient α. In fact, it seems reasonable to reject the hypothesis of a null *shelter effect* when the relative frequency at the category c is significantly greater than the expected totally random proportion $1/m$. Figure 3 depicts this situation in a specific case.

However, when we test a borderline hypothesis (as $H_0 : \delta = 0$) the asymptotic distribution of the *LRT* (4) does not converge to a $\chi^2_{(1)}$ random variable [21, 26, 28] as predicted by the asymptotic standard theory. An acceptable approximate solution is to halve the p-value of a $\chi^2_{(1)}$ distribution and to check this simplification for finite sample sizes by a simulation. For finite sample sizes, this simulation experiment has been performed for varying values of n and m and *LRT* is preferred in regular situations; more evidence of these experiments is reported in [4].

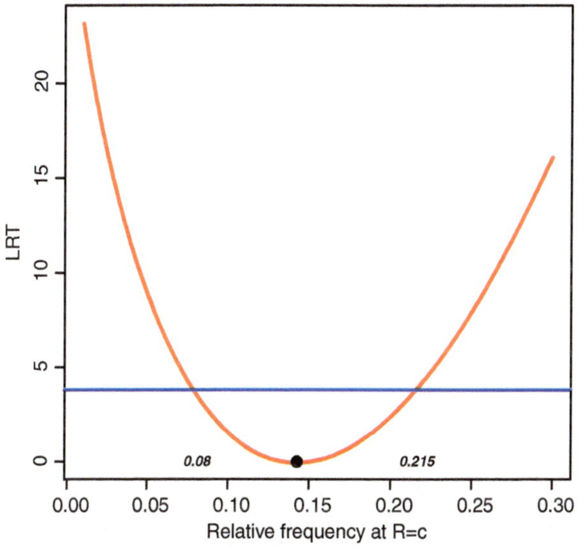

Fig. 3 *LRT* as function of the relative frequency at $R = c$, for a given c ($n = 100$, $m = 7$)

4 A Case Study

Every 5 years, Eurofound carries out the European Working Condition Survey (EWCS) which involves employees and self-employed people with interviews concerning their work and employment conditions across Europe [6].

Data of interest originate from the fifth EWCS and we consider the EU28 Member States only. The sample (a multistage stratified random sampling design) is representative of those aged 15 years and over who are employed and resident in the country being surveyed. The target number of interviews was greater or equal to 1000 in all the countries and global results are based on $n = 31{,}689$ respondents.

The responses to the item: "You have a say in the choice of your working partners" with possible responses: *Never, Rarely, Sometimes, Most-of-the-time, Always* (coded as 1–5) are mostly investigated at an aggregate level. Then, a synthesis of the results for the EU28 Member States is shown. The observed distribution of responses to "*Have a say* ..." is plotted in Fig. 4.

With reference to the selected modelling framework, in Table 1 several alternative and comparable models are presented. First of all, a CUB model without covariates (which is a sort of benchmark for this kind of analysis), then a CUB model

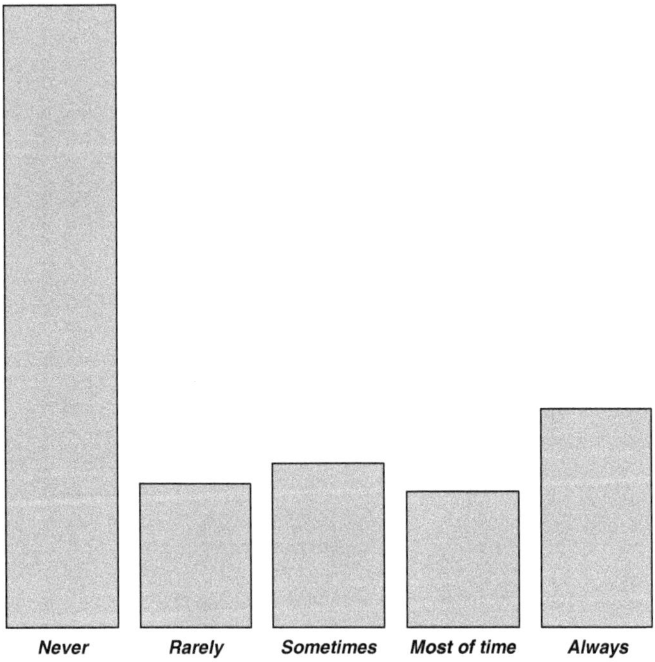

Fig. 4 Frequency distribution of responses to "*Have a say* ..." item

Table 1 Alternative mixture models with uncertainty for responses to *Have a say ...*

Models	Characteristics	*Log − lik*	*BIC*
CUB	*no covariates*	−44595.07	89210.87
CUB +*sh*	*shelter at* R=1	−44595.07	89221.24
CUSH	*no covariates*	−44595.07	89200.50
CUSH +*cov*	**Gender, Lage, Univ, Private**	−44046.38	88154.95

with the inclusion of a shelter effect at $R = 1$. Finally, CUSH models without and with covariates are compared.

The first three models are virtually equivalent since all of them capture a dominant frequency of responses located at $R = 1$ (that is, the proportion of interviewees selecting "Never") and CUSH model should be preferred by a parsimony criterion (as confirmed by the *BIC* index). Then, among the several covariates available in the dataset and able to explain the δ values in subgroups, the following ones have been found significant:

$$\texttt{Lage}_i = \text{deviation from the mean of the logged age in years}$$

$$\texttt{Gender}_i = \begin{cases} 0\,, \text{ if the } i\text{th subject is a man;} \\ 1\,, \text{ if the } i\text{th subject is a woman.} \end{cases}$$

$$\texttt{Univ}_i = \begin{cases} 0\,, \text{ if the } i\text{th subject has not a University education;} \\ 1\,, \text{ if the } i\text{th subject has a University education.} \end{cases}$$

$$\texttt{Private}_i = \begin{cases} 0\,, \text{ if the } i\text{th subject does not work in the private sector;} \\ 1\,, \text{ if the } i\text{th subject works in the private sector.} \end{cases}$$

Given the selected covariates $\boldsymbol{x}_i = (Lage_i, Gender_i, Univ_i, Private_i)$ and within the class of CUSH models, the best result is given by the following *stochastic* and *systematic* components, respectively (standard errors in parentheses):

$$Pr(R = r \mid \boldsymbol{x}_i) = \delta_i D_r^{(1)} + (1 - \delta_i)\,\frac{1}{5}\,; \quad r = 1, 2, \dots, 5; \; i = 1, 2, \dots, n;$$

$$\begin{aligned} logit(\delta_i) = &\underset{(0.038)}{-0.562} + \underset{(0.032)}{0.544}\,Gender_i \underset{(0.039)}{-0.870}\;Univ_i \underset{(0.034)}{-0.295}\;Private_i \\ &\underset{(0.056)}{-0.294}\,Lage_i + \underset{(0.139)}{1.380}\;Lage_i^2\,; \qquad i = 1, \dots, n\,. \end{aligned}$$

The *shelter effect* shows a negative impact (that is, an increasing probability of a "Never" response) when the respondent is a woman, whereas the impact is positive for people working in the private sector as well as for the ones educated at a University level, mostly. The effect of `Lage` is parabolic and globally positive since higher ages correspond to a reduced probability of "Never" responses (Fig. 5).

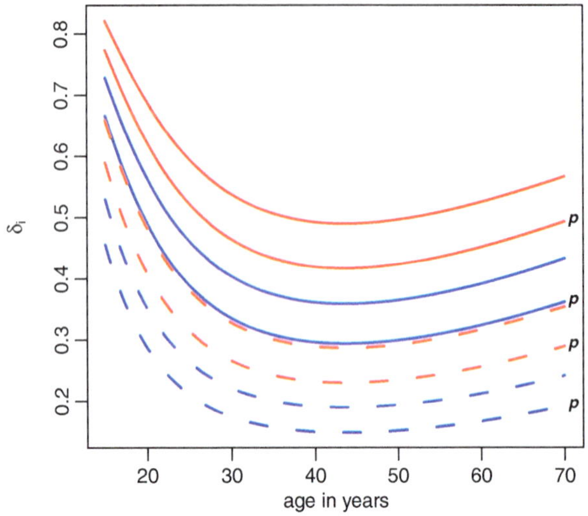

Fig. 5 Estimates of the *shelter effect* for "*Have a say ...*". *Broken lines* denote education at University level; women-lines are systematically above the corresponding men-lines; *p-lines* are for people working in private sector

If we compare the previous estimated models with the standard approach—as the proportional odds model (POM) [17, 18]—we get the same significant covariates and a better result in terms of log-likelihood function caused by an increase in the number of parameters to be estimated (9 instead of 6). Indeed, POM models imply a local fit for each frequency whereas CUSH models involve a global fitting and thus they are by far more parsimonious. In addition, the relationship between the *shelter effect* and the covariates may be sharply depicted by CUSH models as shown in Fig. 5.

The estimated CUSH model has been replicated for every country by considering if and where the impact of the selected covariates was homogeneous and significant. This long modelling exercise may be comparatively presented in a series of panels where the impact of each covariate on the *shelter effect* [according to a logit link as in (1)] is shown with the asymptotic level to check for its significance. Figure 6 presents in four panels the effects of the covariates `Gender`, `Lage`, `University`, and `Private`, respectively, on the item "*Have a say...*".

It is evident that `Gender` and `University` have a substantially homogeneous effect which is almost always significant in all countries and in the same direction; less evident are the influences of `Private` and `Lage` which seem to have an important impact only in some specific countries.

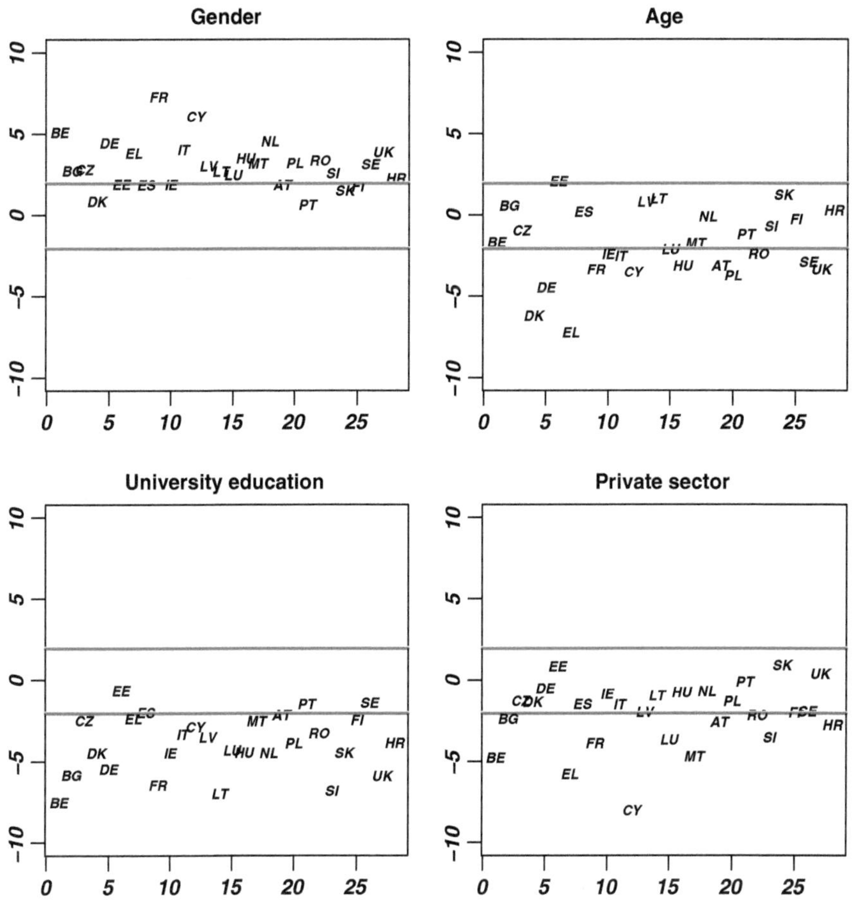

Fig. 6 Differential *shelter effect* on covariates by countries

5 Conclusions

CUSH models are a very simple probability structure introduced to interpret real case studies characterized by large heterogeneity and uncertainty in the responses. Effective methods based on likelihood paradigm can be applied for both estimation and testing. This class of model is particularly parsimonious and offers a sharp graphical representation of the covariates on the inflated category. Some empirical evidence confirms the usefulness of the proposal.

Further open issues worth to be explored include the research of effective tools to select covariates and the investigation of the predictability of these models. Finally, some multivariate suggestions could be very attractive when several items have to be compared and jointly examined.

Acknowledgements This work has been realized within the FIRB2012 project (code RBFR12SHVV) at University of Perugia and the frame of Programme STAR (CUP E68C13000020003) at University of Naples Federico II, financially supported by UniNA and Compagnia di San Paolo.

References

1. Agresti, A.: Analysis of Ordinal Categorical Data, 2nd edn. Wiley, Hoboken (2010)
2. Baumgartner, H., Steenback, J.-B.E.M.: Response styles in marketing research: a cross-national investigation. J. Mark. Res. **38**, 143–156 (2001)
3. Capecchi, S., Iannario, M.: Gini heterogeneity index for detecting uncertainty in ordinal data surveys. METRON (2016). doi:10.1007/s40300-016-0088-5
4. Capecchi, S., Piccolo, D.: Dealing with heterogeneity in ordinal responses. Qual. Quant. (2016). doi:10.1007/s11135-016-0393-3
5. D'Elia, A., Piccolo, D.: A mixture model for preference data analysis. Comput. Stat. Data Anal. **49**, 917–934 (2005)
6. Eurofound: Fifth European Working Conditions Survey. Publications Office of the European Union, Luxembourg (2012)
7. Gini, C.: Variabilità e mutabilità. Studi economico-giuridici, Facoltà di Giurisprudenza, Università di Cagliari, A, III, parte II (1912)
8. Greene, W.H.: Some accounting for excess zeros and sample selection in Poisson and negative binomial regression models. Working Paper EC-94-10, Department of Economics, New York University (1994)
9. Harzing, A.W.K.: Response styles in cross-national survey research: a 26-country study. Int. J. Cross Cult. Manag. **6**, 243–266 (2006)
10. Iannario, M.: Hierarchical CUB models for ordinal variables. Commun. Stat. Theory Methods **41**, 3110–3125 (2012)
11. Iannario, M.: Modelling shelter choices in a class of mixture models for ordinal responses. Stat. Methods Appl. **21**, 1–22 (2012)
12. Iannario, M.: Modelling uncertainty and overdispersion in ordinal data. Commun. Stat. Theory Methods **43**, 771–786 (2014)
13. Iannario, M., Piccolo, D.: CUB models: statistical methods and empirical evidence. In: Kenett, R.S., Salini, S. (eds.) Modern Analysis of Customer Surveys: With Applications Using R, pp. 231–258. Wiley, Chichester (2012)
14. Iannario, M., Piccolo, D., Simone, R.: CUB : a class of mixture models for ordinal data (2016). R package version 0.1. http://CRAN.R-project.org/package=CUB
15. Laakso, M., Taagepera, R.: Effective number of parties: a measure with application to West Europe. Comp. Pol. Stud. **12**, 3–27 (1989)
16. Lambert, D.: Zero-inflated poisson regression, with an application to defects in manufacturing. Technometrics **34**, 1–14 (1992)
17. McCullagh, P.: Regression models for ordinal data (with discussion). J. R. Stat. Soc. Ser. B **42**, 109–142 (1980)
18. McCullagh, P., Nelder, J.A.: Generalized Linear Models, 2nd edn. Chapman & Hall, London (1989)
19. Manisera, M., Zuccolotto, P.: Modeling rating data with Nonlinear CUB models. Comput. Stat. Data Anal. **78**, 100–118 (2014)
20. Manisera, M., Zuccolotto, P.: Modelling "Don't know" responses in rating scales. Pattern Recogn. Lett. **45**, 226–234 (2014)
21. Molenberghs, G., Verbeke, G.: Likelihood ratio, score, and Wald tests in a constrained parameter space. Am. Stat. **61**, 22–27 (2007)

22. Moors, G.: Exploring the effect of a middle response category on response style in attitude measurement. Qual. Quant. **42**, 779–794 (2008)
23. Piccolo, D.: On the moments of a mixture of uniform and shifted binomial random variables. Quaderni di Statistica **5**, 85–104 (2003)
24. Piccolo, D.: Observed information matrix in MUB models. Quaderni di Statistica **8**, 33–78 (2006)
25. Piccolo, D.: Inferential issues on CUBE models with covariates. Commun. Stat. Theory Methods **44**, 5023–5036 (2015)
26. Self, S.G., Liang, K.Y.: Asymptotic properties of maximum likelihood estimators and likelihood ratio test under nonstandard conditions. J. Am. Stat. Assoc. **82**, 605–610 (2003)
27. Tutz, G.: Regression for Categorical Data. Cambridge University Press, Cambridge (2012)
28. Vu, H.T.V., Zhou, S.: Generalization of likelihood ratio tests under nonstandard conditions. Ann. Stat. **25**, 897–916 (1997)

Functional Data Analysis for Optimizing Strategies of Cash-Flow Management

Francesca Di Salvo, Marcello Chiodi, and Pietro Patricola

Abstract The cash management deals with problem of automating and managing cash-flow processes. Optimization of the management processes greatly reduces overall cash handling costs. The present analysis is an empirical study of cash flows, from and to bank branches, deriving an underlying theoretical framework, which can in a reasonable way be connected with the optimal strategy. Functional data analysis is considered an appropriate framework to analyze the dynamics of the time series behavior of cash flows: since the observations are not equally spaced in time and their number is different for each series, they are converted into a collection of random curves in a space spanned by finite dimensional functional bases. A central issue in the analysis is describing specific patterns of the curves, taking into account the temporal dependence, and the dependence between curves. The analysis provides a dynamic cash management model that is applied with alternative strategies for programming a cash in transit for the difference between cash inflows and cash outflows in a fixed interval of time. As the strategies are affected by changes in the behavior of the cash flows, the dynamic model outperforms more traditional approaches in identifying the optimal strategy.

1 Introduction

The cash management deals with problem of automating and managing cash-flow processes in and out of the branches of banks. Nowadays much reliance is placed on forecasting software to solve cash management and the statistical approach brings significant efficiency in predicting how cash is expected to flow. Two figures are important to monitor: cash availability and overall cost. The cash availability is the key objective related to productivity and efficiency. An optimization strategy needs to benchmark its specific requirements, given that higher cash availability produces higher costs: for example, a benchmark of 99% means that no more than 1 in 100 can be out of cash at any one time and its cost may be far higher than 95%. The

F. Di Salvo (✉) • M. Chiodi • P. Patricola
Department of Economics, Management and Statistics, University of Palermo, Palermo, Italy
e-mail: francesca.disalvo@unipa.it; marcello.chiodi@unipa.it; pietro.patricola@alice.it

F. Palumbo et al. (eds.), *Data Science*, Studies in Classification, Data Analysis, and Knowledge Organization, DOI 10.1007/978-3-319-55723-6_17

cost of cash is calculated by adding up the annual costs of cash intake/withdrawals and the marginal increase in cost, that occurs with the level of availability required and with the number of transports; it depends also on factors such as the amount of cash delivered each time, insurance, number of services, number of currencies, number of denominations. There is a natural trade-off between cash availability and the cost of cash. Hence, some basic quantities are supposed to be set, as a minimum threshold at which you do not want to see cash availability fall below, a maximum threshold, beyond which more cash in the supply chain leads to interest costs too high.

Trends in cash management are evolving to include more active methods of forecasting and an emerging topic is the role of new statistical techniques for cash-flow analysis. The statistical models can form the basis of optimization strategies that, programming the frequency of transports, reduce the overall cost. In this paper we look at the cash-flow data as multiple time series, modeling their behavior through a functional data analysis approach in order to have models self-adjusting to demand or supply chain changes. The proposed approach is the methodological proposal of a joint work with Sikelia Service, a consultant and system integrator company, offering information technologies and outsourcing services to Financial Institutions. One of the emerging interests of Sikelia is developing software application to manage cash flows on the cash points of Banking Institutes.

2 A Statistical Approach to the Cash Management

In the past programming cash in transit has been considered an empirical process. Some statistical techniques were considered as logistics analysis or neural networks as statistical methods [8]. These kinds of models use information only from a given static point of time and are informative only if the process remains stable over time, but such assumption is often violated.

In the last year cash management has been supposed to require expert analysis, focusing on adjusting models to changes in the operating environment, and modeling more dynamically the time series behavior. Moreover a shift to high frequency data has occurred due to their huge availability.

Data analysis performed on such datasets for decision-making is generally referred to as big data analysis. In the case of dynamic statistical analysis, data are also potentially unbounded in size and irregularly spaced. For this kind of data, very frequent in Econometrics, new insights are obtained in sequential approach to sensitivity analysis, as described in [6] and [7]: influential analysis is performed with attention to influential data and also to the change point, which is the point in time when the population parameters of the target data change. The recent literature on this topic has an increasing interest in models for functional data (see [9]). High frequency and irregularly spaced data observed in time are considered as samples from a collection of curves and the functional data approach analyzes the temporal dynamics of the curves (see [11]).

In our proposal, the time series of cash flows recorded at each instant are considered discretized observations from functional curves that we estimate through smoothing methods. We synthesize the mode of variations of the curves through the functional principal component analysis. Three different strategies data-based that determine the daily cash need, on the basis of the difference between cash inflows and cash outflows, are applied on the estimated curves. On the basis of the temporal dynamics, the approach allows the identification of the better strategy, optimizing the cash in transit and reducing the overall handling costs.

We illustrate the usefulness of our approach through an application to real data from the branches of a Banking Institution. The performance in overcoming the problem of missing data is evaluated through simulations. R code has been implemented for performing the analysis. Both the simulation studies and applications on empirical data show improvements in the curve reconstruction; the results indicate that a functional data-based approach for the dynamic cash management model improves the process of identification of the optimal strategy.

3 Preliminary Analysis of the Data

Data consists of the amount of money of all the transactions from/to bank, recorded at each instant in the cash points, of two Banking Institutions during 2014; the data were extracted from the data warehouse of Sikelia Services. In Fig. 1 (left) the transactions to the banks have signum plus and those from the banks, that represent payments, have signum minus. The cash flow for each branch is reconstructed by cumulating the values of the transactions over all the period considered (Fig. 1, right). These cumulative functions are the object of our analysis.

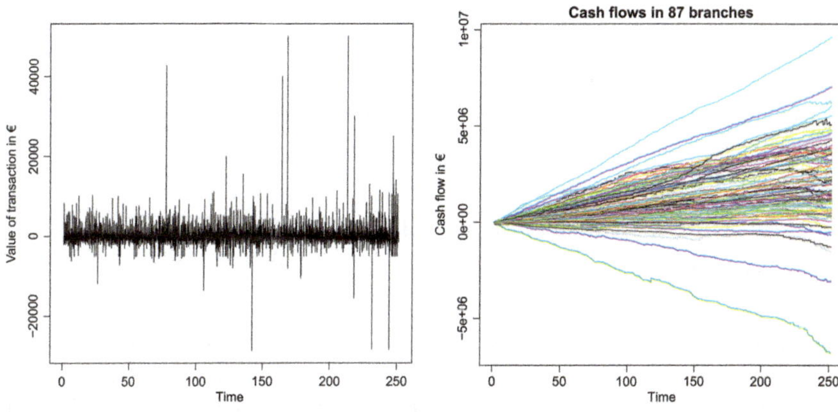

Fig. 1 Observed time series of the values of the transactions for one of the Banks (*left*); Cash flow obtained as cumulative functions of the single transactions (*right*)

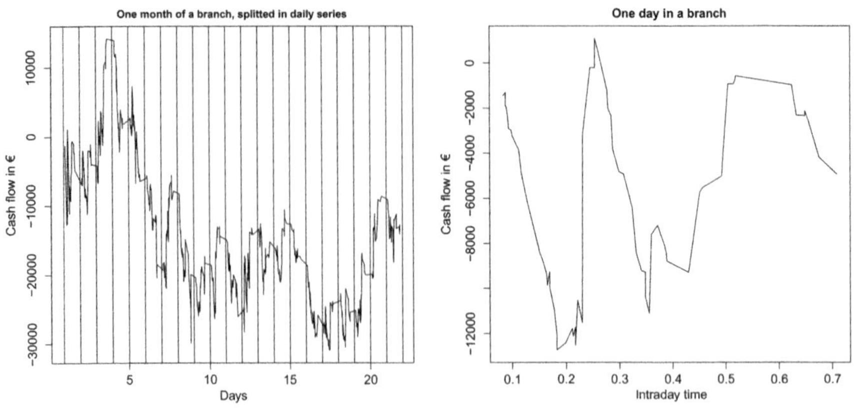

Fig. 2 One-month time series splitted into 1-day time series (*left*); time series spanned by 1 day (*right*)

For a better understanding of dynamics of the temporal patterns, we distinguish the intra-day variations from variations among the days: the time series has been splitted into intra-day series, as described in Fig. 2 (left), where only the first month is represented. So we have multiple time series, one for each day and for each cash point. The time interval considered is the day and it has been normalized in a unit interval [0,1] on the basis of the timing of all the series, as in Fig. 2 (right). This allows comparability among all the time series observed.

An observed time series of cash flow at the pth day and for the branch i is represented as:

$$y_i^p(t_j)$$

where $i = 1, \ldots, N$, are the bank branches; $p = 1, \ldots, P$, P are the work days in the year; the number of transactions of the branch i at day p is n_i^p and $t_1^p, \ldots, t_{n_i}^p$, are the intra-day standardized times of the day p for the branch i;

The intra-day time series are considered as realizations of random functions.

4 The Approach Based on Functional Principal Component Analysis

Our goal is to propose the functional data analysis (FDA) as an appropriate framework for identifying the dynamics of the yearly streamflow patterns. We consider a generalization of the FPCA approach proposed by Ramsay and Silverman [11] based on Generalized Additive Models.

Generalized Additive Models (GAM) [5] are an extension of Generalized Linear Models (GLM), in which the linear predictor is not restricted to be linear in the covariates but is the sum of smoothing functions applied to the covariates. Our aim is to extend these ideas to the functional covariates. There are some previous works in this direction, specially devoted to extend GLM to functional data (see [1]). The GAM models in the contest of functional data are introduced in [10]. In the recent paper [3], FDA is extended to the multivariate linking GAM to Multivariate Functional Data. Wavelet Models are proposed in [9] as solution for the problem of irregularly spaced time series. According to FDA, observed data $\mathbf{y}$, recorded at discrete times $(t_1, \ldots t_j, \ldots t_T)$ in N units and P dimensions, are considered realizations of continuous processes, [11]; in other words, data $\mathbf{y}$ are assumed to be the result of a signal plus a noise:

$$\underbrace{y_i^p(t)}_{data} = \underbrace{x_i^p(t)}_{signal} + \underbrace{\varepsilon_i^p(t)}_{noise} \tag{1}$$

where $t \in [0, 1]$, $i = 1, \ldots, N$, $p = 1, \ldots, P$

The random error › is typically assumed normally distributed, but generically it may follow any exponential family distribution [5]. In the standard model it is also assumed that errors are independently distributed with mean zero and constant variance σ^2.

The data are converted into a functional form by means of linear smooth function (see [11]).

$$x_i^p(t) = \sum_{k=1}^{K} \theta_{i,k}^p \phi_k(t) \tag{2}$$

where: $\Phi(t) = \{\phi_k(t)\}$ is the set of K basis functions and $\boldsymbol{\theta}_i^p = \left[\theta_{i,1}^p, \ldots, \theta_{i,K}^p\right]$ is the vector of the coefficients. In our approach we generalize this framework linking together GAMs model and FPCA, in order to overcome some problems relating to the structures of the data.

Indeed each branch may have different numbers of work days, the length of the vectors of intra-day transactions changes from day to day. The transactions happen at different times and sequences of missing data may occur; in general, we may be interested in smooth models of time and other covariates. We assume a GAM Model with a multidimensional smooth function:

$$x_i^p(t) = f(t, \mathbf{z}) \tag{3}$$

Here $f(t, \mathbf{z})$ is smooth function of several covariates, as tensor product of the two marginal smoothing models $f(t)$ and $f(\mathbf{z})$, [12]. In matrix terms:

$$X^p = f(t, \mathbf{z})$$

$$X^p = \Phi \Theta^p \tag{4}$$

The basis matrix $\Phi_{T\times K}$ is the row tensor product of two marginal basis $\Phi_1(T \times k_1)$, $\Phi_2(T \times k_2)$ for $f(t)$ and $f(\mathbf{z})$, respectively, and $K = k_1k_2$, [4]:

$$\Phi = \Phi_2 \Box \Phi_1 = (\Phi_2 \otimes 1'_{k_1}) \odot (1'_{k_2} \otimes \Phi_1),$$

In order to estimate the $K \times N$ matrix Θ^p, the P-spline (penalized B-spline) approach is here considered, minimizing the penalized sum of squares:

$$PENRSS_{\lambda}(y) = \|y - \Phi\Theta^p\|^2 + \Theta^{p'}\mathbf{H}_{\lambda}\Theta^p$$

P-splines have a number of useful and very attractive properties; a detailed overview is reported in [4]. The penalty term controls the smoothness of the fit applying penalties over adjacent coefficients. $\mathbf{H}_{\lambda}$ is a $(K \times K)$ penalty matrix depending on $\boldsymbol{\lambda}$, a vector of smoothing parameters. The appropriate degree of smoothness can be estimated from data using different criterions such as AIC, BIC, or Generalized Cross Validation (GCV) methods, (see [2]). The penalty matrix $\mathbf{H}_{\lambda}$ is defined as:

$$\mathbf{H}_{\lambda} = \lambda_1 I_{k2} \otimes \mathbf{H}_1 + \lambda_2 \mathbf{H}_2 \otimes I_{k1}$$

$\mathbf{H}_1$ and $\mathbf{H}_2$ are the marginal second order penalty matrices, although first or third order penalties can be used as well, and the parameters $\boldsymbol{\lambda} = (\lambda_1, \lambda_2)$ measure the rate of exchange between fit to the data and smoothness of the functions. The model allows different levels of smoothing for each set of covariates. The estimated coefficients are arranged into a three-mode array $\hat{\Theta}_{(K\times N\times P)}$, where $\hat{\Theta}^p$ is the pth slice $K \times N$ extracted, by fixing the index p:

$$\hat{\Theta}^p = \begin{vmatrix} \hat{\theta}^p_{1,1} & \hat{\theta}^p_{1,2} & \dots & \hat{\theta}^p_{1,N} \\ \dots & \dots & \dots & \dots \\ \hat{\theta}^p_{K,1} & \hat{\theta}^p_{K,2} & \dots & \hat{\theta}^p_{K,N} \end{vmatrix}$$

5 The Smoothed Functional Principal Component Analysis

In this section we aim at linking together results from GAM with the approach proposed by Ramsay and Silverman [11], in order to provide a generalization of the basic idea of reducing random trajectories to a set of functional principal component scores for a three-mode analysis. As the main objective is to understand the dynamic of the temporal variations of the smoothed data, we focus on a variance–covariance function, that properly quantifies the variation observed in functional data.

The Covariance Operator, a continuous function of the time, is the variance–covariance functions $\mathbf{V}^{p_ip_j}(t,t^*) = COV(X^{p_i}(t), X^{p_j}(t^*))$ is a suitable measure of the information retained by the p_i and p_j variables, as described in [11].

A better synthesis of the variance structure may be achieved through a dimension reduction technique: the functional principal component analysis provides dimension reduction for functional data, finding directions in the observation space along which the data have the highest variability and determining uncorrelated linear combinations of the original variables. Functional principal component analysis (FPCA) consists in performing the eigen decomposition of the covariance structure of functional data identifying the dominant modes of variation of the random trajectories and reducing the dimensions of the data. Moreover the functional principal component represents an optimal Empirical Orthonormal Basis system for finding an optimal approximations of the data. This result also allows reconstructions of the curves based on the GAM model, estimated as in the previous section, and provides a perspective of estimating the missing data that frequently occur in this kind of data. The Functional Eigen Analysis in the Multivariate case consists in solving a functional eigensystem of P equations:

$$\int V^{p,p_1}(t,t^*)\xi^{p_1}(t)dt^* + \int V^{p,p_2}(t,t^*)\xi^{p_2}(t)dt^* \ldots + \int V^{p,p_P}(t,t^*)\xi^{p_P}(t)dt^* = \rho\xi^p(t) \quad (5)$$

for $p = 1, \ldots, P$.

We estimate the covariance operator by means of the linear expansion in terms of the bases system Φ and the matrices of the estimated coefficients $\hat{\Theta}^{p_l}$ (for more details, see [3]).

$$V^{p_u,p_l} = \Phi\Theta^{p_u}\mathbf{W}\Theta'^{p_l}\mathbf{W}\Phi' \quad (6)$$

where $\mathbf{W}$ is the product of an order K symmetric matrix $(\Phi'\Phi + \mathbf{H})^{-1}$ and the transpose of the basis matrix $\Phi'_{K\times S}$:

$$\mathbf{W} = \left(\Phi'\Phi + \lambda_1 I_{k2} \otimes H_1 + \lambda_2 H_2 \otimes I_{k1}\right)^{-1}\Phi' \quad (7)$$

By replacing the variance functions with the discretized variances (6) and representing the eigenfunctions in terms of linear combinations of basis Φ and coefficients:

$$\xi^p(t) = \Phi(t)'\mathbf{b}^p, \quad (8)$$

the integral eigenequation system (5) becomes:

$$\Phi\mathbf{W}\Theta_{p_1}\,\Theta'_p\mathbf{W}\Phi'\mathbf{b}^{p_1} + \ldots + \Phi\mathbf{W}\Theta_{p_P}\,\Theta'_p\mathbf{W}\Phi'\mathbf{b}^{p_P} = \rho\mathbf{b}^p \quad (9)$$

for $p = 1, \ldots, P$.

In order to compute the first M eigenfunctions, the eigenequation has to be solved for the hth eigenvalue ρ^h, for $h = 1, \ldots, M$:

$$\tilde{\mathbf{V}}\,[\mathbf{b}^{p_1}, \mathbf{b}^{p_2}, \ldots, \mathbf{b}^{p_P}]' = \rho\,[\mathbf{b}^{p_1}, \mathbf{b}^{p_2}, \ldots, \mathbf{b}^{p_P}]' \quad (10)$$

The hth resulting $(K \times P)$ vector of coefficients $\mathbf{b}_h$ can be splitted into P parts and the eigenfunctions are computed as:

$$\xi_h^p(t) = \Phi(t)\mathbf{b}_h^p$$

The first M harmonics, $\xi_h^p(t)$, for $h = 1, \ldots, M$ can be arranged in a $(T \times P \times M)$ array. By varying t and p, the hth harmonic results in a $(T \times P)$ matrix:

$$\begin{vmatrix} \xi_{1,1,h} & \cdots & \xi_{1,p,h} & \cdots & \xi_{1,P,h} \\ \cdots & & \cdots\cdots & & \cdots\cdots \\ \xi_{t,1,h} & \cdots & \xi_{t,p,h} & \cdots & \xi_{t,P,h} \\ \cdots & & \cdots\cdots & & \cdots\cdots \\ \xi_{T,1,h} & \cdots & \xi_{T,p,h} & \cdots & \xi_{T,P,h} \end{vmatrix}$$

The eigenfunctions or harmonics are functions of time,

$$\xi_h^p(t) = \Phi(t)\mathbf{b}_h^p$$

The proportion of the variability in the hth principal component accounted for by variation in the dimension p is measured by:

$$\left\| \xi_h^p \right\|^2$$

Principal component scores resume variability over time and over the p dimension,

$$f^h(i) = \sum_{p=1}^{P} \sum_{t=1}^{T} x_t^p(i)\mathbf{b}_h^p$$

6 The Application: Validation of the Strategies Through the FPCA

In order to evaluate the performance of FPCA in identifying the optimal strategy, we have considered an application to the set of transaction value for the branches from two Banking Institutes, as described in Sect. 3. Then, for each branch, the values of the cash-flow process are computed summarizing the value of the transactions $z_i^p(t)$ up to the moment t:

$$y_i^p(t) = \sum_{j=1}^{N_{(i,t)}} z_i^p(t_j)$$

where $N_{(i,t)}$ is the number of transactions of the branch i up to the moment t. Then we have P irregularly spaced time series and the lengths are different from a branch to another. The next step is to convert raw data into functional, according to the model (4) introduced in Sect. 4, and in this way we overcome the previous problems of observed time series with different number of observations, as hereafter we deal with the random curves, estimated with a common basis matrix $\Phi(t)$. Then the functional principal components are extracted, the eigen decompositions allowing us to get a small dimension space which exhibits the main modes of variation under the functional model considered. Two main purposes are of interest: first, achieving a synthesis of the variability in and between the curves reducing the dimensionality of the initial problem; second, identifying an optimal strategy. In the following section we analyze the results.

7 Results and Comments

For both the two Banking Institutes, the FPCA has a good performance in classifying the branches according to the optimal strategy.

The branches are projected on the space of the first two principal components, accounting for a 99% of variability (in Figs. 3 and 4). After having classified the branches in according to their optimal strategy (the strategies are indicated with the three different colors in Figs. 3 and 4), we find out that the FPCA reproduces the same classification, as branches with the same optimal strategy are arranged in the same cluster. The functional principal components produce a valuable synthesis of the original data and describe properly the dynamic patterns relevant to the optimization process.

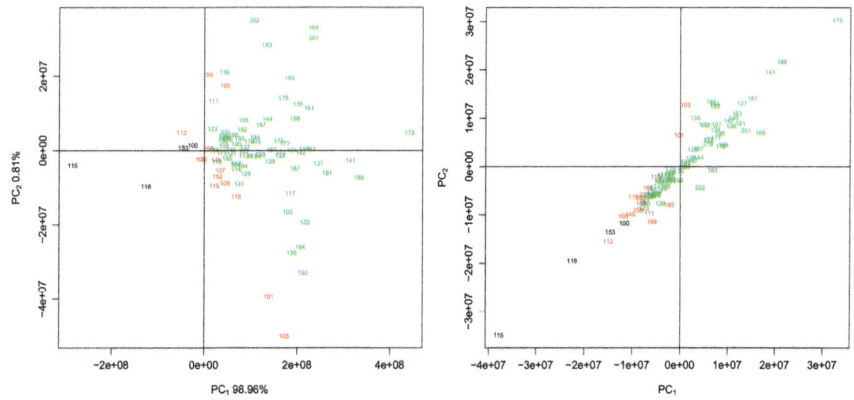

Fig. 3 Bank 1: Projection in the space of the first two principal components (*left*) and VARIMAX rotation (*right*)

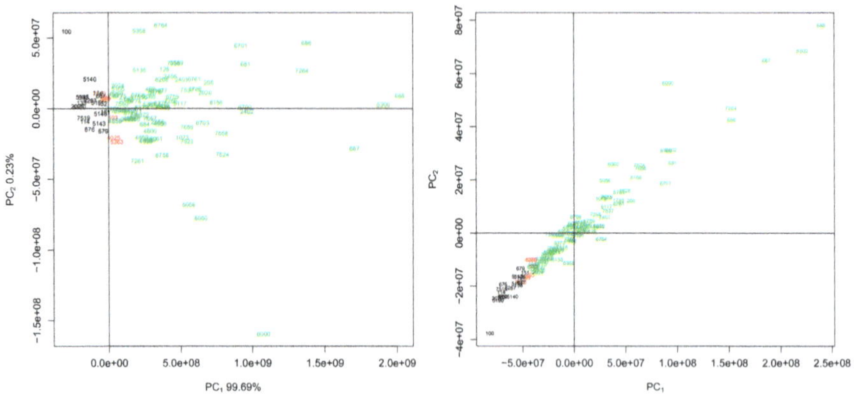

Fig. 4 Bank 2: Projection of the data in the space of the first two principal components (*left*) and VARIMAX rotation (*right*)

8 Performance of FPCA in Filling Missing Data

Missing data may occur for such data, due to a failure of the recording system or when some sequences seem to be unreliable and it is preferred to omit them, with a substantial loss of information rather then to retain them. Using FPCs as Empirical Orthonormal Basis system, we evaluate their performance in filling missing data in two different situations:

1. intervals of 10 consecutive missing days are randomly generated for each branch;
2. after having selected randomly 1 day, the last 5 day are supposed missing.

For both the situations, the following steps are implemented:

- The artificial gaps are reconstructed by mean of Empirical Orthonormal Bases according to:

$$\hat{x}_i^p(t) = \hat{\mu}(t) + \sum_{h=1}^{M} f^h(i)\xi_h^p(t).$$

- The alternative optimization strategies are implemented with the reconstructed data

The results in Tables 1 and 2 show that the reconstructions through FPCA, in presence of missing data, do not invalidate the results of the strategies neither for the availability, in terms of number of day with a negative balance, neither for the costs, in terms of increasing number of programmed cash transits.

Table 1 Simulation of the random sequences of 10 days of missing data

	Differences in the number of programmed cash in transit (%)	Occurrences of negative balance (%)
Strategy 1	0.00	0.00
Strategy 2	+10.11	0.00
Strategy 3	+3.18	−7.69

Table 2 Simulation of the random sequences of the missing data for the last five missing days

	Differences in the number of programmed cash in transit (%)	Occurrences of negative balance (%)
Strategy 1	+1.22	0.00
Strategy 2	−1.69	0.00
Strategy 3	0.00	−7.69

9 Conclusions

The present analysis is based on an empirical study of Data Stream of cash flow from/to bank branches. Three alternative strategies for optimizing the cash in transit and reducing overall handling costs are considered. The goal is to propose the functional data analysis (FDA) as an appropriate framework for identifying the dynamics of the yearly streamflow patterns. We consider a generalization of the traditional FPCA approach proposed by Ramsay [11] based on Generalized Additive Models for the estimation of the functional data and the variance function. The results indicate that a functional data-based approach for the dynamic cash management model improves the process of identification of the optimal strategy. The idea of merging the GAMs Model with FPCA is suggested and the procedure has good performance in describing the dynamic of the temporal patterns, reducing the dimensionality of the multivariate time series data, reconstructing sequences of missing values through Empirical Orthogonal Functions, validating the process of identification of the optimal strategy.

Acknowledgements This study was supported with funding by Research Grant Modelli statistici di previsione dei movimenti di cassa, 2015, Sikelia Service S.p.A.

References

1. Cardot, H., Sarda, P.: Estimation in generalized linear models for functional data via penalized likelihood. J. Multivar. Anal. **92**, 24–41 (2005)
2. Craven, P., Wahba, G.: Smoothing noisy data with spline functions: estimating the correct degree of smoothing by the method of generalized cross-validation. Numer. Math. **31**, 377–403 (1979)

3. Di Salvo, F., Ruggieri, M., Plaia, A.: Functional principal component analysis for multivariate multidimensional environmental data. Environ. Ecol. Stat. **22**(4), 739–757 (2015)
4. Eilers, P.H.C., Marx, B.D.: Flexible smoothing with B-splines and penalties. Stat. Sci. **11**(2), 89–102 (1996)
5. Hastie, T., Tibshirani, R.: Generalized Additive Models. Chapman & Hall, London (1990)
6. Hayashi, K., Kurihara, K.: Detecting a change point using statistical sensitivity analysis based on the influence function. In: SCIS&ISIS 2014, Kitakyushu (2014)
7. Hayashi K., Tanaka, Y.: Sensitivity analysis for multiple similarity method and its application. In: International Statistical Institute Proceedings of 58th World Statistical Congress, Dublin (2011)
8. Laitinen E.K., Laitinen T.: Cash management behavior and failure prediction. J. Bus. Fin. Account. **25**(7–8), 893–919 (1998)
9. Laukaitis, A.: Functional data analysis for cash flow and transactions intensity continuous-time prediction using Hilbert-valued autoregressive processes. Eur. J. Oper. Res. **185**, 1607–1614 (2008)
10. McLean, M.W., Hooker, G., Staicu, A., Scheipl, F., Ruppert, D.: Functional generalized additive models. Comput. Graph. Stat. **23**(1), 249–269, (2014)
11. Ramsay, J.O., Silverman, B.W.: Functional Data Analysis. 2nd edn. Springer, Berlin (2005)
12. Wood, S.N.: Generalized Additive Models: An Introduction with R. Chapman & Hall, London (2006)

The Five Factor Model of Personality and Evaluation of Drug Consumption Risk

Elaine Fehrman, Awaz K. Muhammad, Evgeny M. Mirkes, Vincent Egan, and Alexander N. Gorban

Abstract The problem of evaluating an individual's risk of drug consumption and misuse is highly important and novel. An online survey methodology was employed to collect data including personality traits (NEO-FFI-R), impulsivity (BIS-11), sensation seeking (ImpSS), and demographic information. The data set contained information on the consumption of 18 central nervous system psychoactive drugs. Correlation analysis using a relative information gain model demonstrates the existence of a group of drugs (amphetamines, cannabis, cocaine, ecstasy, legal highs, LSD, and magic mushrooms) with strongly correlated consumption. An exhaustive search was performed to select the most effective subset of input features and data mining methods to classify users and non-users for each drug. A number of classification methods were employed (decision tree, random forest, k-nearest neighbours, linear discriminant analysis, Gaussian mixture, probability density function estimation, logistic regression, and naïve Bayes) and the most effective method selected for each drug. The quality of classification was surprisingly high. The best results with sensitivity and specificity being greater than 75% were achieved for cannabis, crack, ecstasy, legal highs, LSD, and volatile substance abuse. Sensitivity and specificity greater than 70% were achieved for amphetamines, amyl nitrite, benzodiazepines, chocolate, caffeine, heroin, ketamine, methadone, and nicotine. The poorest result was obtained for prediction of alcohol consumption.

E. Fehrman (✉)
Men's Personality Disorder and National Women's Directorate, Rampton Hospital, Retford, Nottinghamshire DN22 0PD, UK
e-mail: Elaine.Fehrman@nottshc.nhs.uk

A.K. Muhammad • E.M. Mirkes • A.N. Gorban
Department of Mathematics, University of Leicester, Leicester LE1 7RH, UK
e-mail: akm40@le.ac.uk; em322@le.ac.uk; ag153@le.ac.uk

V. Egan
Department of Psychiatry and Applied Psychology, University of Nottingham, Nottingham NG8 1BB, UK
e-mail: Vincent.Egan@nottingham.ac.uk

F. Palumbo et al. (eds.), *Data Science*, Studies in Classification, Data Analysis, and Knowledge Organization, DOI 10.1007/978-3-319-55723-6_18

1 Introduction

Drug consumption and addiction constitutes a serious problem globally. Drug use is a risk behaviour that does not happen in isolation. It includes numerous risk factors, which are defined as any attribute, characteristic, or event in the life of an individual that increases the probability of drug consumption. Psychological, social, environmental, economic, and individual factors are correlated with initial drug use [8, 43]. These factors are likewise associated with several *personality traits* [5]. Legal drugs such as alcohol, and tobacco are probably responsible for far more premature deaths than illegal recreational drugs [2]. Psychologists have mostly agreed that the personality traits of the Five Factor Model (FFM) are the most comprehensive and adaptable technique for understanding human individual differences [9]. The FFM comprises Neuroticism (N), Extraversion (E), Openness to Experience (O), Agreeableness (A), and Conscientiousness (C).

Previous studies demonstrate that high N and O, and low A and C are associated with higher risk of drug use (including cocaine, cannabis, tobacco, heroin, and alcohol) [41]. Our findings improve the knowledge concerning the pathways leading to drug consumption. Sensation seeking (SS) is also higher for users of recreational drugs [24]. The problem of risk evaluation for individuals is much more complex. This was explored very recently [6, 42, 44].

The goal of the study was to predict the risk of drug consumption for each individual according to their personality traits. For this purpose, several data mining approaches were used: decision tree, random forest, k-nearest neighbours, linear discriminant analysis, Gaussian mixture, probability density function estimation, logistic regression, and naïve Bayes. For each drug, the most effective subset of input attributes was selected to provide the highest level of accuracy. Unexpectedly good classifiers were found for all drugs (see Table 1). Successful construction of a classifier provides an instrument for the evaluation of the risk of drug consumption for each individual, along with the creation of a map of risk [12, 31, 32].

The database was collected by an anonymous online survey methodology by Elaine Fehrman yielding 2051 respondents. It included personality traits, impulsivity (Imp), SS, and demographic information including country of location, ethnicity, level of education, gender, and age. The data set contains information on the consumption of 18 central nervous system psychoactive drugs including alcohol, amphetamines, amyl nitrite, benzodiazepines, cannabis, chocolate, cocaine, caffeine, crack, ecstasy, heroin, ketamine, legal highs, LSD, methadone, magic mushrooms, nicotine, and Volatile Substance Abuse (VSA).

Correlation analysis on base of Relative Information Gain (RIG) [33] demonstrates that the consumption of three legal drugs (alcohol, chocolate, and caffeine) is not correlated with other drugs. The consumption of seven illicit drugs (amphetamines, cannabis, cocaine, ecstasy, legal highs, LSD, and magic mushrooms) is symmetrically correlated. There are many strongly asymmetric correlations (see Fig. 3b). For example, knowledge about amphetamines consumption is useful for the evaluation of ketamine usage, but knowledge about ketamine consumption is significantly less useful for the evaluation of amphetamines usage.

2 Materials and Methods

The Database was collected by Elaine Fehrman between 2011 and 2012. An online survey tool from Survey Gizmo was employed to gather data which maximized anonymity, this being particularly relevant to canvassing respondents' views, given the sensitive nature of drug use. All participants were required to declare themselves at least 18 years of age prior to informed consent being given. The study recruited 2051 participants over an 18-month recruitment period. One thousand eight hundred and eighty-five participants (male/female = 943/942) were included following data cleansing. The snowball sampling methodology recruited a primarily (93.7%) native English-speaking sample, with participants from the UK (55.4%), the USA (29.5%), Canada (4.6%), Australia (2.9%), New Zealand (0.3%), and Ireland (1.1%). A total of 6.3% came from a diversity of other countries, none of whom individually met 1% of the sample or did not declare the country of location. Further optimizing anonymity, persons reported their age band, rather than their exact age; 18–24 years (34.1%), 25–34 years (25.5%), 35–44 years (18.9%), 45–54 years (15.6%), 55–64 (4.9%), and over 65 (1%). This indicates that although the biggest age cohort band was in the 18–24 range, some 40% of the cohort was 35 or above, which is an age range often missed in studies of this kind. The sample recruited was highly educated, with 59.5% educated to, at least, degree or professional certificate level: 14.4% reported holding a professional certificate or diploma, 25.5% an undergraduate degree, 15% a master's degree, and 4.7% a doctorate. Approximately 26.8% of the sample had received some college or university tuition although they did not hold any certificates; lastly, 13.6% had left school at the age of 18 or younger. Participants were asked to indicate which racial category was broadly representative of their cultural background. An overwhelming majority (91.2%) reported being White, (1.8%) stated they were Black, and were (1.4%) Asian. The remainder of the sample (5.6%) described themselves as 'Other' or 'Mixed' categories. This small number of persons belonging to specific non-white ethnicities precludes any analyses involving racial categories.

Personality Measurements In order to assess personality traits of the sample, the Revised NEO-Five Factor Inventory (NEO-FFI-R) questionnaire was employed. The NEO-FFI-R was developed and validated by Costa and McCrae [30]. The reliability of its component was studied in [11]. The scale is a 60-item inventory comprised of five personality domains. The five factors are: N, E, O, A, and C with 12 items per domain. Participants were asked to read the 60 NEO-FFI-R statements and indicate on a five-point Likert scale how much a given item applied to them. The second used measure was the *Barratt Impulsiveness Scale* (BIS-11) [40]. The BIS-11 is a 30-item self-report questionnaire, which measures the behavioural

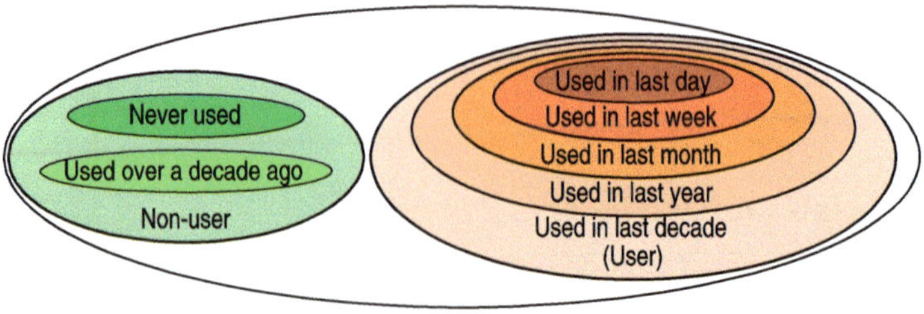

Fig. 1 Categories of drug users

construct of impulsiveness and comprises three subscales: motor impulsiveness, attentional impulsiveness, and non-planning. The scale's items are scored on a four-point Likert scale. The third measurement tool employed was the Impulsiveness Sensation Seeking Scale (ImpSS). Although the ImpSS combines the traits of Imp and SS, it is regarded as a measure of a general SS trait [45]. The scale consists of 19 statements in true–false format, comprising eight items measuring Imp, and 11 items gauging SS.

Drug Use Participants were questioned concerning their use of 18 legal and illegal drugs (alcohol, amphetamines, amyl nitrite, benzodiazepines, cannabis, chocolate, cocaine, caffeine, crack, ecstasy, heroin, ketamine, legal highs, LSD, methadone, magic mushrooms, nicotine, and VSA), and one fictitious drug (Semeron) which was introduced to identify over-claimers (analogously to [21]). Finer distinctions concerning the measurement of drug use have been deployed, due to the potential for the existence of qualitative differences amongst individuals with varying usage levels. It has seven categories of drug users that are depicted in Fig. 1. There are two special categories (see Fig. 1): 'Never used' and 'Used over a decade ago'. These two categories were placed into the class of 'Non-user' and all other five categories into the class 'User', as the simplest version of binary classification.

Input Feature Transformation Many data mining methods were developed to work with continuous data. To use these methods, it is necessary to quantify correctly all the discrete scales and categorical features (for example, 48 categories for each personality trait).

Ordinal Features Quantification The calculation of Polychoric Correlation (PolC) [26] is one of the widely used procedures to analyse categorical data. The matrix of PolC further is used to calculate Principal Components (PCs), etc. The technique of PolC is based on suggestion that categories of ordinal feature are the result of discretization of normally distributed random variable with fixed thresholds. There are two limitations of PolC techniques: it defines the thresholds of discretization but not the values for each category and the defined thresholds are different for different pairs of attributes. Let us have the ordinal feature o with categories o_1,o_2,...,o_k and with number of cases n_i of category o_i. The empirical estimation of probability of category o_i is $p_i = n_i/N$, where $N = \sum n_i$. The sample estimation of thresholds t_i

and average value of i category q_i are evaluating as:

$$t_i = \phi^{-1}\left(\sum_{j=1}^{i} p_j\right), q_i = \phi^{-1}\left(\sum_{j=1}^{i-1} p_j + \frac{p_i}{2}\right).$$

where ϕ is the cumulative normal distribution function.

The correlation coefficients, calculated on base of quantification q_i, have less likelihood than PolC. The merit of this approach is the usage of the same thresholds for all pairs of attributes and explicit formula for calculating the categories' values.

Nominal Feature Quantification We applied the procedure of nonlinear CatPCA [28] to quantify nominal features. This technique includes four steps: exclude nominal features from the set of input features and calculate the informative PCs [14–16, 35] in space of retained input features (to select informative PCs we use Kaiser's rule [18, 23]); calculate the centroid of each category in projection on selected PCs; and calculate the first PC of centroids; put the difference between projection of category centroid and projection of coordinate origin onto calculated PC as the numerical value for this category. As an alternative variant of nominal feature quantification we used dummy coding [17] of nominal variables: country and ethnicity. Each of these attributes was transformed into seven binary features with values 1 or 0.

Input Feature Ranking We used three different procedures of input feature ranking. First, we exploited *principal variables* [29]. Second, we applied Principal Component Analysis (PCA) with *double Kaiser's selection*: Calculated PCs and selected informative PCs by Kaiser's rule [18, 23]. This rule states the all PCs which are corresponding to eigenvalues greater than average are informative and all other PCs are uninformative. We applied the covariance based PCs. For this type of PCs, the Kaiser rule threshold is equal to the sum of diagonal elements of covariance matrix divided by the number of attributes. The importance of an attribute is defined as the maximum of absolute value of the corresponding coordinates in the important PCs represented by unit length vectors. For attribute selection we defined the threshold of importance as $1/\sqrt{n}$ for a unit length vector of dimension n, where n is the number of attributes. If the attribute importance is greater than the threshold of importance, then this attribute is informative. Otherwise the attribute is trivial. If there are trivial attributes, then the worst attribute is the attribute with the minimal value of importance. We removed the worst attribute and repeated the procedure. This procedure stops if there are no trivial attributes. This algorithm ranks attributes from the worst to the best one. Third, we used *sparse PCA* [34]. We applied the simplest threshold for sparse PCA. The detailed description of implemented sparse PCA version is presented in [12]. When using PCA for preselection of the important input features without considering their relationships with drug usage variables, some useful information may be lost. To overcome this limitation of the classical PCA, the goal-oriented tools should be used, like *supervised PCA* [25].

Classification Methods For this study, we applied several classification methods which provided risk evaluation as well.

1. K-Nearest Neighbours (KNN). The basic concept of KNN is: the class of an object is the class of the majority of its k-nearest neighbours [7]. We tested the KNN versions, which differed by: the number of NN, which was varied between 1 and 20; the set of input features; distances such as Euclidean distance, adaptive distance [19], and Fishers distance [13]; the kernel function for adaptive distance transformation; and the kernel functions [27] for voting.
2. Decision Tree (DT) [36, 39] is a method that constructs a tree like structure, which can be used to choose between several courses of action. The binary decision trees were used. We tested the decision trees, which differed by: the three split criterion (information gain, Gini gain, or DKM gain); linearly combined or separately used input features; the set of the input features; and the minimal number of cases in the leaf, which varied between 3 and 30.
3. Linear Discriminant Analysis (LDA). We used Fisher's linear discriminant [13]. We tested the LDA which differed by RIG, Gini gain, DKM gain, or accuracy as criterion for threshold defining.
4. Gaussian mixture is a method to estimate the probability under assumption that each category of target feature has the multivariate normal distribution [10].
5. Probability Density Function Estimation (PDFE). We implemented the radial-basis functions method for it [38]. We tested the PDFE versions which differed by: the number of the nearest neighbours which was varied between 5 and 30; the set of the input features; and the kernel function [27] which was placed in each data points.
6. Logistic regression. We implemented the weighted version of logistic regression [22]. The log likelihood estimation of the regression coefficients was used.
7. Naïve Bayes. We implemented the standard version of naïve Bayes [37]. All attributes which contained less than or equal to 20 different values were interpreted as categorical and the standard contingency tables were calculated for such attributes. All other attributes were considered as normally distributed.
8. Random forest was used for building a predictor ensemble with a set of decision trees that grow in randomly selected subspaces of data [4].

Maximal value of minimum of sensitivity and specificity were employed as the criteria. If minimum of sensitivity and specificity was the same, then we selected classifier with maximal sum of the sensitivity and specificity. This criterion selects the classifier furthest from the 'completely random guess' classifier. Classifiers with sensitivity or specificity less than 50% were not considered. *Leave-One-Out Cross Validation* (LOOCV) [1] was used for all tests.

3 Results

The study's sample was found to be significantly biased in comparison with the general population [11, 30]. Such a bias is usual for clinical cohorts [41]. For each drug average values of personality traits for groups of users are shifted from the sample means in the same direction as the sample means deviate from the population norm. For all drugs the average values of N and O for groups of users are neutral or moderately high. Groups of user of illicit drugs have moderately low A and C. Groups of licit drug users have neutral scores of A and C, apart from nicotine users, who have moderately low scores of C. For the groups of users of crack, heroin, VSA, and methadone, the score of E is moderately low. For groups of users of other drugs, the score of E is neutral. Detailed exploratory analysis of personality traits is presented in [12].

All three feature ranking approaches in Sect. 2 show that ethnicity and country are not important for evaluation of drug consumption risk. The reason for this is the high imbalance of frequency of different categories: more than 91% of cases in the 'White' category of ethnicity and 55% of UK and 29% of USA for country. These two features were removed from further study.

The Pearson's correlation coefficients (PCC) r is not appropriate for general categorical attributes but for the Boolean random variables (with 0,1 values) it gives a reasonable measure of dependence because for them $\text{cov}(X, Y) = P(X = 1 \& Y = 1) - P(X = 1)P(Y = 1)$. The majority of the PCCs between the usage of different drugs are significant. One hundred twenty-four pairs of drug usages from a total of 153 pairs have p-values less than 0.01 (p-value is the probability to observe by chance the same or greater correlation coefficient for independent variables). A multi-testing approach is necessary when testing 153 pairs of drug usages in order to estimate the significance of the correlation [3]. We applied the most conservative technique, Bonferroni correction, and used Benjamini-Hochberg (BH) step-up procedure [3] to control False Discovery Rate (FDR) for estimation of the genuine significance of these correlations. One hundred and fifteen correlation coefficients were significant with Bonferroni corrected p-value 0.001. BH step-up procedure with threshold of FDR equals 0.01 defined 127 significant correlation coefficients. We consider correlations with absolute values of PCC $|r| \geq 0.4$. Figure 2 sets out all identified significant correlations greater than 0.4. The correlation can be considered weak if $|r| < 0.4$; medium if $0.45 > |r| \geq 0.4$; strong if $0.5 > |r| \geq 0.45$; and very strong if $|r| \geq 0.5$. Figure 2 shows that the usage of amphetamines, cannabis, cocaine, ecstasy, ketamine, legal highs, LSD, and magic mushrooms correlates with

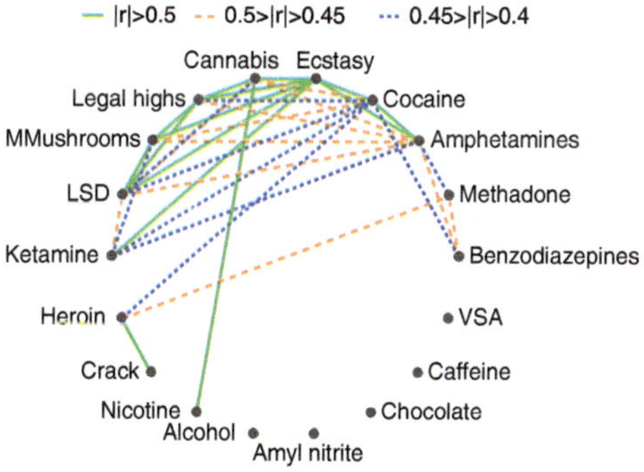

Fig. 2 Strong drug usage correlations

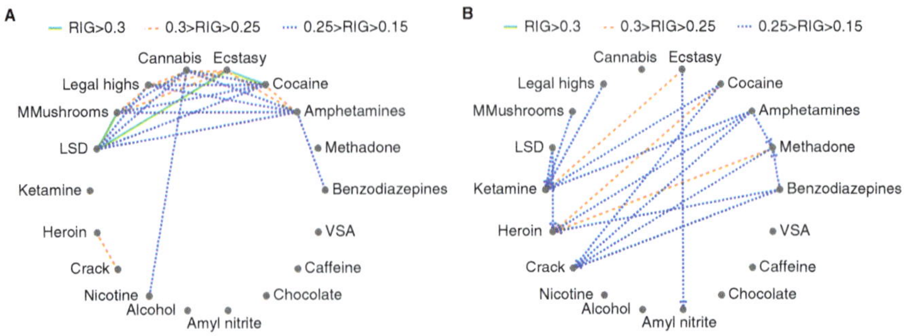

Fig. 3 Pairs of drug usages with high relative information gain: (**a**) more or less symmetric RIG and (**b**) significantly asymmetric RIG. In figure (**b**) *arrow* from LSD usage to heroin usage, for example, means that knowledge of LSD usage can decrease uncertainty in heroin usage

all other drugs in the same group, excluding correlations between cannabis and ketamine usage (r=0.302) and between legal highs and ketamine usage (r=0.393).

RIG is widely used in data mining to measure dependence between categorical attributes [33]. The greater the value of RIG, the stronger is the indicated correlation. RIG is zero for independent attributes. RIG is not symmetric. It is a measure of mutual information. Figure 3 presents all pairs with RIG >0.15. Figure 3a shows "symmetric" RIGs (we call RIG($X|Y$) symmetric if $\frac{|\mathrm{RIG}(X|Y)-\mathrm{RIG}(Y|X)|}{\min\left(\mathrm{RIG}(X|Y),\mathrm{RIG}(Y|X)\right)} < 0.2$). Figure 3b shows asymmetric RIG. It can be seen that in Fig. 3a the usage of amphetamines, cannabis, cocaine, ecstasy, legal highs, LSD, and magic mushrooms are correlated with each other. This group is the same that in Fig. 2 except for ketamine usage. Asymmetric RIGs illustrate pattern significantly different from Fig. 2.

To find the best classifier we used eight different types of classifiers and selected the best one for each drug. Results of the best classifier selection are presented in Table 1. Table 1 shows that for all drugs except alcohol, cocaine, and magic mushrooms, the sensitivity and specificity are greater than 70%. It is an unexpectedly high accuracy. After initial feature selection data contain 10 input features. Each of them is important at least for five drugs. There is no single most effective classifier employing all input features. The maximum number of used features is 6 out of 10 and the least number is 2. Table 1 shows that the best choice of the input attributes is different for different drugs. Age is used in the best classifiers for 14 drugs (the most universal attribute). Gender is used in the best classifiers for 10 drugs.

It is important to note that attributes which are unused in the best classifiers are not non-informative or redundant. For example, for ecstasy the best classifier is based on Age, SS, and Gender and has sensitivity 76.17% and specificity 77.16%. There exist a DT for the same drug based on Age, Edu., O, C, and SS with sensitivity 77.23% and specificity 75.22%, a DT based on Age, Edu., E, O, and A with sensitivity 73.24% and specificity 78.22%, and a KNN classifier based on Age, Edu.,

Table 1 The best results of the drug users classifiers

Target feature	Method	Age	Edu.	N	E	O	A	C	Imp.	SS	Gender	Sens. (%)	Spec. (%)	Sum (%)
Alcohol	LDA	X	X	X						X	X	75.34	63.24	138.58
Amphetamines	DT	X		X		X		X	X	X		81.30	71.48	152.77
Amyl nitrite	DT			X		X		X		X		73.51	87.86	161.37
Benzodiazepines	DT	X		X	X				X	X	X	70.87	71.51	142.38
Cannabis	DT	X	X			X	X	X	X			79.29	80.00	159.29
Chocolate	KNN	X			X			X			X	72.43	71.43	143.86
Cocaine	DT	X				X	X		X	X		68.27	83.06	151.32
Caffeine	KNN	X	X			X	X		X			70.51	72.97	143.48
Crack	DT				X			X				80.63	78.57	159.20
Ecstasy	DT	X								X	X	76.17	77.16	153.33
Heroin	DT	X							X		X	82.55	72.98	155.53
Ketamine	DT	X			X		X		X	X		72.29	80.98	153.26
Legal highs	DT	X				X	X	X		X	X	79.53	82.37	161.90
LSD	DT	X		X	X	X			X		X	85.46	77.56	163.02
Methadone	DT	X	X		X	X					X	79.14	72.48	151.62
MMushrooms	DT				X						X	65.56	94.79	160.36
Nicotine	DT			X	X			X			X	71.28	79.07	150.35
VSA	DT	X	X		X		X	X		X		83.48	77.64	161.12

Symbol 'X' means used input feature. Results are calculated by LOOCV

N, E, O, C, Imp., SS, and Gender with sensitivity 75.63% and specificity 75.75%. It means that for ecstasy users risk evaluation all input attributes are informative but required information can be extracted from part of attributes.

The results presented in Table 1 were calculated by LOOCV. It should be stressed that different methods of testing give different sensitivity and specificity. For example, a decision tree formed for the entire sample can have accuracy, sensitivity, and specificity different from LOOCV [20]. For illustration we can use the decision tree for ecstasy, depicted in Fig. 4. It has sensitivity 78.56% and specificity 71.16%, calculated using the whole sample. Results of LOOCV for a tree with the same options presented in the Table 1 show sensitivity 76.17% and specificity 77.16%.

4 Discussion

We evaluated the individual drug consumption risk for each drug. We analysed interrelations between the individual drug consumption risks for different drugs. We applied eight data mining approaches and selected the best one for each drug. Classifiers with sensitivity and specificity greater than 70% were found for all drugs

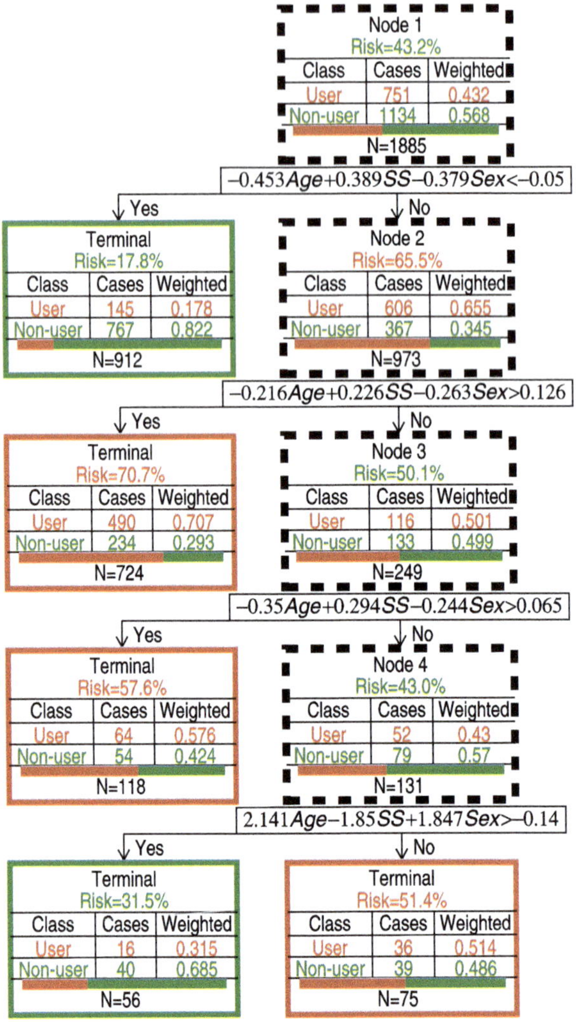

Fig. 4 Decision tree for ecstasy. Input features are: Age, SS, and Gender. Non-terminal nodes are depicted with *dashed border*. Values of Age, SS, and Gender are calculated by quantification procedures described in Sect. 2. Weight of each case of user class is 1.15 and of non-user class is 1. Column 'Weighted' presents normalized weights: weight of each class is divided by sum of weights

except magic mushrooms, alcohol, and cocaine. This accuracy is unexpectedly high for this type of problem. Correlation analysis using a RIG model demonstrated the existence of a group of drugs (see Fig. 3a) with strongly correlated consumption. There are limitations of this study. The collected sample is biased with respect to the general population, but it can still be used for risk evaluation. A further limitation concerns the fact that a number of the findings may be culturally specific.

References

1. Arlot, S., Celisse, A.: A survey of cross-validation procedures for model selection. Stat. Surv. **4**, 40–79 (2010)
2. Beaglehole, R., Bonita, R., Horton, R., Adams, C., Alleyne, G., Asaria, P., et al.: Priority actions for the non-communicable disease crisis. Lancet **377**(9775), 1438–1447 (2011)
3. Benjamini, Y., Hochberg, Y.: Controlling the false discovery rate: a practical and powerful approach to multiple testing. J. Royal Stat. Soc. **57**(1), 289–300 (1995)
4. Biau, G.: Analysis of a random forests model. J. Mach. Learn. Res. **13**(1), 1063–1095 (2012)
5. Bogg, T., Roberts, B.W.:. Conscientiousness and health-related behaviors: a meta-analysis of the leading behavioral contributors to mortality. Psychol. Bull. **130**(6), 887–919 (2004)
6. Bulut, F., Bucak, İ.Ö.: An urgent precaution system to detect students at risk of substance abuse through classification algorithms. Turk. J. Electr. Eng. Comput. Sci. **22**(3), 690–707 (2014)
7. Clarkson, K.L.: Nearest-neighbor searching and metric space dimensions. In: Shakhnarovich, G., Darrell, T., Indyk, P. (eds.) Nearest-Neighbor Methods for Learning and Vision: Theory and Practice, pp. 15–59. MIT, Cambridge (2005)
8. Cleveland, M.J., Feinberg, M.E., Bontempo, D.E., Greenberg, M.T.: The role of risk and protective factors in substance use across adolescence. J. Adolesc. Health **43**(2), 157–164 (2008)
9. Costa, P.T., MacCrae, R.R.: Revised NEO-Personality Inventory (NEO-PI-R) and the NEO-Five Factor Inventory (NEO-FFI): Personality manual. Psychological Assessment Resources, Odessa, FL (1992)
10. Dinov, I.D.: Expectation maximization and mixture modeling tutorial. UCLA, Statistics Online Computational Resource (2008). http://escholarship.org/uc/item/1rb7097
11. Egan, V., Deary, I., Austin, E.: The NEO-FFI: emerging British norms and an item-level analysis suggest N, A and C are more reliable than O and E. Personal. Individ. Differ. **29**(5), 907–920 (2000)
12. Fehrman, E., Muhammad, A.K., Mirkes, E.M., Egan, V., Gorban, A.N.: The five factor model of personality and evaluation of drug consumption risk. arXiv preprint arXiv:1506.06297
13. Fisher, R.A.: The use of multiple measurements in taxonomic problems. Ann. Eugenics **7**(2), 179–188 (1936)
14. Gorban, A.N., Zinovyev, A.Y.: Principal graphs and manifolds. In Olivas, E.S., Guerrero, J.D.M., Sober, M.M., Benedito, J.R.M., Lόpez, A.J.S. (eds.) Handbook of Research on Machine Learning Applications and Trends: Algorithms, Methods, and Techniques, pp 28–59. IGI Global, Hershey, NY (2009)
15. Gorban, A.N., Zinovyev, A.Y.: Principal manifolds and graphs in practice: from molecular biology to dynamical systems. Int. J. Neural Syst. **20**(3), 219–232 (2010)
16. Gorban, A.N., Kégl, B., Wunsch, D.C., Zinovyev, A.Y. (eds.): Principal Manifolds for Data Visualisation and Dimension Reduction. Lecture Notes in Computer Science and Engineering, vol. 58. Springer, Berlin, Heidelberg (2008)
17. Gujarati, D.N.: Basic Econometrics, 4th edn. McGraw-Hill, New York (2003)
18. Guttman, L.: Some necessary conditions for common-factor analysis. Psychometrika **19**(2), 149–161 (1954)
19. Hastie, T., Tibshirani, R.: Discriminant adaptive nearest neighbor classification. IEEE Trans. Pattern Anal. Mach. Intell. **18**(6), 607–616 (1996)
20. Hastie, T., Tibshirani, R., Friedman, J.: The Elements of Statistical Learning. Springer, New York (2009)
21. Hoare, J., Moon, D.: Drug misuse declared: findings from the 2009/10 British Crime Survey Home Office Statistical Bulletin 13/10 (2010)
22. Hosmer, D.W. Jr., Lemeshow, S.: Applied Logistic Regression. Wiley, New York (2004)
23. Kaiser, H.F.: The application of electronic computers to factor analysis. Educ. Psychol. Meas. **20**, 141–151 (1960)

24. Kopstein, A.N., Crum, R.M., Celentano, D.D., Martin, S.S.: Sensation seeking needs among 8th and 11th graders: characteristics associated with cigarette and marijuana use. Drug Alcohol Depend. **62**(3), 195–203 (2001)
25. Koren, Y., Carmel, L.: Robust linear dimensionality reduction. IEEE Trans. Vis. Comput. Graph. **10**(4), 459–470 (2004)
26. Lee, S.Y., Poon, W.Y., Bentler, P.M.: A two-stage estimation of structural equation models with continuous and polytomous variables. Br. J. Math. Stat. Psychol. **48**(2), 339–358 (1995)
27. Li, Q., Racine, J.S.: Nonparametric Econometrics: Theory and Practice. Princeton University Press, Princeton, NJ (2007)
28. Linting, M., van der Kooij, A.: Nonlinear principal components analysis with CATPCA: a tutorial. J. Pers. Assess. **94**(1), 12–25 (2012)
29. McCabe, G.P.: Principal variables. Technometrics **26**(2), 137–144 (1984)
30. McCrae, R.R., Costa, P.T.: A contemplated revision of the NEO Five-Factor Inventory. Personal. Individ. Differ. **36**(3), 587–596 (2004)
31. Mirkes, E.M., Alexandrakis, I., Slater, K., Tuli, R., Gorban, A.N.: Computational diagnosis and risk evaluation for canine lymphoma. Comput. Biol. Med. **53**, 279–290 (2014)
32. Mirkes, E.M., Alexandrakis, I., Slater, K., Tuli, R., Gorban, A.N.: Computational diagnosis of canine lymphoma. J. Phys. Conf. Ser. **490**(1), 012135 (2014). http://stacks.iop.org/1742--6596/490/i=1/a=012135
33. Mitchell, T.M.: Machine learning. 1997. Burr Ridge, IL: McGraw Hill **45** (1997).
34. Naikal, N., Yang, A.Y., Sastry, S.S.: Informative feature selection for object recognition via sparse PCA. In: 2011 IEEE International Conference on Computer Vision (ICCV), pp. 818–825. IEEE, New York (2011)
35. Pearson, K.: On lines and planes of closest fit to systems of points in space. Philos. Mag. **2**(6), 559–572 (1901)
36. Quinlan, J.R.: Simplifying decision trees. Int. J. Man Mach. Stud. **27**(3), 221–234 (1987)
37. Russell, S., Norvig, P.: Artificial Intelligence: A Modern Approach, Prentice Hall, NJ (1995)
38. Scott, D.W.: Multivariate Density Estimation: Theory, Practice and Visualization. Wiley, New York (1992)
39. Sofeikov, K.I., Tyukin, I.Y., Gorban, A.N., Mirkes, E.M., Prokhorov, D.V., Romanenko, I.V.: Learning optimization for decision tree classification of non-categorical data with information gain impurity criterion. In: 2014 International Joint Conference on Neural Networks (IJCNN), pp. 3548–3555. IEEE, New York (2014)
40. Stanford, M.S., Mathias, C.W., Dougherty, D.M., Lake, S.L., Anderson, N.E., Patton, J.H.: Fifty years of the Barratt impulsiveness scale: an update and review. Personal. Individ. Differ. **47**(5), 385–395 (2009)
41. Terracciano, A., Löckenhoff, C.E., Crum, R.M., Bienvenu, O.J., Costa, P.T.: Five factor model personality profiles of drug users. BMC Psych. **8**(1), 22 (2008)
42. Valeroa, S., Daigre, C., Rodríguz-Cintas, L., Barral C., Gomà-i-Freixanet, M., Ferrer, M., Casasa, M., Roncero, C.R.: Neuroticism and impulsivity: Their hierarchical organization in the personality characterization of drug-dependent patients from a decision tree learning perspective. Compr. Psychiatry **55**(5), 1227–1233 (2014)
43. Ventura, C.A., de Souza, J., Hayashida, M., Ferreira, P.S.: Risk factors for involvement with illegal drugs: opinion of family members or significant others. J. Subst. Use **20**(2), 136–142 (2014)
44. Yasnitskiy, L., Gratsilev, V., Kulyashova, J., Cherepanov, F.: Possibilities of artificial intellect in detection of predisposition to drug addiction. Perm University Herald Series "Philosophy Psychology Sociology" **1**(21), 61–73 (2015)
45. Zuckerman, M.: Behavioral Expressions and Biosocial Bases of Sensation Seeking. Cambridge University Press, New York (1994)

Correlation Analysis for Multivariate Functional Data

Tomasz Górecki, Mirosław Krzyśko, and Waldemar Wołyński

Abstract The relationship between two sets of real variables defined for the same individuals can be evaluated by few different correlation coefficients. For the functional data we have only one important tool: the canonical correlations. It is not immediately straightforward to extend other similar measures to the context of functional data analysis. In this work we show how to use commonly known measures of correlation for two sets of variables: rV coefficient and distance correlation coefficient for multivariate functional case. Finally, these three different coefficients are compared and their use is demonstrated on two real examples.

1 Introduction

In recent years methods for data representing functions or curves have received much attention. Such data are known in the literature as the functional data [11, 19]. Examples of functional data can be found in several application domains, such as medicine, economics, meteorology, and many others. In many applications there is need for using statistical methods for objects characterized by many features observed in many time points (double multivariate data). In this such data are called the multivariate functional data. The pioneer theoretical paper were works of Besse [2], where random variables have values into a general Hilbert space. Saporta [24], Leurgans et al. [16], He et al. [10], Berrenderoa et al. [1], Górecki et al. [9], Jacques and Preda [13] present an analysis of multivariate functional data from the point of view of factorial methods (principal components analysis and canonical analysis).

In this paper we focused on correlation analysis for multivariate functional data. In the literature, there are different strategies to explore the association between two sets of variables (p dimensional X and q dimensional Y). Historically, the first approach was put forward by Hotelling [12] who proposed the canonical correlation in the framework of Canonical Correlation Analysis (CCA). One of the most important works in the functional data analysis context has been made by

T. Górecki (✉) • M. Krzyśko • W. Wołyński
Faculty of Mathematics and Computer Science, Adam Mickiewicz University, Umultowska 87, 61-614 Poznań, Poland
e-mail: tomasz.gorecki@amu.edu.pl; mkrzysko@amu.edu.pl; wolynski@amu.edu.pl

F. Palumbo et al. (eds.), *Data Science*, Studies in Classification, Data Analysis, and Knowledge Organization, DOI 10.1007/978-3-319-55723-6_19

Leurgans et al. [16]. They showed that smoothing is necessary in order to define the functional canonical correlations meaningfully. CCA is a reference tool concerned with describing linear dependencies between two sets of variables; it seeks a linear combination of the variables of the first group which is maximally correlated with a linear combination of the variables of the second group. The correlation coefficient thus obtained is said to be canonical and the linear combinations are called canonical variables. Other pairs of canonical variables can be recursively defined, maximizing the canonical correlation subject to corresponding variables being orthogonal to previously found pairs. CCA provides thus a sequence of canonical correlation coefficients. Consequently, the association between the two sets viewed by the canonical correlation is not a global measurement since the intensity of the relationship is expressed component by component. However, in CCA, other global measures of correlation exist, but the core of CCA intrinsically relies on the canonical coefficient.

Another approach is to consider each set of variables trough its individual cloud, and to compare the structures (i.e., the shapes) of the two point clouds. In this way, the rV coefficient [5–7, 22] provides insight into the global association between the two sets of variables. However, as for many correlation coefficients, it is difficult to evaluate the magnitude of the relationship just by considering its values, hence the relevance of proposing a test.

Székely et al. [25], Székely and Rizzo [26–28] defined a measure of dependence between random vectors: the distance correlation (dCor) coefficient. The authors showed that for all random variables with finite first moments, the dCor coefficient generalizes the idea of correlation in two ways. Firstly, this coefficient can be applied when $\boldsymbol{X}$ and $\boldsymbol{Y}$ are of any dimensions and not only for the simple case where $p = q = 1$. They constructed their coefficient as a generalization of the simple correlation coefficient without reference to the earlier rV literature. Secondly, the dCor coefficient is equal to zero, if and only if there is independence between the random vectors. Indeed, a correlation coefficient measures linear relationships and can be equal to 0 even when the variables are related. This can be seen as a major shortcoming of the correlation coefficient and the rV coefficient.

The rest of this paper is organized as follows. We first review the concept of transformation of discrete data to multivariate functional data (Sect. 2). Section 3 contains the functional version of canonical correlation coefficients analysis. Sections 4 and 5 describe our extension of the ρV and distance correlation coefficients to the functional case. Section 6 contains the results of our experiments on two real data sets. Conclusions are given in Sect. 7.

2 Smoothing of Stochastic Processes

Let us assume that $\boldsymbol{X} \in L_2^p(I_1)$ and $\boldsymbol{Y} \in L_2^q(I_2)$ are random processes, where $L_2(I)$ is a Hilbert space of square integrable functions on the interval I. We also assume that $\mathrm{E}(\boldsymbol{X}(s)) = \boldsymbol{0},\ s \in I_1$ and $\mathrm{E}(\boldsymbol{Y}(t)) = \boldsymbol{0},\ t \in I_2$. This fact does not cause loss of generality, because functional correlation coefficients are calculated based on the covariance functions of processes $\boldsymbol{X}$ and $\boldsymbol{Y}$ of the form

$$\mathrm{Cov}\begin{bmatrix}\boldsymbol{X}\\ \boldsymbol{Y}\end{bmatrix} = \boldsymbol{\Sigma}(s,t) = \begin{bmatrix}\boldsymbol{\Sigma}_{XX}(s,t) & \boldsymbol{\Sigma}_{XY}(s,t)\\ \boldsymbol{\Sigma}_{YX}(s,t) & \boldsymbol{\Sigma}_{YY}(s,t)\end{bmatrix},\ s \in I_1,\ t \in I_2,$$

where

$$\boldsymbol{\Sigma}_{XX}(s,t) = \mathrm{E}[\boldsymbol{X}(s)\boldsymbol{X}'(t)],\ s,t \subset I_1,$$

$$\boldsymbol{\Sigma}_{XY}(s,t) = \mathrm{E}[\boldsymbol{X}(s)\boldsymbol{Y}'(t)],\ s \in I_1,\ t \in I_2,$$

$$\boldsymbol{\Sigma}_{YX}(s,t) = \mathrm{E}[\boldsymbol{Y}(s)\boldsymbol{X}'(t)],\ s \in I_2,\ t \in I_1,$$

$$\boldsymbol{\Sigma}_{YY}(s,t) = \mathrm{E}[\boldsymbol{Y}(s)\boldsymbol{Y}'(t)],\ s,t \in I_2.$$

We will further assume that each component X_g of the process $\boldsymbol{X}$ and Y_h of the process $\boldsymbol{Y}$ can be represented by a finite number of orthonormal basis functions $\{\varphi_e\}$ and $\{\varphi_f\}$ of space $L_2(I_1)$ and $L_2(I_2)$, respectively:

$$X_g(s) = \sum_{e=0}^{E_g} \alpha_{ge}\varphi_e(s), s \in I_1, g = 1,2,\ldots,p,$$

$$Y_h(t) = \sum_{f=0}^{F_h} \beta_{hf}\varphi_f(t), t \in I_2, h = 1,2,\ldots,q.$$

The degree of smoothness of functions X_g and Y_h depends on the values E_g and F_h respectively (a small values cause more smoothing). We introduce the following notation:

$$\boldsymbol{\alpha} = (\alpha_{10},\ldots,\alpha_{1E_1},\ldots,\alpha_{p0},\ldots,\alpha_{pE_p})',$$

$$\boldsymbol{\beta} = (\beta_{10},\ldots,\beta_{1F_1},\ldots,\beta_{q0},\ldots,\beta_{qF_q})',$$

$$\boldsymbol{\varphi}_{E_g}(s) = (\varphi_0(s),\ldots,\varphi_{E_g}(s))', s \in I_1, g = 1,2,\ldots,p,$$

$$\boldsymbol{\varphi}_{F_h}(t) = (\varphi_0(t),\ldots,\varphi_{F_h}(t))', t \in I_2, h = 1,2,\ldots,q,$$

$$\boldsymbol{\Phi}_1(s) = \begin{bmatrix} \boldsymbol{\varphi}'_{E_1}(s) & \mathbf{0} & \dots & \mathbf{0} \\ \mathbf{0} & \boldsymbol{\varphi}'_{E_2}(s) & \dots & \mathbf{0} \\ \dots & \dots & \dots & \dots \\ \mathbf{0} & \mathbf{0} & \dots & \boldsymbol{\varphi}'_{E_p}(s) \end{bmatrix}, \quad \boldsymbol{\Phi}_2(t) = \begin{bmatrix} \boldsymbol{\varphi}'_{F_1}(t) & \mathbf{0} & \dots & \mathbf{0} \\ \mathbf{0} & \boldsymbol{\varphi}'_{F_2}(t) & \dots & \mathbf{0} \\ \dots & \dots & \dots & \dots \\ \mathbf{0} & \mathbf{0} & \dots & \boldsymbol{\varphi}'_{F_q}(t) \end{bmatrix}.$$

Using the above matrix notation the processes $\boldsymbol{X}$ and $\boldsymbol{Y}$ can be represented as:

$$\boldsymbol{X}(s) = \boldsymbol{\Phi}_1(s)\boldsymbol{\alpha}, \quad \boldsymbol{Y}(t) = \boldsymbol{\Phi}_2(t)\boldsymbol{\beta}.$$

This means that the realizations of processes $\boldsymbol{X}$ and $\boldsymbol{Y}$ are in finite dimensional subspaces of $L_2^p(I_1)$ and $L_2^q(I_2)$, respectively. We will denote these subspaces by $\mathscr{L}_2^p(I_1)$ and $\mathscr{L}_2^q(I_2)$. For random vectors $\boldsymbol{\alpha}$ and $\boldsymbol{\beta}$ we have:

$$\mathrm{E}(\boldsymbol{\alpha}) = \mathbf{0}, \ \mathrm{E}(\boldsymbol{\beta}) = \mathbf{0}$$

and

$$\mathrm{Cov}\begin{bmatrix} \boldsymbol{\alpha} \\ \boldsymbol{\beta} \end{bmatrix} = \boldsymbol{\Sigma} = \begin{bmatrix} \boldsymbol{\Sigma}_{\alpha\alpha} & \boldsymbol{\Sigma}_{\alpha\beta} \\ \boldsymbol{\Sigma}_{\beta\alpha} & \boldsymbol{\Sigma}_{\beta\beta} \end{bmatrix},$$

where $\boldsymbol{\Sigma}_{\alpha\alpha} = \mathrm{E}(\boldsymbol{\alpha}\boldsymbol{\alpha}')$, $\boldsymbol{\Sigma}_{\alpha\beta} = \mathrm{E}(\boldsymbol{\alpha}\boldsymbol{\beta}')$, $\boldsymbol{\Sigma}_{\beta\alpha} = \mathrm{E}(\boldsymbol{\beta}\boldsymbol{\alpha}')$ and $\boldsymbol{\Sigma}_{\beta\beta} = \mathrm{E}(\boldsymbol{\beta}\boldsymbol{\beta}')$. Note that

$$\boldsymbol{\Sigma}_{XX}(s,t) = \mathrm{E}[\boldsymbol{\Phi}_1(s)\boldsymbol{\alpha}\boldsymbol{\alpha}'\boldsymbol{\Phi}_1'(t)] = \boldsymbol{\Phi}_1(s)\,\mathrm{E}(\boldsymbol{\alpha}\boldsymbol{\alpha}')\boldsymbol{\Phi}_1'(t) = \boldsymbol{\Phi}_1(s)\boldsymbol{\Sigma}_{\alpha\alpha}\boldsymbol{\Phi}_1'(t).$$

Similarly

$$\boldsymbol{\Sigma}_{XY}(s,t) = \boldsymbol{\Phi}_1(s)\boldsymbol{\Sigma}_{\alpha\beta}\boldsymbol{\Phi}_2'(t),$$

$$\boldsymbol{\Sigma}_{YX}(s,t) = \boldsymbol{\Phi}_2(s)\boldsymbol{\Sigma}_{\beta\alpha}\boldsymbol{\Phi}_1'(t),$$

$$\boldsymbol{\Sigma}_{YY}(s,t) = \boldsymbol{\Phi}_2(s)\boldsymbol{\Sigma}_{\beta\beta}\boldsymbol{\Phi}_2'(t).$$

In fact correlation analysis for the random processes is based on the matrices $\boldsymbol{\Sigma}_{\alpha\alpha}$, $\boldsymbol{\Sigma}_{\beta\beta}$, and $\boldsymbol{\Sigma}_{\alpha\beta}$ which are unknown. We have to estimate them on the basis of n independent realizations $\boldsymbol{X}_1, \boldsymbol{X}_2, \dots, \boldsymbol{X}_n$ and $\boldsymbol{Y}_1, \boldsymbol{Y}_2, \dots, \boldsymbol{Y}_n$ of random processes $\boldsymbol{X}$ and $\boldsymbol{Y}$. We have $\boldsymbol{X}_i(s) = \boldsymbol{\Phi}_1(s)\boldsymbol{\alpha}_i$ and $\boldsymbol{Y}_i(t) = \boldsymbol{\Phi}_2(t)\boldsymbol{\beta}_i$, $i = 1, 2, \dots, n$. The coefficients $\boldsymbol{\alpha}_i$ and $\boldsymbol{\beta}_i$ are estimated by the least squares method. Let us denote these estimates by $\boldsymbol{a}_i$ and $\boldsymbol{b}_i$, $i = 1, 2, \dots, n$.

Typically data are recorded at discrete moments in time. The process of transformation of discrete data to functional data is performed for each realization and each variable separately. As a result of the transformation process, we obtain functional data of the form:

$$\boldsymbol{x}_i(s) = \boldsymbol{\Phi}_1(s)\boldsymbol{a}_i,\ \boldsymbol{y}_i(t) = \boldsymbol{\Phi}_2(t)\boldsymbol{b}_i, \tag{1}$$

where $s \in I_1$, $t \in I_2$ and $i = 1, 2, \ldots, n$.

Let $\boldsymbol{A} = (\boldsymbol{a}\prime_1, \boldsymbol{a}\prime_2, \ldots, \boldsymbol{a}\prime_n)'$, and $\boldsymbol{B} = (\boldsymbol{b}\prime_1, \boldsymbol{b}\prime_2, \ldots, \boldsymbol{b}\prime_n)'$. Then

$$\hat{\boldsymbol{\Sigma}}_{\alpha\alpha} = \frac{1}{n}\boldsymbol{A}'\boldsymbol{A},\quad \hat{\boldsymbol{\Sigma}}_{\beta\beta} = \frac{1}{n}\boldsymbol{B}'\boldsymbol{B},\quad \hat{\boldsymbol{\Sigma}}_{\alpha\beta} = \frac{1}{n}\boldsymbol{A}'\boldsymbol{B}. \tag{2}$$

3 Functional Canonical Correlation Coefficient

Functional canonical variables U and V for stochastic processes $\boldsymbol{X} \in \mathcal{L}_2^p(I_1)$ and $\boldsymbol{Y} \in \mathcal{L}_2^q(I_2)$ are defined as follows:

$$U(s) = \boldsymbol{l}'(s)\boldsymbol{X}(s),\ \boldsymbol{l} \in \mathcal{L}_2^p(I_1),$$
$$V(t) = \boldsymbol{m}'(t)\boldsymbol{Y}(t),\ \boldsymbol{m} \in \mathcal{L}_2^q(I_2),$$

where $\boldsymbol{l}$ and $\boldsymbol{m}$ are the weight functions. We have

$$\mathrm{E}(U(s)) = \mathrm{E}(V(t)) = 0,\ s \in I_1,\ t \in I_2.$$

Let us denote the covariance matrix of the processes U and V by

$$\boldsymbol{\Sigma}_{UV}(s,t) = \begin{bmatrix} \sigma_{UU}(s,t) & \sigma_{UV}(s,t) \\ \sigma_{VU}(s,t) & \sigma_{VV}(s,t) \end{bmatrix}.$$

Because $\boldsymbol{l} \in \mathcal{L}_2^p(I_1)$ and $\boldsymbol{m} \in \mathcal{L}_2^q(I_2)$ we have

$$\boldsymbol{l}(s) = \boldsymbol{\Phi}_1(s)\boldsymbol{\lambda},\ \boldsymbol{m}(t) = \boldsymbol{\Phi}_2(t)\boldsymbol{\mu},$$

where $\boldsymbol{\lambda} \in \mathbb{R}^{K_1+p}$, $\boldsymbol{\mu} \in \mathbb{R}^{K_2+q}$ and $K_1 = E_1 + \ldots + E_p$, $K_2 = F_1 + \ldots + F_q$. Moreover

$$\begin{aligned}\sigma_{UU}(s,t) &= \mathrm{E}[U(s)U'(t)] = \mathrm{E}[\boldsymbol{l}'(s)\boldsymbol{X}(s)\boldsymbol{X}'(t)\boldsymbol{l}(t)] \\ &= \mathrm{E}[\boldsymbol{\lambda}'\boldsymbol{\Phi}_1'(s)\boldsymbol{\Phi}_1(s)\boldsymbol{\alpha}\boldsymbol{\alpha}'\boldsymbol{\Phi}_1'(t)\boldsymbol{\Phi}_1(t)\boldsymbol{\lambda}] \\ &= \boldsymbol{\lambda}'\boldsymbol{\Phi}_1'(s)\boldsymbol{\Phi}_1(s)\boldsymbol{\Sigma}_{\alpha\alpha}\boldsymbol{\Phi}_1'(t)\boldsymbol{\Phi}_1(t)\boldsymbol{\lambda}.\end{aligned}$$

Similarly

$$\sigma_{UV}(s,t) = \boldsymbol{\lambda}'\boldsymbol{\Phi}_1'(s)\boldsymbol{\Phi}_1(s)\boldsymbol{\Sigma}_{\alpha\beta}\boldsymbol{\Phi}_2'(t)\boldsymbol{\Phi}_2(t)\boldsymbol{\mu},$$
$$\sigma_{VU}(s,t) = \boldsymbol{\mu}'\boldsymbol{\Phi}_2'(s)\boldsymbol{\Phi}_2(s)\boldsymbol{\Sigma}_{\beta\alpha}\boldsymbol{\Phi}_1'(t)\boldsymbol{\Phi}_1(t)\boldsymbol{\lambda},$$
$$\sigma_{VV}(s,t) = \boldsymbol{\mu}'\boldsymbol{\Phi}_2'(s)\boldsymbol{\Phi}_2(s)\boldsymbol{\Sigma}_{\beta\beta}\boldsymbol{\Phi}_2'(t)\boldsymbol{\Phi}_2(t)\boldsymbol{\mu}.$$

The functional canonical coefficient $\rho_{X,Y}$ is defined as

$$\rho_{X,Y} = \max_{l,m}\int_{I_1}\int_{I_2}\sigma_{UV}(s,t)dsdt = \max_{l,m}\int_{I_1}\int_{I_2}\mathrm{E}[\boldsymbol{l}'(s)\boldsymbol{X}(s)\boldsymbol{Y}'(t)\boldsymbol{m}(t)]dsdt,$$

subject to the constraint

$$\int_{I_1}\int_{I_2}\sigma_{UU}(s,t)dsdt = \int_{I_1}\int_{I_2}\sigma_{VV}(s,t)dsdt = 1.$$

Because

$$\int_{I_1}\int_{I_2}\sigma_{UV}(s,t)dsdt = \boldsymbol{\lambda}'\boldsymbol{\Sigma}_{\alpha\beta}\boldsymbol{\mu},$$
$$\int_{I_1}\int_{I_2}\sigma_{UU}(s,t)dsdt = \boldsymbol{\lambda}'\boldsymbol{\Sigma}_{\alpha\alpha}\boldsymbol{\lambda},$$
$$\int_{I_1}\int_{I_2}\sigma_{VV}(s,t)dsdt = \boldsymbol{\mu}'\boldsymbol{\Sigma}_{\beta\beta}\boldsymbol{\mu}$$

we have

$$\rho_{X,Y} = \max_{\boldsymbol{\lambda},\boldsymbol{\mu}}\boldsymbol{\lambda}'\boldsymbol{\Sigma}_{\alpha\beta}\boldsymbol{\mu} = \rho_{\alpha,\beta},$$

subject to the restriction

$$\boldsymbol{\lambda}'\boldsymbol{\Sigma}_{\alpha\alpha}\boldsymbol{\lambda} = \boldsymbol{\mu}'\boldsymbol{\Sigma}_{\beta\beta}\boldsymbol{\mu} = 1.$$

Theorem 1 *Functional canonical correlation coefficient $\rho_{X,Y}$ of the pair of random processes $\boldsymbol{X} \in \mathcal{L}_2^p(I_1)$ and $\boldsymbol{Y} \in \mathcal{L}_2^q(I_2)$ is equivalent to the canonical correlation coefficient $\rho_{\alpha,\beta}$ of the pair of the random vectors $\boldsymbol{\alpha}$ and $\boldsymbol{\beta}$.*

Canonical correlation coefficient $\rho_{\alpha,\beta}$ of the pair of the random vectors $\boldsymbol{\alpha}$ and $\boldsymbol{\beta}$ is based on the matrices $\boldsymbol{\Sigma}_{\alpha\alpha}$, $\boldsymbol{\Sigma}_{\beta\beta}$, and $\boldsymbol{\Sigma}_{\alpha\beta}$. If they are not known, we have to use their estimators (2). Hence

$$\hat{\rho}_{\alpha,\beta} = \max_{\boldsymbol{\lambda},\boldsymbol{\mu}}\boldsymbol{\lambda}'\hat{\boldsymbol{\Sigma}}_{\alpha\beta}\boldsymbol{\mu},$$

under the condition

$$\boldsymbol{\lambda}'\hat{\boldsymbol{\Sigma}}_{\alpha\alpha}\boldsymbol{\lambda} = \boldsymbol{\mu}'\hat{\boldsymbol{\Sigma}}_{\beta\beta}\boldsymbol{\mu} = 1.$$

As with the ordinary correlation coefficient, a high value of the $\hat{\rho}_{\alpha,\beta}$ coefficient does not necessarily mean that is a significant relationship between the two random processes. Hence the need for a valid inferential procedure for testing the significance of the association. Note that the problem of testing $H_0: \rho_{X,Y} = 0$ vs. $H_1: \rho_{X,Y} \neq 0$ for the pair of random processes $\boldsymbol{X} \in \mathcal{L}_2^p(I_1)$ and $\boldsymbol{Y} \in \mathcal{L}_2^q(I_2)$ is equivalent to the problem of testing $H_0: \rho_{\alpha,\beta} = 0$ vs. $H_1: \rho_{\alpha,\beta} \neq 0$ for the pair of random vectors $\boldsymbol{\alpha}$ and $\boldsymbol{\beta}$. Typically we use the test statistic of the form (Wilk's Lambda)

$$\Lambda = \prod_{i=1}^{K_1+p} (1 - r_i^2),$$

where r_i are the eigenvalues of the matrix

$$\hat{\boldsymbol{\Sigma}}_{\alpha\alpha}^{-1}\hat{\boldsymbol{\Sigma}}_{\alpha\beta}\hat{\boldsymbol{\Sigma}}_{\beta\beta}^{-1}\hat{\boldsymbol{\Sigma}}_{\beta\alpha}.$$

Under the null hypothesis, and when the joint distribution of the random variables belongs to the class of normal distribution the expression (Bartlett's approximation)

$$-[n - 0.5(K_1 + p + K_2 + q + 3)] \ln \Lambda$$

converges to the chi-square distribution with $(K_1 + p)(K_2 + q)$ degrees of freedom.

Instead, it is possible to use a permutation test.

4 Functional ρV Coefficient

The squared functional ρV coefficient is defined as

$$(\rho V_{X,Y})^2 = \frac{\|\boldsymbol{\Sigma}_{XY}\|^2}{\|\boldsymbol{\Sigma}_{XX}\|\|\boldsymbol{\Sigma}_{YY}\|},$$

where

$$\|\boldsymbol{\Sigma}_{XY}\| = \sqrt{\int_{I_1}\int_{I_2} \mathrm{tr}(\boldsymbol{\Sigma}'_{XY}(s,t)\boldsymbol{\Sigma}_{XY}(s,t))dsdt}.$$

Because

$$\|\boldsymbol{\Sigma}_{XY}\|^2 = \int_{I_1}\int_{I_2} \operatorname{tr}(\boldsymbol{\Sigma}'_{XY}(s,t)\boldsymbol{\Sigma}_{XY}(s,t))dsdt =$$
$$= \int_{I_1}\int_{I_2} \operatorname{tr}(\boldsymbol{\Phi}_2(t)\boldsymbol{\Sigma}'_{\alpha\beta}\boldsymbol{\Phi}'_1(s)\boldsymbol{\Phi}_1(s)\boldsymbol{\Sigma}_{\alpha\beta}\boldsymbol{\Phi}'_2(t))dsdt$$
$$= \operatorname{tr}(\boldsymbol{\Sigma}'_{\alpha\beta}\boldsymbol{\Sigma}_{\alpha\beta})$$

and

$$\|\boldsymbol{\Sigma}_{XX}\|^2 = \int_{I_1}\int_{I_2} \operatorname{tr}(\boldsymbol{\Sigma}'_{XX}(s,t)\boldsymbol{\Sigma}_{XX}(s,t))dsdt = \operatorname{tr}(\boldsymbol{\Sigma}'_{\alpha\alpha}\boldsymbol{\Sigma}_{\alpha\alpha})$$
$$\|\boldsymbol{\Sigma}_{YY}\|^2 = \int_{I_1}\int_{I_2} \operatorname{tr}(\boldsymbol{\Sigma}'_{YY}(s,t)\boldsymbol{\Sigma}_{YY}(s,t))dsdt = \operatorname{tr}(\boldsymbol{\Sigma}'_{\beta\beta}\boldsymbol{\Sigma}_{\beta\beta})$$

then

$$(\rho V_{X,Y})^2 = \frac{\operatorname{tr}(\boldsymbol{\Sigma}'_{\alpha\beta}\boldsymbol{\Sigma}_{\alpha\beta})}{\sqrt{\operatorname{tr}(\boldsymbol{\Sigma}'_{\alpha\alpha}\boldsymbol{\Sigma}_{\alpha\alpha})\operatorname{tr}(\boldsymbol{\Sigma}'_{\beta\beta}\boldsymbol{\Sigma}_{\beta\beta})}} = (\rho V_{\alpha,\beta})^2.$$

Theorem 2 *Functional ρV correlation coefficient $\rho V_{X,Y}$ of the pair of random processes $\boldsymbol{X} \in \mathcal{L}_2^p(I_1)$ and $\boldsymbol{Y} \in \mathcal{L}_2^q(I_2)$ is equivalent to the ρV correlation coefficient $\rho V_{\boldsymbol{\alpha},\boldsymbol{\beta}}$ of the pair of the random vectors $\boldsymbol{\alpha}$ and $\boldsymbol{\beta}$.*

Functional correlation coefficient $\rho_{\boldsymbol{X},\boldsymbol{Y}}$ has the following properties:

- $0 \le \rho_{\boldsymbol{X},\boldsymbol{Y}} \le 1$.
- $\rho_{\boldsymbol{X},\boldsymbol{Y}} = 0$ if and only if for all $g = 1,2,\ldots,p$ and $h = 1,2,\ldots,q$ such that $g \neq h$, X_g component of the process $\boldsymbol{X}$ and Y_h component of the process $\boldsymbol{Y}$ are uncorrelated.

Using the data we can estimate the $\rho V_{X,Y}$ coefficient [6]. Let

$$\boldsymbol{W}_{\alpha\alpha} = \boldsymbol{A}\boldsymbol{A}', \ \boldsymbol{W}_{\beta\beta} = \boldsymbol{B}\boldsymbol{B}'.$$

Then

$$(\widehat{\rho V}_{X,Y})^2 = (rV_{X,Y})^2 = \frac{\operatorname{tr}(\boldsymbol{W}_{\alpha\alpha}\boldsymbol{W}_{\beta\beta})}{\sqrt{\operatorname{tr}(\boldsymbol{W}^2_{\alpha\alpha})\operatorname{tr}(\boldsymbol{W}^2_{\beta\beta})}}.$$

The problem of testing $H_0\colon \rho V_{X,Y} = 0$ vs. $H_1\colon \rho V_{X,Y} \neq 0$ for the pair of random processes $\boldsymbol{X} \in \mathcal{L}_2^p(I_1)$ and $\boldsymbol{Y} \in \mathcal{L}_2^q(I_2)$ is equivalent to the problem of testing $H_0\colon \rho V_{\boldsymbol{\alpha},\boldsymbol{\beta}} = 0$ vs. $H_1\colon \rho V_{\boldsymbol{\alpha},\boldsymbol{\beta}} \neq 0$ for the pair of random vectors $\boldsymbol{\alpha}$ and $\boldsymbol{\beta}$. Under the null hypothesis, and when the joint distribution of the random vectors belongs to the class of elliptical distribution, $nrV_{\alpha,\beta}$ converges to

$$\frac{1+k}{\mathrm{tr}(\boldsymbol{\Sigma}^2_{\alpha\alpha})\mathrm{tr}(\boldsymbol{\Sigma}^2_{\beta\beta})}\sum_{l=1}^{K_1+p}\sum_{m=1}^{K_2+q}\lambda_l\gamma_m Z^2_{lm},$$

where k is the kurtosis parameter of the elliptical distribution; λ_l are the eigenvalues of the covariance matrix $\boldsymbol{\Sigma}_{\alpha\alpha}$, γ_m are the eigenvalues of the covariance matrix $\boldsymbol{\Sigma}_{\beta\beta}$, and Z_{lm} are i.i.d $N(0, 1)$ [3, 23].

Because the asymptotic tests fail in many situations, it is a reasonable strategy to use a permutation test [15].

5 Functional Distance Correlation

First, let's define the joint characteristic function of the pair of random processes $(\boldsymbol{X}, \boldsymbol{Y})$. If for all functions $\boldsymbol{l} \in L_2^p(I_1)$ the integral $\int_{I_1} \boldsymbol{l}'(s)\boldsymbol{X}(s)ds$ converges for almost all realization of $\boldsymbol{X}$, and for all functions $\boldsymbol{m} \in L_2^q(I_1)$ the integral $\int_{I_2} \boldsymbol{m}'(t)\boldsymbol{Y}(t)dt$ converges for almost all realization of $\boldsymbol{Y}$, then the characteristic function of the pair of random processes $(\boldsymbol{X}, \boldsymbol{Y})$ has the following form:

$$\varphi_{X,Y}(\boldsymbol{l}, \boldsymbol{m}) = \mathrm{E}\{\exp[i <\boldsymbol{l}, \boldsymbol{X}>_p + i <\boldsymbol{m}, \boldsymbol{Y}>_q]\},$$

where

$$<\boldsymbol{l}, \boldsymbol{X}>_p = \int_{I_1} \boldsymbol{l}'(s)\boldsymbol{X}(s)ds, \quad <\boldsymbol{m}, \boldsymbol{Y}>_q = \int_{I_2} \boldsymbol{m}'(t)\boldsymbol{Y}(t)dt$$

and $i^2 = -1$. Moreover, we define the marginal characteristic function of $\boldsymbol{X}$ and $\boldsymbol{Y}$ as follows: $\varphi_X(\boldsymbol{l}) = \varphi_{X,Y}(\boldsymbol{l}, \boldsymbol{0})$ and $\varphi_Y(\boldsymbol{m}) = \varphi_{X,Y}(\boldsymbol{0}, \boldsymbol{m})$.

Now, let's assume that $\boldsymbol{X} \in \mathscr{L}_2^p(I_1)$ and $\boldsymbol{Y} \in \mathscr{L}_2^q(I_2)$. Then the processes $\boldsymbol{X}$ and $\boldsymbol{Y}$ can be represented as:

$$\boldsymbol{X}(s) = \boldsymbol{\Phi}_1(s)\boldsymbol{\alpha}, \quad \boldsymbol{Y}(t) = \boldsymbol{\Phi}_2(t)\boldsymbol{\beta},$$

where $\boldsymbol{\alpha} \in \mathbb{R}^{K_1+p}$ and $\boldsymbol{\beta} \in \mathbb{R}^{K_2+q}$. In this case, we may assume [19] that the vector function $\boldsymbol{l}$ and the process $\boldsymbol{X}$ are in the same space, i.e., the function $\boldsymbol{l}$ can be written in the form

$$\boldsymbol{l}(s) = \boldsymbol{\Phi}_1(s)\boldsymbol{\lambda},$$

where $\boldsymbol{\lambda} \in \mathbb{R}^{K_1+p}$. The same we may assume for the vector function $\boldsymbol{m}$ and the process $\boldsymbol{Y}$. Then, we have

$$\boldsymbol{m}(t) = \boldsymbol{\Phi}_2(t)\boldsymbol{\mu},$$

where $\boldsymbol{\mu} \in \mathbb{R}^{K_2+q}$. Hence

$$< \boldsymbol{l}, \boldsymbol{X} >_p = \int_{I_1} \boldsymbol{l}'(s)\boldsymbol{X}(s)ds = \boldsymbol{\lambda}'[\int_{I_1} \boldsymbol{\Phi}_1'(s)\boldsymbol{\Phi}_1(s)ds]\boldsymbol{\alpha} = \boldsymbol{\lambda}'\boldsymbol{\alpha},$$

$$< \boldsymbol{m}, \boldsymbol{Y} >_q = \int_{I_2} \boldsymbol{m}'(t)\boldsymbol{Y}(t)dt = \boldsymbol{\mu}'[\int_{I_2} \boldsymbol{\Phi}_2'(t)\boldsymbol{\Phi}_2(t)dt]\boldsymbol{\beta} = \boldsymbol{\mu}'\boldsymbol{\beta}$$

then

$$\varphi_{\boldsymbol{X},\boldsymbol{Y}}(\boldsymbol{l}, \boldsymbol{m}) = \mathrm{E}\{\exp[i\boldsymbol{\lambda}'\boldsymbol{\alpha} + i\boldsymbol{\mu}'\boldsymbol{\beta}]\} = \varphi_{\boldsymbol{\alpha},\boldsymbol{\beta}}(\boldsymbol{\lambda}, \boldsymbol{\mu}),$$

where $\varphi_{\boldsymbol{\alpha},\boldsymbol{\beta}}(\boldsymbol{\lambda}, \boldsymbol{\mu})$ is the joint characteristic function of the pair of the random vectors $(\boldsymbol{\alpha}, \boldsymbol{\beta})$. Based on the idea of the distance covariance between two random vectors [25], we may introduced the functional distance covariance between random processes $\boldsymbol{X}$ and $\boldsymbol{Y}$ as the nonnegative number $\nu_{\boldsymbol{X},\boldsymbol{Y}}$ defined by

$$\nu_{\boldsymbol{X},\boldsymbol{Y}} = \nu_{\boldsymbol{\alpha},\boldsymbol{\beta}},$$

where

$$\nu^2_{\boldsymbol{\alpha},\boldsymbol{\beta}} = \frac{1}{C_{K_1+p}C_{K_2+q}} \int_{\mathbb{R}^{K_1+K_2+p+q}} \frac{|\phi_{\boldsymbol{\alpha},\boldsymbol{\beta}}(\boldsymbol{\lambda}, \boldsymbol{\mu}) - \phi_{\boldsymbol{\alpha}}(\boldsymbol{\lambda})\phi_{\boldsymbol{\beta}}(\boldsymbol{\mu})|^2}{\|\boldsymbol{\lambda}\|_{K_1+p}^{K_1+p+1} \|\boldsymbol{\mu}\|_{K_2+q}^{K_2+q+1}} d\boldsymbol{\lambda} d\boldsymbol{\mu},$$

and $|z|$ denotes the modulus of $z \in \mathbb{C}$, $\|\boldsymbol{\lambda}\|_{K_1+p}$, $\|\boldsymbol{\mu}\|_{K_2+q}$ the standard Euclidean norms on the corresponding spaces, and

$$C_r = \frac{\pi^{\frac{1}{2}(r+1)}}{\Gamma(\frac{1}{2}(r+1))}.$$

The functional distance correlation between random processes $\boldsymbol{X}$ and $\boldsymbol{Y}$ is the nonnegative number defined by

$$\mathcal{R}_{\boldsymbol{X},\boldsymbol{Y}} = \frac{\nu_{\boldsymbol{X},\boldsymbol{Y}}}{\sqrt{\nu_{\boldsymbol{X},\boldsymbol{X}}\nu_{\boldsymbol{Y},\boldsymbol{Y}}}}$$

if both $\nu_{\boldsymbol{X},\boldsymbol{X}}$ and $\nu_{\boldsymbol{Y},\boldsymbol{Y}}$ are strictly positive, and zero otherwise. For distributions with finite first moments, the distance correlation characterizes independence in that $0 \le \mathcal{R}_{\boldsymbol{X},\boldsymbol{Y}} \le 1$ with $\mathcal{R}_{\boldsymbol{X},\boldsymbol{Y}} = 0$ if and only if $\boldsymbol{X}$ and $\boldsymbol{Y}$ are independent.

We can estimate the functional distance covariance using the functional data of the form (1). Based on the result of Székely et al. [25], we have

$$\hat{\nu}^2_{\boldsymbol{X},\boldsymbol{Y}} = \frac{1}{n^2} \sum_{k,l=1}^{n} A_{kl}B_{kl},$$

where $a_{kl} = \|\boldsymbol{a}_k - \boldsymbol{a}_l\|_{K_1+p}$, $\bar{a}_{k\cdot} = \frac{1}{n}\sum_{l=1}^{n} a_{kl}$, $\bar{a}_{\cdot l} = \frac{1}{n}\sum_{k=1}^{n} a_{kl}$, $\bar{a}_{\cdot\cdot} = \frac{1}{n^2}\sum_{k,l=1}^{n} a_{kl}$ and $A_{kl} = a_{kl} - \bar{a}_{k\cdot} - \bar{a}_{\cdot l} + \bar{a}_{\cdot\cdot}$, and similarly for $b_{kl} = \|\boldsymbol{b}_k - \boldsymbol{b}_l\|_{K_2+q}$, $\bar{b}_{k\cdot}$, $\bar{b}_{\cdot l}$, $\bar{b}_{\cdot\cdot}$, and B_{kl}, where $k, l = 1, \ldots, n$. Thus, the squared sample distance covariance equals the average entry in the component-wise or Schur product of the centered distance matrices for the two variables.

The sample functional distance correlation then is defined by

$$\hat{\mathscr{R}}_{X,Y} = \frac{\hat{\mathcal{V}}_{X,Y}}{\sqrt{\hat{\mathcal{V}}_{X,X}\hat{\mathcal{V}}_{Y,Y}}}$$

if both $\hat{\mathcal{V}}_{X,X}$ and $\hat{\mathcal{V}}_{Y,Y}$ are strictly positive, and zero otherwise.

The problem of testing the independence between the random processes $\boldsymbol{X} \in \mathscr{L}_2^p(I_1)$ and $\boldsymbol{Y} \in \mathscr{L}_2^q(I_2)$ is equivalent to the problem of testing $H_0\colon \mathscr{R}_{X,Y} = 0$. Székely et al. [25] showed that under the null hypothesis of independence, $n\hat{\mathscr{R}}_{X,Y}$ converges to $\sum_{j=1}^{\infty} \nu_j Z_j^2$, where Z_j are i.i.d $N(0, 1)$, and ν_j depend on the distribution of $(\boldsymbol{X}, \boldsymbol{Y})$. In practice, permutation tests are used to assess the significance of the functional distance correlation [14].

6 Examples

6.1 Univariate Case

As a first example we used average daily temperature (in Celsius degrees) for each day of the year and average daily rainfall (in mm) for each day of the year rounded to 0.1 mm at 35 different weather stations in Canada from 1960 to 1994. Each station belongs to one of the four climate zones (Fig. 1): Arctic (3 stations), Atlantic (15), Continental (12), or Pacific (5) zone. This data set comes from Ramsay and Silverman [19].

In the first step we smoothed data. We used Fourier basis with various value of smoothing parameter (number of basis functions) from 3 to 15. We can observe the effect of smoothing on Figs. 2 and 3 (for Fourier basis with 15 basis functions). We decided to Fourier basis because of it has excellent computational properties, especially if the observations are equally spaced and it is natural for describing periodic data, such as the annual weather cycles.

From the plots we can observe that the level of smoothness seems big enough. Additionally, we can observe some relationship between average temperature and precipitation. Namely, for weather stations with large average temperature we observe relatively bigger average precipitation. While for Arctic stations with lowest average temperatures we observe the smallest average precipitation. So we can expect some correlation between average temperature and average precipitation for Canadian weather stations.

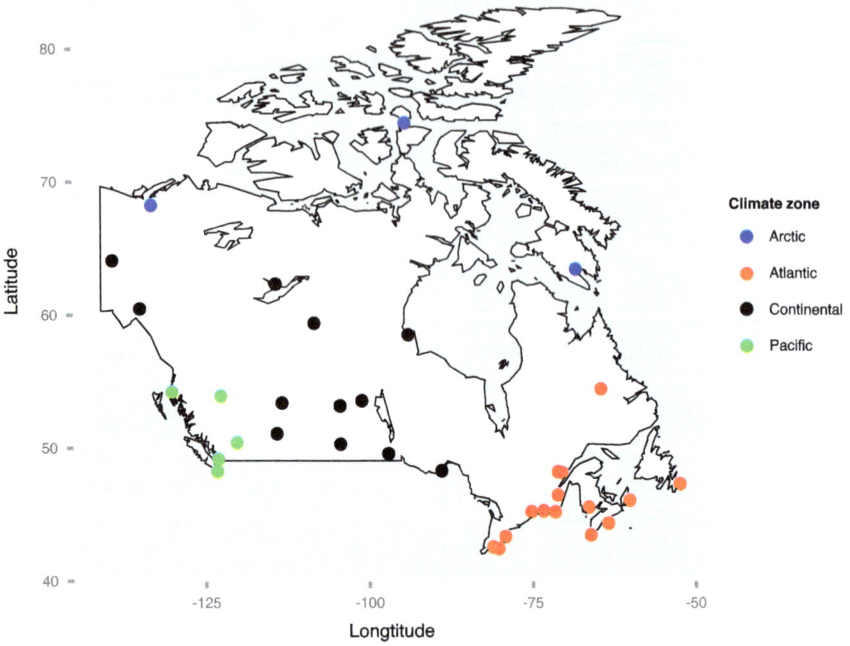

Fig. 1 Location of Canadian weather stations

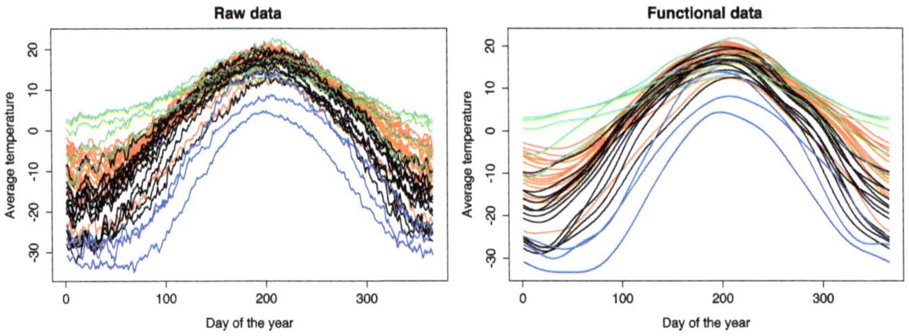

Fig. 2 Raw and functional temperature for Canadian weather stations

In the next step we calculated the values of described earlier coefficients of correlation. To find the parameters of regularization for CC we used leave-one-out criterion on the grid from 1.0×10^{-10} to 1.0×10^{-1}. The values of correlation coefficients are presented in Fig. 4. We observe quite big values of correlation, but it is impossible to inference about correlation from these values. We see that the values of rV and dCor coefficients are stable (both do not depend on basis size). However, the value of canonical correlation coefficient increase a little with the size of basis despite the regularization.

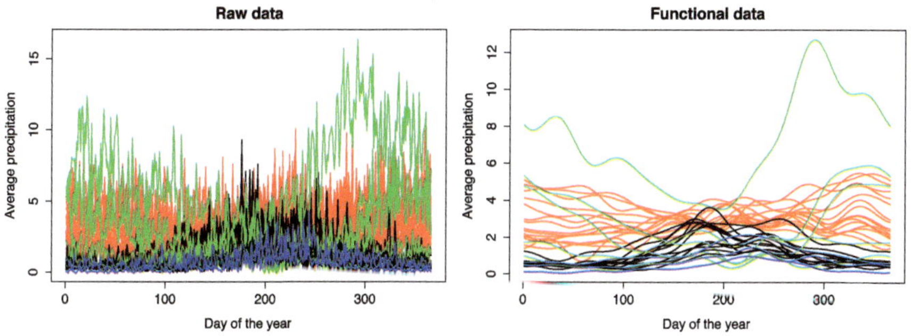

Fig. 3 Raw and functional precipitation for Canadian weather stations

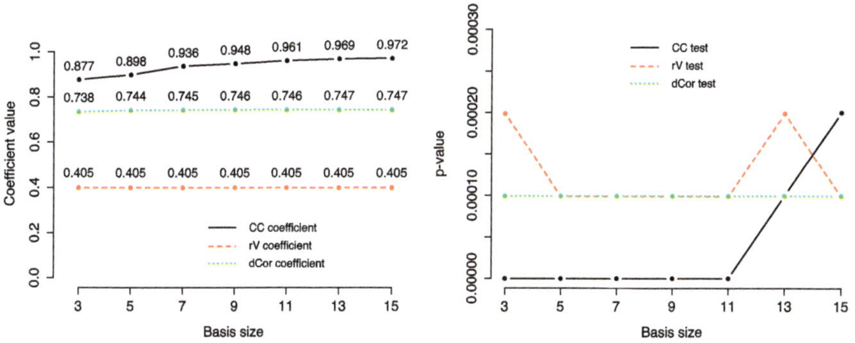

Fig. 4 Correlation coefficients and p-values (permutation tests) for Canadian weather data

Finally we performed permutation tests for correlation coefficients. We used 100,000 permutation resamples. The results are presented in Fig. 4. All tests rejected H_0 (p-values 0 or close to 0) for all basis sizes, so we can inference that we have some relationship between average temperature and average precipitation for Canadian weather stations. Unfortunately we know nothing about the strength and direction of correlation. Only from plots we can suppose that we have strong and positive correlation.

6.2 *Multivariate Case*

As a second example we used data obtained from Professor R. Nadulski of Department of Food Machinery and Engineering, University of Life Sciences, Lublin, Poland. In the experiment apples of six varieties were under consideration. The apples were obtained from a specialist orchard, and were picked in the first half of October 2010. They were stored in two types of refrigeration chamber: Ultra Low Oxygen and ordinary refrigeration chamber (CH). The textural properties of apples were tested by five different methods (in each of them, different sets of traits were measured):

Table 1 Correlation coefficients for apples data set

	CC				rV				dCor			
	M2	M3	M4	M5	M2	M3	M4	M5	M2	M3	M4	M5
M1	0.913	0.967	0.964	0.918	0.647	0.771	0.719	0.607	0.828	0.905	0.879	0.808
M2		0.935	0.929	0.852		0.566	0.507	0.293		0.804	0.802	0.696
M3			0.995	0.946			0.771	0.660			0.924	0.846
M4				0.955				0.475				0.756

- M1—penetrometric test—3 traits,
- M2—puncture test—4 traits,
- M3—TPA test—6 traits,
- M4—single compression test—5 traits,
- M5—apple flesh tension test—4 traits.

Apples were tested immediately following harvesting and then at 15-day intervals for a period of 180 days. The experimental results were transformed into functional data using the method described earlier. Again we used Fourier basis with different values of basis size. We calculated the values of correlation coefficients for each pair of methods M1-M5 for both types of refrigeration. Then, we tested the significance of the obtained results with permutation tests. For each size of basis and both methods of refrigeration we obtained similar results. For example, we present correlation coefficients for CH refrigeration method and basis size 3: Table 1. We can observe rather big values of correlation coefficients.

From permutation tests we obtained p-values equal to 0 for both method of refrigeration and all basis sizes. We conclude that methods are correlated (rather heavily). As a hint for the future we can recommend the cheapest method of testing the textural properties of apples.

During the numerical calculation process we used R software [18] and packages: `ade4` [4], `CCA` [8], `CCP` [17], `energy` [21], and `fda` [20].

7 Conclusions

We proposed an extension of two classical correlation coefficients for two sets of variables for multivariate functional data. We proposed to use permutation test to examine the significance of results because the values of proposed coefficients are rather hard to interpret. The proposed method has proved useful, on two real examples, in investigating the correlation between two sets of variables.

During the study of proposed coefficients we discovered that the size of basis (smoothing parameter) is rather unimportant, the values (and p-values for permutation tests) do not depend on the basis size (Fourier). We also discovered that despite regularization the CC coefficient increase with the size of basis. Hence, we recommend to use in practice rV or dCor correlation coefficients.

Of course, the performance of the methods needs to be further evaluated on additional real and artificial data sets. Additionally we can examine the behavior of coefficients (and tests) for different basis like B-splines or wavelets (when data are not periodic Fourier basis could fail). This could be direction of our future research.

References

1. Berrenderoa, J.R., Justela, A., Svarcb, M.: Principal components for multivariate functional data. Comput. Stat. Data Anal. **55**(9), 2619–2634 (2011)
2. Besse, P.: Étude descriptive d'un processus: aproximation et interpolation. Ph.D. thesis, Université Paul Sabatier, Toulouse III (1979)
3. Cléroux, R., Ducharme, G.R.: Vector correlation for elliptical distributions. Comput. Stat. Theory Methods **18**(4), 1441–1454 (1989)
4. Dray, S., Dufour, A.B.: The ade4 package: implementing the duality diagram for ecologists. J. Stat. Softw. **22**(4), 1–20 (2007)
5. Escoufier, Y.: Echantillonnage dans une population de variables aléatoires réelles. Ph.D thesis, Université des sciences et techniques du Languedoc, Montpellier (1970)
6. Escoufier, Y.: Le traitement des variables vectorielles. Biometrics **29**(4), 751–760 (1973)
7. Escoufier, Y., Robert, P.: Choosing variables and metrics by optimizing the RV coefficient. In: Rustagi, J.S. (ed.) Optimizing Methods in Statistics, pp. 205–219. Academic, New York (1979)
8. González I., Déjean, S.: CCA: Canonical correlation analysis. R package version 1.2 (2012). http://CRAN.R-project.org/package=CCA
9. Górecki, T, Krzyśko, M., Waszak, Ł., Wołyński, W.: Methods of reducing dimension for functional data. Stat. Transit. New Ser. **15**(2), 231–242 (2014)
10. He, G., Muller, H.G., Wang, J.L.: Methods of canonical analysis for functional data. J. Stat. Plan. Infer. **122**(1–2), 141–159 (2004)
11. Horváth, L., Kokoszka, P.: Inference for Functional Data with Applications. Springer, Berlin (2012)
12. Hotelling, H.: Relation between two sets of variables. Biometrika **28**(3/4), 321–377 (1936)
13. Jacques, J., Preda, C.: Model-based clustering for multivariate functional data. Comput. Stat. Data Anal. **71**(C), 92–106 (2014)
14. Josse, J., Holmes, S.: Tests of independence and beyond (2014). arXiv:1307.7383v3
15. Josse, J., Pagès, J., Husson, F.: Testing the significance of the RV coefficient. Comput. Stat. Data Anal. **53**(1), 82–91 (2008)
16. Leurgans, S.E., Moyeed, R.A., Silverman, B.W.: Canonical correlation analysis when the data are curves. J. R. Stat. Soc. Ser. B **55**(3), 725–740 (1993)
17. Menzel, U.: CCP: Significance Tests for Canonical Correlation Analysis (CCA). R package version 1.1 (2012). http://CRAN.R-project.org/package=CCP
18. R Core Team: R: A language and environment for statistical computing. R Foundation for Statistical Computing, Vienna (2015). https://www.R-project.org/
19. Ramsay, J.O., Silverman, B.W.: Functional Data Analysis. Springer, Berlin (2005)
20. Ramsay, J.O., Wickham, H., Graves, S., Hooker, G.: fda: Functional Data Analysis. R package version 2.4.4 (2014). http://CRAN.R-project.org/package=fda
21. Rizzo, M.L., Székely, G.J.: energy: E-statistics (energy statistics). R package version 1.6.2 (2014). http://CRAN.R-project.org/package=energy
22. Robert, P., Escoufier, Y.: A unifying tool for linear multivariate statistical methods: the RV coefficient. J. R. Stat. Soc. Ser. C **25**(3), 257–265 (1976)
23. Robert, P., Cléroux, R., Ranger, N.: Some results on vector correlation. Comput. Stat. Data Anal. **3**(1), 25–32 (1985)

24. Saporta, G.: Méthodes exploratoires d'analyse de données temporelles. Ph.D. thesis, Université Pierre et Marie Curie, Paris 6 (1981)
25. Székely, G.J., Rizzo, M.L., Bakirov, N.K.: Measuring and testing dependence by correlation of distances. Ann. Stat. **35**(6), 2769–2794 (2007)
26. Székely, G.J., Rizzo, M.L.: Brownian distance covariance. Ann. Appl. Stat. **3**(4), 1236–1265 (2009)
27. Székely, G.J., Rizzo, M.L.: On the uniqueness of distance covariance. Stat. Probab. Lett. **82**(12), 2278–2282 (2012)
28. Székely, G.J., Rizzo, M.L.: The distance correlation t-test of independence in high dimension. J. Multivar. Anal. **117**, 193–213 (2013)

Multi-Dimensional Scaling of Sparse Block Diagonal Similarity Matrix

Tadashi Imaizumi

Abstract Similarity matrix represents the relationship of n objects and gives us a useful information about these objects. Several models for analyzing this data assume that each object of n objects is embedded as a point or a vector in t dimensional "common space" of n objects in general. However, these models are not appropriate for analyzing a sparse block diagonal similarity matrix as each block diagonal matrix indicates us that each member of the set of objects in a block is represented as a point or a vector in not "common space," but, "sub-space." And a model is proposed to analyze this type of a sparse block diagonal similarity matrix. And application to a real data set will be shown.

1 Introduction

Similarity matrix represents the relationship of n objects. And asymmetric similarity matrix has much information about objects than symmetric similarity matrix has. We find many asymmetric relations from an object to another object of n objects, for example, a trading data between two nations, frequency of talking of n persons, journal citation data, etc. Tversky [14] reported the systematic asymmetry in Morse code data. Then we want to explore the complex relationship of n objects in asymmetric similarity matrix by some models and methods. Several models and methods have been proposed to analyze asymmetric (dis)similarity matrix. One is the decomposition of similarity matrix, and another is distance-based model. All of these models and methods try to give us the geometric representation of relationship among n objects in similarity matrix. These models assume that each object is represented as a point or a vector of t dimensional common space. When the number of objects n is small, for example, less than 30, this geometrical representation is useful to understand the global relationship of n objects and gain new insight from the results. However, it will be difficult for us to understand a hidden structure of

T. Imaizumi
School of Management and Information Sciences, Tama University, 4-1-1 Hijirigaoka, Tama-shi, Tokyo 206-0022, Japan
e-mail: imaizumi@tama.ac.jp

F. Palumbo et al. (eds.), *Data Science*, Studies in Classification, Data Analysis, and Knowledge Organization, DOI 10.1007/978-3-319-55723-6_20

these objects as the number of objects as larger one, for example, more than 50, 100, or 200.

On other hand, more essential question arises as to whether these objects have common characteristics or not as n is large. A typical example of this case is a sparse block diagonal similarity matrix as in Eq. (1).

$$\mathbf{S} = \begin{pmatrix} \mathbf{S}_{11} & \mathbf{S}_{12} & \cdots & \mathbf{S}_{1G} \\ \mathbf{S}_{21} & \mathbf{S}_{22} & \cdots & \mathbf{S}_{2G} \\ \vdots & \vdots & \ddots & \vdots \\ \mathbf{S}_{G1} & \mathbf{S}_{G2} & \cdots & \mathbf{S}_{GG} \end{pmatrix}, \tag{1}$$

where G is the number of blocks, and the values of element of $S_{gg'}$ for $g \neq g'$ will be relatively small than that of the S_{gg}. This type of the sparse block diagonal similarity matrix informs us that unknown characteristics will not be common for all objects, but some weak connections exist between blocks as the similarity matrix is not a perfect block diagonal matrix. These connections between blocks will be induced from the characteristics within block as these blocks of similarity matrix came out as a result. From these point of views, the sparse block diagonal similarity matrix will be represented the relationship within block and the relationship between blocks. And these two relationship will be induced from the characteristics of each object of n objects. So, a model and a method which integrates these two aspects will be proposed in this paper. A sparse block diagonal matrix will be explained in the next section. And the model for asymmetric similarity matrix will be explained. Then new model and its method to obtain the model parameters will also be described. And finally, an application of the proposed model to a real data set will be shown.

2 Data and Models

2.1 Data

Let $\mathbf{S} = [s_{jk}]$ be a asymmetric similarity matrix of the size $n \times n, j = 1, 2, \cdots, n; k = 1, 2, \cdots, n$. And $\{s_{jk}\}$ is a ranking number on the degree of similarity of n objects, that is, the more similar object j and object k is, the larger value of s_{jk} is.

$$s_{jk} \geq 0, \quad s_{jk} \neq s_{kj}. \tag{2}$$

We also assume

- If j is similar to k, then s_{jk} is larger.
- Though we do not know the number of blocks, G, in advance, a given similarity matrix is approximately represented as the form of block diagonal matrix after rearrangement of rows and columns as in Eq. (1).

- We also assume the values of the elements of off-diagonal sub-matrix $\mathbf{S}_{gh}, g \neq h$ are relatively small to the values of the elements of diagonal sub-matrix $\mathbf{S}_{gg}$.
- Each object belongs to one and only one block.

Then we want to represent the unknown relationship of **S** as a geometrical representation of n objects and G clusters.

2.2 Asymmetric Scaling Models

Models for analyzing asymmetric similarity matrix are classified into two types, one is a distance-based modeling and the other is non-distance-based modeling. Borg and Groenen [2] reviewed on asymmetric scaling models in the chapter "Big Data Scaling Through Metric Mapping: Exploiting the Remarkable Simplicity of Very High Dimensional Spaces Using Correspondence Analysis." Constantine and Gower [4] proposed a decomposition of asymmetric matrix to symmetric matrix and skew-symmetric matrix. A simple decomposition of **S** matrix will be

$$\mathbf{S} = \mathbf{A} + \mathbf{B}, \mathbf{A} = (\mathbf{S} + \mathbf{S}')/2, \mathbf{B} = (\mathbf{S} - \mathbf{S}')/2.$$

Chino [3] also proposed ASYMSCAL model in which he tried to represent **A** and **B** simultaneously.

$$s_{jk} = a(x_{j1}x_{k1} + x_{j2}x_{k2}) + b(x_{j1}x_{k2} - x_{j2}x_{k1}) + c.$$

This model represents similarity from object j to object k by an inner product between two points j and k, $(x_{j1}x_{k1} + x_{j2}x_{k2})$ and by an outer-product between these two points $(x_{j1}x_{k2} - x_{j2}x_{k1})$. Harshman [8] proposed DEDICOM model,

$$\mathbf{S} = \mathbf{YAY}',$$

where **A** is a asymmetric matrix of $t \times t$. Kiers [9] proposed an algorithm to find a **Y** and **A** from observed **S**. This approach is based on decomposition of a similarity matrix.

And the other approach to analyze an asymmetric similarity matrix is using the distance-based model such as a variants of unfolding model. Let δ_{jk} be dissimilarity from object j to object k. An unfolding model assumes

$$\delta_{jk} = \sqrt{\sum_{p=1}^{t}(x_{jp} - y_{kp})^2},$$

where x_j and y_k be two points in R^t. Young [15] proposed a weighted unfolding model,

$$\delta_{ij} = \left(\sum_{p=1}^{t} w_{ip}(x_{ip} - y_{jp})^2 \right)^{\frac{1}{2}}.$$

And Zielman and Heiser [16] proposed the slide-vector model,

$$\delta_{jk} = \left(\sum_{p=1}^{t} (x_{jp} - x_{kp} + z_p)^2 \right)^{1/2},$$

where $\mathbf{z} = [z_p]$ is a so-called slide vector. Krumhansl [10] proposed distance-density model for analyzing similarity matrix.

$$s_{jk} = f^{-1}(\delta_{jk})$$

$$\delta_{jk} = \sqrt{\sum_{p=1}^{t} (x_{jp} - x_{kp})^2 + a\iota_j + b\iota_k},$$

where ι_j is a term of the density of object j, $\iota_j \geq 0$. Okada and Imaizumi [12] proposed a radius-distance model (Fig. 1),

$$m_{jk} = \left(\sum_{p=1}^{t} (x_{jp} - x_{kp})^2 \right)^{1/2} - r_j + r_k. \tag{3}$$

The radii $r_j, j = 1, 2, \cdots, n$ show the relative dominance of object j over the other objects. The larger the dominant object j is, the smaller the radius of object is. So, the object j is less dominant than the object k, and

$$\delta_{jk} \geq \delta_{kj}$$

As the distance-based model assumes the relation between δ_{jk} and d_{jk} directly, we can capture overall relation among objects by configuration of points. de Rooij and Heiser [6] proposed a model for analyzing the contingency table(s) $F = [f_{jk}]$,

$$f_{jk} = \bar{g} + r_j + c_k - d_{jk}^2.$$

where $\bar{g}$ is grand mean, r_j with $r_j > 0$ is row-effect, and c_k with $c_k > 0$ is column-effect. The distance-based models have some advantage than the decomposition models and methods of asymmetric similarity model in this point. So, the model proposed in Sect. 3 will be based on this distance-based model.

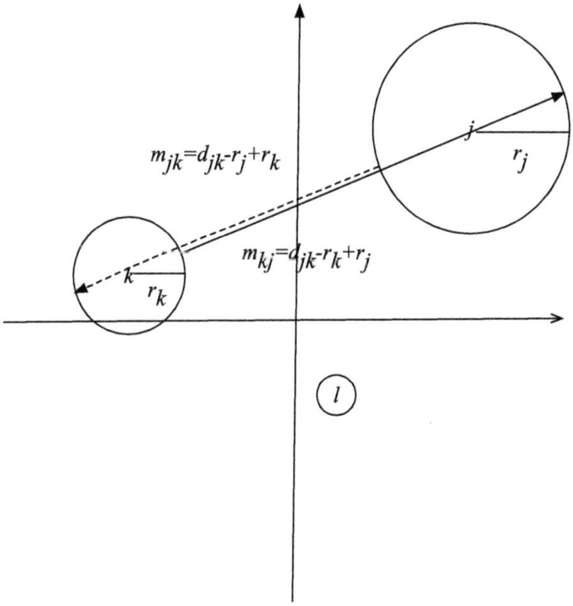

Fig. 1 The radius-distance model: radius of each circle represents $r_j - minr_j$

2.3 *Estimation of Number of Blocks*

How to estimate the number of blocks may be solved by applying one of the subspace clustering methods, for example, density-based spatial clustering of applications with noise (DBSCAN) [7]. However, our data matrix is not object configuration, but, matrix of the rank numbers of pairs, and we need to drive object configuration of higher dimensionality in general. When similarity matrix is a covariance matrix or a correlation matrix, a simple approach is to use factor-analysis approach.

$$\mathbf{S} = \Lambda \Delta \Lambda' + \Theta.$$

Then an estimated number of common factors G will be also an estimated number of blocks G. If this method was adopted, then main problem will be how to derive the covariance matrix between objects. Another method to estimate G is to use cluster analysis and determine the number of clusters, which will be an estimated number of blocks.

3 A New Model and Its Algorithm

For analyzing a given similarity data, it is assumed the radius-distance model Eq. (3) for with-in blocks and between blocks, respectively. And it is also assumed that s_{jk} is monotonically decreasingly related to m_{jk}. Then the diagonal sub-matrix $\mathbf{S}_{gg}, g = 1, 2, \cdots, G$ will be represented by a matrix of distance between objects in same cluster g, and the off-diagonal sub-matrix $\mathbf{S}_{gh}, g \neq h$ will be represented by a matrix of distance between clusters g and h. Let $g(j), g(j) = 1, 2, \cdots, G$ be an index vector in which cluster object j belongs to. When m_{jk} for the case that j and k are members of same cluster, $g(j) = g(k)$, is, then

$$m_{jk} = d_{jk} - r_j + r_k. \tag{4}$$

In the case that j does not belong to the same cluster of k, the distance between clusters needs to be defined. Let $M(g) = \{j' : g(j') = g\}$ be a set of members of the cluster g, then the cluster mean of g,$x^*_{g(j)p}, p = 1, 2, \cdots, t$ is defined by

$$x^*_{g(j)p} = \sum_{j \in M(g)} x_{jp} / \chi(M(g)), p = 1, 2, \cdots, t, \tag{5}$$

where $\chi(\cdot)$ is the number of elements of the set $M(g)$. And the distance between clusters is defined by

$$d^*_{g(j)g(k)} = \left(\sum_{p=1}^{t} (x^*_{g(j)p} - x^*_{g(k)p})^2 \right)^{1/2}, \tag{6}$$

and m_{jk} when $g(j) \neq g(k)$ is defined by

$$m_{jk} = d^*_{g(j)g(k)} - r^*_{g(j)} + r^*_{g(k)}, \tag{7}$$

where $r_{g(j)}$ is the radius of the cluster $g(j)$. Our aims are to find an optimal allocation of each object to a block(cluster) and obtain a geometrical representation of n object in lower dimensional space, for example, two or three dimensional space.

3.1 Monotonic Disparity

The $\{m_{jk}\}$ need to satisfy such that

$$\textit{if} \quad s_{jk} \leq s_{j'k'} \quad \textit{then} \quad m_{jk} \leq m_{j'k'}. \tag{8}$$

Unfortunately, a given $\{m_{jk}\}$ did not satisfy the order relation in $\{s_{jki}\}$ in general. Then we need to improve $\{m_{jk}\}$ as they satisfy it. We introduce the disparities $\{\hat{m}_{jk}\}$ which satisfy

$$\textit{if} \quad s_{jk} \leq s_{j'k'} \quad \textit{then} \quad \hat{m}_{jk} \leq \hat{m}_{j'k'}. \tag{9}$$

As $\{\hat{m}_{jk}\}$, we adopt $\{\hat{m}_{jk}\}$ which minimize

$$S(\hat{m}, m) = \sum_{j=1}^{n}\sum_{k=1}^{n}(\hat{m}_{jk} - m_{jk})^2. \tag{10}$$

Then it is very important how to derive those disparities. We employ $\hat{m}_{jki}$ as values given by PAVA Isotonic Regression algorithm [5]. The original algorithm was given by Kruskal [11], and Barlow et al. [1] formulated those isotonic regression mathematically.

3.2 Allocation of Each Object to a Block (Cluster)

For a given t, $\mathbf{X} = [x_{jp}]$, $\mathbf{r} = [r_j]$, $\mathbf{r}^* = [r_{g(j)}]$, and $\mathbf{g}^* = [g(j)]$, m_{jk} and monotonic disparity $\hat{m}_{jk}$ are calculated. Then the squares of the average of $\hat{m}_{jk}$ and $\hat{m}_{kj}$, $\left((\hat{m}_{jk} + \hat{m}_{kj})/2\right)^2$, are supplied as distance to the Ward method, and the updated number of clusters G^* and its allocation $\mathbf{g}^* = [g(j)]$ are determined using Rousseeuw's silhouette index [13].

3.3 Loss Function for Geometrical Representation

For the updated G^* and its allocation $\mathbf{g}^*$, the radii of clusters, $\mathbf{r}_g^* = [r_{g(j)}^*]$, the object configuration $\mathbf{X}$, and $\mathbf{r} = [r_j]$ need to be improved. The LQ criterion in Eq. (11) is adopted to update them,

$$LSQ(\mathbf{X}, \mathbf{r}, \mathbf{r}^* | G^*, \mathbf{g}^*) = \sum_{j=1}^{n}\sum_{k=1}^{n}(\hat{m}_{jk} - m_{jk})^2 / \sum_{j=1}^{n}\sum_{k=1}^{n}\hat{m}_{jk}^2. \tag{11}$$

And the flow of computational procedure is shown in Fig. 2.

The dimensionality of configuration is limited to 2 or 3 as visualization of the results.

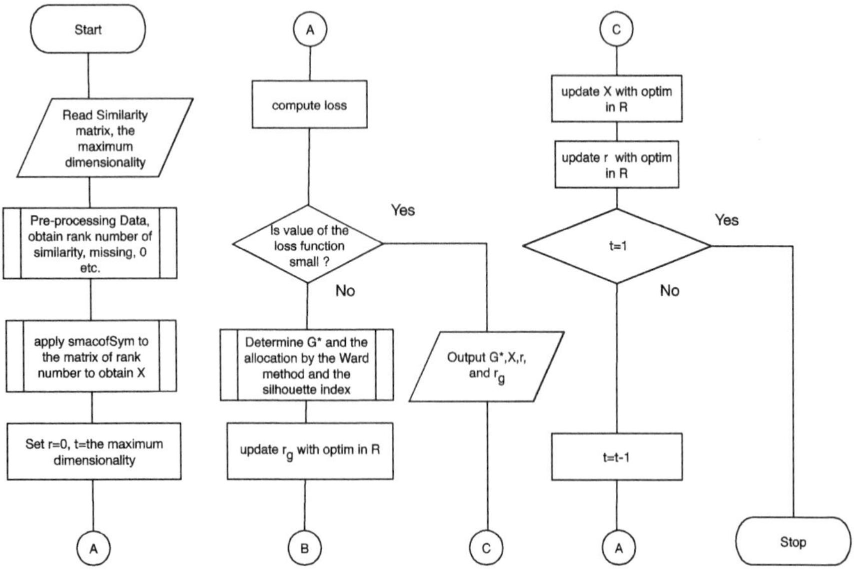

Fig. 2 Flowchart of the computational procedure

4 Application

The Japanese Government has been promoting to increase the number of visitors for raising an economic effects. To do that, it needs for us to develop the infrastructure at many regions in Japan. And the analysis of OD table of the visitors will be important. An application of the proposed model and method to traveling records data of the visitors to Japan. This data was collected by the ministry of Land, Infrastructure, Transport, and Tourism on a week in the autumn season of the year 2010. There are 47 prefectures in Japan as the administrative partition. The location of each prefectures in Japan is shown in the left panel of Fig. 3. The rows and columns of this transaction table of visitors are basically corresponding to the 47 prefectures in Japan, but Hokkaido prefecture is divided into four regions, and Chiba prefecture also divided into three regions. No visitors were in the Eastern Hokkaido region and Saga prefecture, these two regions were omitted from table. And the data table was 50 × 50. The sub-matrix of data table was shown in Table 1, for example. The values in Table 1 are the number of persons moved to the row ordination to the column destination, and 0 frequency indicates no visitors observed in the surveyed week.

Fig. 3 The location of 47 prefectures in Japan

As each sums of the row frequency is different, the data table was normalized as to $\sum_{k=1}^{n} s_{jk}$, and this normalized table was supplied as the data to be analyzed. The normalized data matrix of Table 1 is also shown in Table 2.

A heatmap of this normalized data (Table 2) is shown in Fig. 4. This heatmap indicates asymmetric tendency in this data, and the data is the block-diagonal matrix.

We applied the proposed model with setting from $t = 3$ to $t = 1$, and the maximum number of clusters being 10. The value of $LSQ(\cdot)$ was 0.297 and 0.340, and 0.556 for $t = 3$, $t = 2$, and $t = 1$, respectively. And we adopted two dimensional representation. The final number of clusters when $t = 2$ was 4. The list of members in each clusters is shown in Table 3.

As the member of Cluster 1 are the regions of the northern part of Japan, this Cluster shows the "geographical" characteristics. The characteristics of Cluster 4 is also "geographical" as the members are the prefectures of the southern part of Japan. The member of Cluster 3 are the regions near to Tokyo, the regions near to

Table 1 Example of the original data

Ordination(row) To destination(column)	Northern Hokkaido	Central Hokkaido	Southern Hokkaido	Aomori	Iwate	Miyagi	Akita	Yamagata
Northern Hokkaido	266	598	272	13	0	0	10	0
Central Hokkaido	729	4858	979	72	0	0	60	0
Southern Hokkaido	287	1051	382	85	0	0	74	0
Aomori	22	4	221	463	50	44	171	0
Iwate	0	0	0	39	0	0	6	0
Miyagi	0	0	0	45	10	277	205	36
Akita	8	40	42	157	45	157	183	0
Yamagata	0	0	0	18	53	8	34	0
Fukushima	0	0	0	0	0	60	4	28
Ibaragi	22	0	0	0	0	0	0	0

Table 2 Example of the normalized data

Ordination (row) to destination (column)	Northern Hokkaido	Central Hokkaido	Southern Hokkaido	Aomori	Iwate	Miyagi	Akita	Yamagata
Northern Hokkaido	0.21	0.09	0.12	0.02	0.00	0.00	0.01	0.00
Central Hokkaido	0.47	0.60	0.45	0.00	0.00	0.00	0.04	0.00
Southern Hokkaido	0.21	0.12	0.16	0.16	0.00	0.00	0.04	0.00
Aomori	0.01	0.01	0.04	0.33	0.12	0.04	0.15	0.08
Iwate	0.00	0.00	0.00	0.04	0.00	0.01	0.04	0.24
Miyagi	0.00	0.00	0.00	0.03	0.00	0.25	0.15	0.04
Akita	0.01	0.01	0.03	0.12	0.02	0.18	0.17	0.15
Yamagata	0.00	0.00	0.00	0.00	0.00	0.03	0.00	0 .00
Fukushima	0.00	0.00	0.00	0.04	0.00	0.00	0.01	0.00
Ibaragi	0.00	0.00	0.00	0.00	0.00	0.00	0.00	0.00

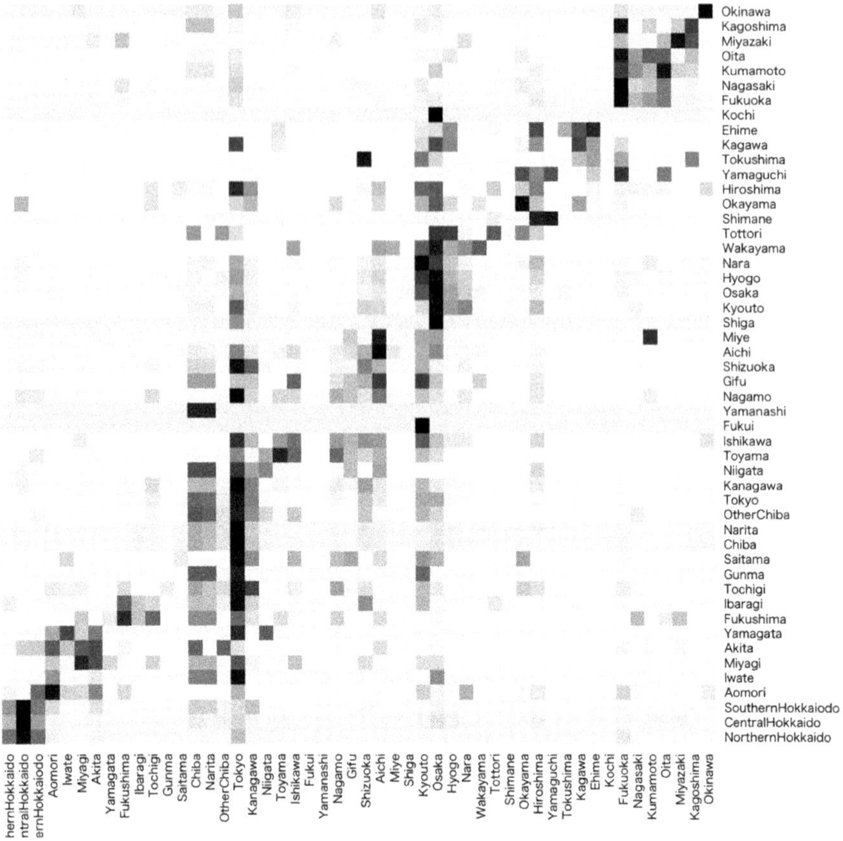

Fig. 4 Heat map of visitors to Japan, *row* is the ordination region and *column* is the destination region

Kyouto and Osaka, and the regions near to Aichi. As the big international airports in Japan are located in each of Narita, Tokyo, Osaka, and Aichi, the characteristics of this Cluster will be "accessibility." As the radius of each cluster is very small, the visitors would move with-in cluster. The object configuration with with-in radii is shown in Fig. 5. The radius of the regions, Chiba, Narita, Fukioka, and Yamagata is larger, respectively. So, the visitors would move from one of these ordinations to the near destinations.

Table 3 Clustering results and $r_{g(j)}$

Cluster	$r_{g(j)}$	Prefecture				
1	0.0000	N. Hokkaido	C. Hokkaido	S. Hokkaido	Aomori	Iwate
		Miyagi	Akita	Yamagata		
2	0.0004	Fukushima	Ibaragi	Tochigi	Gunma	Saitama
		Fukui	Yamanashi	Miye	Shiga	Wakayama
		Tottori	Shimane	Okayama	Hiroshima	Yamaguchi
		Tokushima	Kagawa	Ehime	Kochi	Okinawa
3	0.0043	Chiba	Narita	Chiba reg.	Tokyo	Kanagawa
		Niigata	Toyama	Ishikawa	Nagano	Gifu
		Shizuoka	Aichi	Kyouto	Osaka	Hyogo
		Nara				
4	0.0006	Fukuoka	Nagasaki	Kumamoto	Oita	Miyazaki
		Kagoshima				

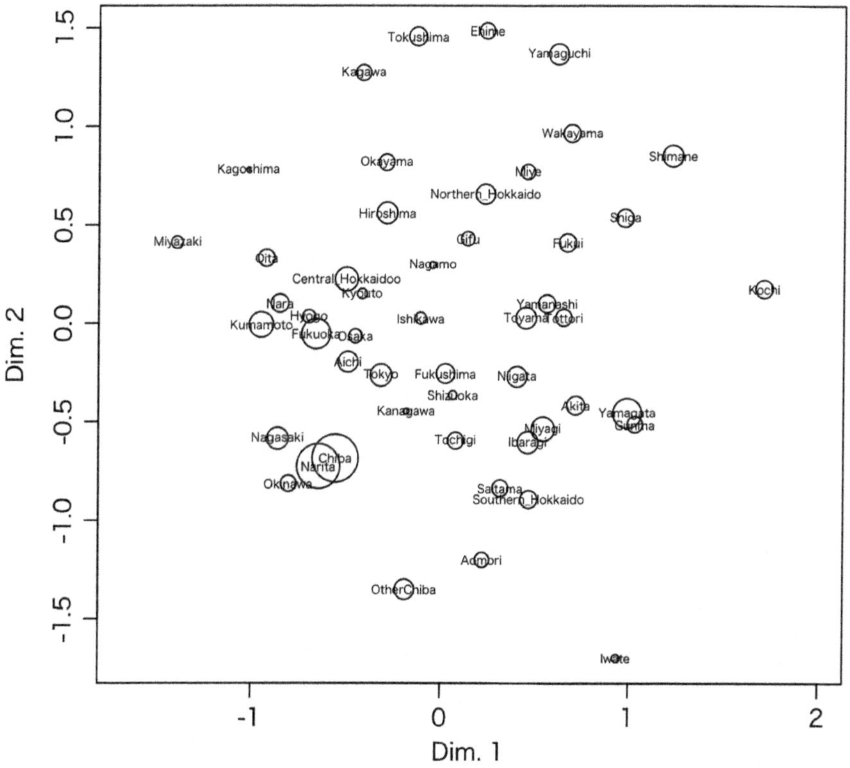

Fig. 5 Object configuration with with-in radii

5 Conclusion

The proposed model revealed how the tourists were moving in Japan. We can make a decision on how to maintain the infrastructure of each prefectures as to fit the visitors purpose.

The proposed model is a model for one mode two way data. More dynamical model for two mode three way data will be more valuable, specially for the case that one mode is time. Then clusters will vary on time-line.

References

1. Barlow, R.E., Bartholomew, J.M., Bremner, J.M., Brunk, H.D.: Statistical Inference under Order Restrictions, the Theory and Application of Isotonic Regression. Wiley, New York (1972)
2. Borg, I., Groenen, P.J.F.: Modern Multidimensional Scaling. Springer, New York (2005)
3. Chino, N.: A graphical technique for representing the asymmetric relationships between N objects. Behaviormetrika **5**, 23–40 (1978)
4. Constantine, A.G., Gower, J.C.: Graphic representations of asymmetric matrices. Appl. Stat. **27**, 297–304 (1978)
5. de Leeuw, J., Hornik, K., Mair, P.: Isotone optimization in R: Pool-Adjacent-Violators Algorithm (PAVA) and active set methods. J. Stat. Softw. **32**, 1–24 (2009)
6. de Rooij, M., Heiser, W.: A distance representation of the quasi-symmetry model and related distance models. In: Yanai, H., Okada, A., Shigemasu, K., Kano, Y., Meulman, J. (eds.) New Developments in Psychometrics, pp. 487–494. Springer, Tokyo (2002)
7. Ester, M., Kriegel, H.-P., Sander, J., Xu, X.: A density-based algorithm for discovering clusters in large spatial databases with noise. In: Simoudis, E., Han, J., Fayyad, U.M. (eds.) Proceedings of the Second International Conference on Knowledge Discovery and Data Mining (KDD-96), pp. 226–231. AAAI Press, Portland (1996)
8. Harshman, R.: Models for analysis of asymmetrical relationships among N objects or stimuli. Paper presented at the First Joint Meeting of the Psychometric Society and the Mathematical Psychology, McMaster University, Hamilton, ON, August 1978
9. Kiers, H.A.L.: An alternating least squares algorithm for fitting the two- and three-way DEDICOM model and the IDIOSCAL model. Psychometrika **54**, 515–521 (1989)
10. Krumhansl, C.L.: Concerning the applicability of geometric models to similarity data: the interrelationship between similarity and spatial density. Psychol. Rev. **85**, 445–463 (1978)
11. Kruskal, J.B.: Nonmetric multidimensional scaling: a numerical method. Psychometrika **29**, 115–129 (1964)
12. Okada, A., Imaizumi, T.: Geometric models for asymmetric similarity data. Behaviormetrika **21**, 81–96 (1987)
13. Rousseeuw, P.: Silhouettes: a graphical aid to the interpretation and validation of cluster analysis. J. Comput. Appl. Math. **20**, 53–65 (1987)
14. Tversky, A.: Features of similarity. Psychol. Rev. **84**, 327–352 (1977)
15. Young, F.W.: An asymmetric Euclidean model for multi-process asymmetric data. Paper presented at U.S.-Japan Seminar on MDS. San Diego (1975)
16. Zielman, B., Heiser, W.J.: Models for asymmetric proximities. Br. J. Math. Stat. Psychol. **49**, 127–147 (1996)

The Application of Classical and Positional TOPSIS Methods to Assessment Financial Self-sufficiency Levels in Local Government Units

Agnieszka Kozera, Aleksandra Łuczak, and Feliks Wysocki

Abstract The aim of this study was to conduct a comparative analysis of the results provided by rank ordering of objects and identification of their development types using the TOPSIS (*Technique for Order Preference by Similarity to an Ideal Solution*) method in two approaches—classical and positional. The classical approach aiming at the construction of a synthetic measure of development is based on the classical standardisation of characteristics values and Euclidean distances of each object from model objects (the positive and negative ideals of development), while the positional approach applies standardisation of the Weber medians and median absolute deviation. The proposed approaches are presented based on an example concerning the construction of a synthetic measure and identification of types of financial self-sufficiency in rural communes of the Wielkopolska Province, Poland, in 2013. The study is based on statistical data supplied by the Central Statistical Office of Poland (*Local Data Bank*) and the Ministry of Finances (*Indicators to assessment the financial situation of local government*). Analyses showed that the application of the positional approach has led (in comparison to the classical approach) to such a value of the synthetic measure, which covered a greater part of the potential range of its variation and as a result—to a more accurate and more comprehensive identification of types of financial self-sufficiency in rural communes.

1 Introduction

In the assessment of the financial situation of local government units we frequently find features characterised by outliers or strong asymmetry. In such cases the classical methods of construction applied to a synthetic measure of development may fail, since the incidence of even one outlier for a given object (either very high or very small) may have a significant effect on the excessively high (or low)

A. Kozera (✉) • A. Łuczak • F. Wysocki
Poznań University of Life Sciences, ul. Wojska Polskiego 28, 60-637 Poznań, Poland
e-mail: akozera@up.poznan.pl; luczak@up.poznan.pl; wysocki@up.poznan.pl

F. Palumbo et al. (eds.), *Data Science*, Studies in Classification, Data Analysis, and Knowledge Organization, DOI 10.1007/978-3-319-55723-6_21

rank ascribed in the final classification. Thus in some typological studies concerning object ranking the focus should be on the methods not affected by this phenomenon, particularly those using the *Weber median* (see Rousseeuw and Leroy [7], Vardi and Zhang [10], Lira et al. [5], Wysocki [13]). This statistic shows considerable resistance to the effect of atypical observations—outlying and extreme.

The aim of this paper is to present a comparative analysis of results of object linear ordering and identification of their development types using the TOPSIS (*Technique for Order Preference by Similarity to an Ideal Solution*) methods in two approaches—classical and positional. The proposed approaches are presented based on an example concerning the construction of a synthetic measure and identification of types of financial self-sufficiency in rural communes of the Wielkopolska[1] Province, Poland, in 2013.

2 Methods

In proposed approach was applied multi-criteria decision method (MCDM). MCDM allows to refer to many criteria simultaneously during the problem solving. There are many methods to solve the multi-criteria problems, although some of them have been criticised as "ad hoc" and used only because of the lack of better alternatives, because their use was not fully theoretically or empirically justified [8]. Each method not only has its advantages, but also some limitations. However, as one of the most interesting to construct a synthetic characteristic may be considered the TOPSIS (*Technique for Order Preference by Similarity to an Ideal Solution*). TOPSIS method is a popular approach and has been widely used in many issues. It is useful in creating the ranking of objects describing by many features. The method allows us to find the shortest distance from the positive ideal solution and the farthest distance from the negative ideal solution for solving multiple-criteria decision making problems. On the base of this distances the synthetic measure is constructed. The higher values of synthetic measure denote the better position of objects in ranking.

The procedure to construct a synthetic measure, based on TOPSIS method, is a multi-stage process, in which we may distinguish six basic stages [13]. The first stage includes the selection (based on substantive and statistical criteria) of features describing selected objects, as well as determination of the direction of their preferences in relation to the investigated general criterion. Characteristics

[1]The Wielkopolska Province is one of the 16 provinces in Poland. It is situated on the central-western part of Poland.

considered to be destimulant[2] may be transformed into stimulants[3] using the following transformation:

$$x_{ik} = a - b \cdot x_{ik}^{D} \tag{1}$$

where: x_{ik}^{D}—value of the kth feature which is a destimulant ($k \in I_D$) in the ith object ($i = 1, \ldots, N$),

I_D—set of destimulant indexes,
N—number of objects,
x_{ik}—value of the kth feature ($k = 1, \ldots, K$) transformed into a stimulant,
a, b—arbitrary constants, here $a = 0$ and $b = 1$.

In the next stage (II), consisting in the standardisation of values of simple characteristics in the classical approach, standardisation is applied as follows:

$$z_{ik} = \frac{x_{ik} - \bar{x}_k}{s_k} \tag{2}$$

while in the positional approach it is the Weber median standardisation [5]:

$$z_{ik} = \frac{x_{ik} - \tilde{med}_k}{1,4826\, \tilde{mad}_k} \tag{3}$$

where: x_{ik}—value of the kth feature in the ith object,
$\tilde{med}_k$—L_1-median[4] vector component corresponding to the kth feature,
$\tilde{mad}_k = \tilde{med}_i |x_{ik} - \tilde{med}_k|$—median absolute deviation of kth feature values from the median component of the kth feature,
1.4826—a constant scaling factor corresponding to normally distributed data, $\sigma \approx E(1.4826 \cdot \tilde{mad}_k(X_1, X_2, \ldots X_k))$, σ—standard deviation.

The distribution of thus standardised feature values is considered close to the normal distribution of zero expectation and unitary standard deviation [6].

In stage III the coordinates of ideal and negative ideals were computed according to the following formulae:

$$A^{+} = \left(\max_i(z_{i1}), \max_i(z_{i2}), \ldots, \max_i(z_{iK})\right) = \left(z_1^{+}, z_2^{+}, \ldots, z_K^{+}\right) \tag{4}$$

[2]Destimulant is a kind of characteristic where the smaller values are desirable and high values are undesirable from the viewpoint of the considered synthetic property. It is negatively correlated with the synthetic characteristic.

[3]Stimulant is a kind of characteristic where the higher values are desirable and small values are undesirable from the viewpoint of the considered synthetic property. It is positively correlated with the synthetic characteristic.

[4]The L_1-median is also called the Weber median, geometric median, median centre, spatial median or the Weber point [5].

for the positive ideal solution, and:

$$A^- = \left(\min_i(z_{i1}), \min_i(z_{i2}), \ldots, \min_i(z_{iK})\right) = \left(z_1^-, z_2^-, \ldots, z_K^-\right) \quad (5)$$

for the negative ideal solution.

The positive ideal solution[5] (PIS) are the best values of features. PIS includes the maximum (ideal) values of each feature. Whereas, the negative ideal solution (NIS) are the worst values of features. NIS contains the minimum (nadir) values of each feature which are stimulant or are transformed into stimulant.

Coordinates of model objects constitute the basis for the calculation of distances for each assessed object from the positive ideal (A^+) and the negative ideal of development (A^-) (stage IV), i.e. in the classical approach using Euclidean distances [4]:

$$d_i^+ = \sqrt{\sum_{k=1}^{K}\left(z_{ik} - z_k^+\right)^2}, \qquad d_i^- = \sqrt{\sum_{k=1}^{K}\left(z_{ik} - z_k^-\right)^2}, \quad (6)$$

while in the ordering approach based on the median absolute deviation [13]:

$$d_i^+ = med_k\left(\left|z_{ik} - z_k^+\right|\right), \qquad d_i^- = \left(\left|z_{ik} - z_k^-\right|\right), \quad (7)$$

where: med_k—marginal median for the kth feature.

The construction of a synthetic measure (stage V) used the TOPSIS method (*Technique for Order Preference by Similarity to an Ideal Solution*) based on the idea introduced by Hellwig [2] (see also Hellwig [3], Hwang and Yoon [4]):

$$S_i = \frac{d_i^-}{d_i^- + d_i^+}, (i = 1, \ldots, N), \quad (8)$$

where $0 \leq S_i \leq 1$.

The smaller the distance of a given model object—the positive ideal of development, and thus greater from the other extreme (the negative ideal of development), the closer to 1 is the value of the synthetic measure.

Established values of the synthetic measure are used in rank ordering of communes and—on this basis—identification of their typological classes (stage VI). Identification of classes for the entire range of variation of a synthetic measure may be performed using statistical methods based on the mean and standard deviation from values of the synthetic measure, or in an arbitrary manner, assuming, e.g. numerical ranges of values for measure S_i [13]:

[5]The positive ideal solution is also called the ideal solution.

$< 0.00, 0.20)$—very low, $< 0.20, 0.40)$—low, $< 0.40, 0.60)$—medium, $< 0.60, 0.80)$—high, and $< 0.80, 1.00)$ very high levels of financial self-sufficiency of local government units.

The identified typological classes of financial self-sufficiency of communes were evaluated statistically, applying for this purpose measures of homogeneity (H_0), heterogeneity (H_e) and correctness of classes (PS) based on the concept of the centre of gravity of a class (which constitutes the Weber median of its elements) and distances from it [6]:

$$H_0 = med_{i:\Gamma_i \in P_c} \delta(\Gamma_i, \Gamma_{kc}), \tag{9}$$

$$H_e = med_{i=1,2,\ldots,c} \delta(\Gamma_{ki}, \Gamma_{kc}), \tag{10}$$

$$PS = \frac{H_0}{H_e}, \tag{11}$$

where:

c—the number of typological class,

Γ_{kc}—centre of gravity of class P_c (the Weber median of its elements).

The homogeneity index measures the compactness of the identified class (the lower the value of the index, the more homogeneous the class), while the heterogeneity index investigates its distinctness from other classes (the higher its value, the greater the inter-class differences). Finally, the lower value of measure of correctness of classes (PS), the typology is more correct in terms of statistical criteria.

3 Results of Research

Analyses concerning the level of financial self-sufficiency in rural communes of the Wielkopolska Province (N=117) are based on statistical data from 2013 coming from the Central Statistical Office of Poland (Local Data Bank) and the Ministry of Finances (*Indicators to assessment the financial situation of local government*). In the first stage of the evaluation of the financial self-sufficiency, based on substantive criteria, nine indicators were preliminarily selected, describing the level of financial self-sufficiency of communes (Table 1).

As a result of conducted statistical verification the WSFW II index was eliminated, due to the very low variation of its value (6.9%) (Table 2), as well as WSFD and WBF due to their high correlation with the other indicators. Finally in this study six indicators were adopted, of which one, i.e. WFIP, was considered to be a destimulant, while the others were stimulants of financial self-sufficiency of commune local government units.

It may be observed that the application of the positional approach, based on the Weber median and median absolute deviations from the positive and negative

Table 1 Indicators adopted in the evaluation of financial self-sufficiency levels of communes

	Characteristic	The formula of indicator	Unit of measure
WSFD	Indicator of financial self-sufficiency (income)	$WSFD = \frac{DW-PD}{DO} \cdot 100\%$	%
WSFW I	Indicator of financial self-sufficiency (expenditure) of the first degree	$WSFW_1 = \frac{DW}{DO} \cdot 100\%$	%
WSFW II	Indicator of financial self-sufficiency (expenditure) of the second degree	$WSFW_2 = \frac{DW+SO}{DO} \cdot 100\%$	%
WDWM	Indicator of level of own income	$WDWM = \frac{DW}{LM} \cdot 100\%$	PLN *per capita*
WBF	Indicator of level of tax income	$WBF = \frac{DP}{LM} \cdot 100\%$	PLN *per capita*
WAP	Share of ratio of tax income to current income	$WAP = \frac{DP}{DB} \cdot 100\%$	%
WFIP	Share of general and targeted subsidies in total income	$WFIP = \frac{DT}{DO} \cdot 100\%$	%
WS	Self-financing rate	$WS = \frac{NO+DM}{WM} \cdot 100\%$	%
WIWO	Share of investment expenditures in total expenditures	$WWI = \frac{WI}{WO} \cdot 100\%$	%

Symbols: DW—own income, DO—income, SO—general subsidy, DB—current income (budgetary revenue other than income property), DM—capital income, the DP tax income (tax bill of agricultural, forestry, real estate, from funds transport of civil law, income from taxation, income from mining fee), DT—transfer income (including specific grants and the general subsidy), NO—operating surplus, PD—participation in income taxes of legal and natural persons, WM—capital expenditure, LM—the number of inhabitants
Source: Own calculations based on Wang et al. [11], Trussel and Patrick [9], *Indicators to assess the financial situation of local government in 2011–2013* (2014) [12]

Table 2 Descriptive statistics of indicators describing the level of financial self-sufficiency in rural communes of the Wielkopolska Province in 2013 (N=117)

Characteristic	WSFW I	WDWM	WAP	WFIP	WS	WIWO
Classical statistics						
Min	19.5	616.7	5.5	19.9	29.7	2.1
Max	80.1	4662.8	57.1	78.6	773.1	34.7
Average	40.2	1329.0	18.5	56.1	139.9	14.9
Standard deviation	13.4	688.4	8.3	12.3	89.3	7.1
Coefficient of variation (%)	33.3	51.8	44.7	22.0	63.8	47.8
Skewness	0.9	2.9	1.5	−0.9	3.7	0.6
Positional statistics						
Vector of Weber median (L1-median)[a]	37.8	1174.5	17.1	58.8	137.6	13.4
Median	37.0	1176.9	17.2	58.7	116.6	13.3
Coefficient of variation (%)	25.2	25.3	25.3	12.5	31.1	36.6
Skewness	0.3	0.2	0.0	−0.2	0.3	0.3

Source: Own calculations based on data from the Central Statistical Office (*Local Data Bank* [1]) and the Ministry of Finance (*Indicators for assessing the financial position of local government units*)
[a]The calculations are performed using R program

ideals of development, is justified here. In the set of indicators we have some, which show a relatively strong asymmetry, high variation of their values and outliers. Such properties of distribution were found particularly in two indicators adopted in the construction of the synthetic measure of development (Table 2), i.e. indicator of level of commune income (PLN *per capita*) and the self-financing rate. These indicators were characterised by a relatively high right-hand asymmetry (the skewness index was 2.9 and 3.7, respectively) and its high variations of values in the set of investigated rural communes in the Wielkopolska Province (the coefficient of variation for WDWM and WS was 51.8% and 63.8%, respectively). This was influenced by outliers, observed, e.g. for communes of Przykona, Suchy Las and Mieleszyn. For example, in the communes of Przykona and Suchy Las income exceeded over threefold the mean for all the rural communes, which amounted to 1329.0 PLN per capita. In turn, in the Mieleszyn commune the self-financing rate exceeded over sixfold the average for all the rural communes of the Wielkopolska Province, which was 139.9 in 2013.

In the next stage (II) values of the indicators[6] were standardised using formula (2) in the case of the classical approach, and formula (3) in the positional approach. Standardised values of the indicators made it possible to determine distances of each investigated rural commune from the positive and negative ideals of development (stage IV) applying formulas (6) in the classical approach and formulas (7) in the positional approach. Values of the ideal (stage III) were established assuming maximum values of characteristics, while those of the negative ideal—based on their minimum values in the set of all rural communes in the Wielkopolska Province (N=117). Next values of the synthetic measure of development were calculated using the TOPSIS method (formula (1), stage V). Ranges of variation of the obtained values of synthetic measures are given in Fig. 1. On their basis types of the financial self-sufficiency levels in communes were identified assuming arbitrarily five numerical ranges of the synthetic measure.

These analyses showed that the application of TOPSIS method in the positional approach provides a greater range of values of the synthetic measure in relation to the classical approach (Fig. 1). The range between the maximum and minimum values of the synthetic measure in the positional approach was 0.85, while in the classical approach it was by 50% smaller, amounting to 0.44. This situation results from the atypical values—outliers—in the set of data. In such cases the assumption that maximum and minimum values of characteristics in ideal object rank-ordering methods (e.g. TOPSIS) constitute model objects leads to their excessive distance from typical values of investigated indexes and as a consequence—to a reduction of variation of the constructed measure of development and problems with providing on its basis a reliable identification of development types in the investigated objects. Obtained values of the synthetic measure using the classical TOPSIS method as a result may cover only a small part of the potential range of its variation specified in the range of $< 0, 1 >$.

[6]WFIP was transformed into a destimulant by Eq. (1).

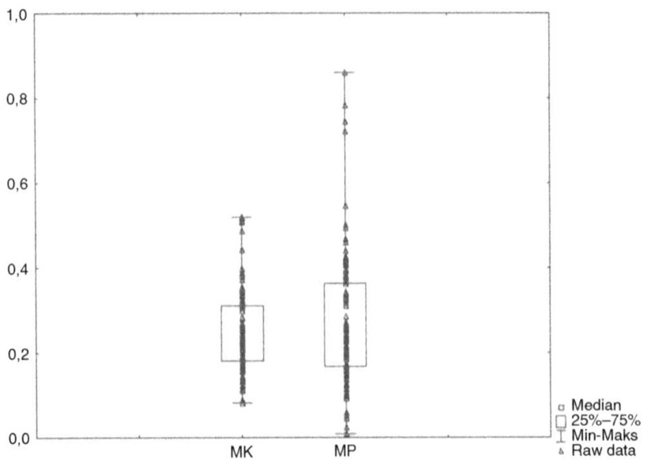

Fig. 1 The range of variation in values of synthetic measures (TOPSIS method: MK—classical approach, MP—positional approach) of financial self-sufficiency in rural communes of the Wielkopolska Province in 2013. Source: Own calculations based on data from the Central Statistical Office (*Local Data Bank*) and the Ministry of Finance (*Indicators for assessing the financial position of local government units*)

Investigated rural communes in the Wielkopolska Province were ordered from the best to the worst based on values of the synthetic measure of financial self-sufficiency obtained by positional TOPSIS method (Table 3). We need to stress here the differences between the orderings of communes in the Wielkopolska Province based on values of synthetic measure provided by the positional and classical approaches. Consistency of ordering of the analysed communes according to the values of the synthetic measure was relatively high, as the Spearman rank correlation coefficient was 0.75. However, considerable discrepancies were found between the ordering of the analysed communes, such as the communes of Mieleszyn, Lisków and Wijewo. The Mieleszyn commune, in the approach using the classical method, was ranked very high, i.e. 6th, while in the approach using the positional method it was classified as low as 86th (Table 3). Similarly, the communes of Lisków and Wijewo in the ranking of communes provided by the classical approach to construct the synthetic index were ranked much higher (48th and 46th) than in the ranking based on the positional approach (107th and 90th, respectively). Such a situation in the case of Mieleszyn commune was the effect of the very high value of the WS indicator (which in this commune was 773.1 at the mean for all the communes amounting to 139.9), while in the case of the Lisków and Wijewo communes it was by very high WIWO values (32.0% and 30.5% at the mean for all the communes of 14.9%).

Thus the conducted studies showed that when the investigated commune is characterised by only one outlying value (either very high or very low) this may have a significant effect on the excessively high or low rank ascribed in the final

Table 3 Values of the synthetic measure according to TOPSIS and the rank of communes[a] according to the level of financial self-sufficiency of communes in the Wielkopolska Province in 2013

		Values of synthetic measure		Rank of commune		
No.	Communes	MP	MK	MP	MK	Level of development
1	Przykona	0.861	0.521	1	1	Very high
2	Tarnowo Podgórne	0.804	0.510	2	4	Very high
3	Suchy Las	0.783	0.519	3	2	Medium
4	Komorniki	0.756	0.512	4	3	Medium
5	Powidz	0.722	0.489	5	5	Medium
6	Brudzew	0.547	0.389	6	9	Medium
7	Baranów	0.503	0.373	7	11	Medium
8	Czerwonak	0.496	0.397	8	8	Medium
9	Rokietnica	0.467	0.350	9	15	Medium
...	...	...	...	...	...	...
86	Mieleszyn	0.170	0.444	86	6	Very low
...	...	...	...	...	...	...
90	Wijewo	0.161	0.259	90	46	Very low
...	...	...	...	...	...	...
107	Lisków	0.114	0.252	107	48	Very low
...	...	...	...	...	...	...
115	Rzgów	0.045	0.120	115	113	Very low
116	Chocz	0.025	0.083	116	117	Very low
117	Grodziec	0.008	0.090	117	116	Very low
	Min	0.008	0.083			
	Max	0.861	0.521			
	Range	0.853	0.438			
	Coefficient of variation (%)	57.8	37.9			

Source: Own calculations based on data from the Central Statistical Office (*Local Data Bank*) and the Ministry of Finance (*Indicators for assessing the financial position of local government units*)
[a]Linear ordering of communes based on the values of synthetic measure obtained by: MP—positional approach, MK—classical approach

classification, which is manifested particularly at the application of the positional method. Observed differences are connected first of all with the adopted methods of measurement for the distances between objects and the positive and negative ideals of development in the TOPSIS method.

In comparison to the positional method, the classical TOPSIS method supplied a much lower range of variation in values of the synthetic measure, which as a result leads to a less substantive identification of development types of financial self-sufficiency in the analysed communes. Based on values of the synthetic measure obtained using the classical method, assuming the arbitrary method of identification of development types, only three typological classes may be distinguished, repre-

Table 4 Typological classification of rural communes in the Wielkopolska Province in terms of the level of financial self-sufficiency in 2013

Typology classes (level of financial self-sufficiency)	S_i	TOPSIS method—approach			
		Classical		Positional	
		N_c	%	N_c	%
I (very high)	$< 0.80, 1.00 >$	0	0.0	2	1.7
II (high)	$< 0.60, 0.80)$	0	0.0	3	2.6
III (medium)	$< 0.40, 0.60)$	6	5.1	14	12.0
IV (low)	$< 0.20, 0.40)$	71	60.7	57	48.7
V (very low)	$< 0.00, 0.20)$	40	34.2	41	35.0

N_c—the number of object in c-th classes ($c = 1, 2, \ldots, 5$)
Source: Own calculations based on data from the Central Statistical Office (*Local Data Bank*) and the Ministry of Finance (*Indicators for assessing the financial position of local government units*)

senting a medium, low and very low level of financial self-sufficiency of communes (Table 4). However, there is no ground to identify classes of communes characterised by very high and high levels of financial self-sufficiency, although it results from the conducted analyses that the rural communes located in the first ring around the city of Poznań (within the so-called Poznań agglomeration[7]), such as Suchy Las, Tarnowo Podgórne and Komorniki, in which their residential and service functions are rapidly gaining in importance (these communes are no longer typically rural, agricultural in character) should represent either a high (Suchy Las, Komorniki) or very high level of financial self-sufficiency (Tarnowo Podgórne).

Values of the synthetic measure provided using the classical TOPSIS method did not sufficiently reflect inter-class differences in the level of financial self-sufficiency of communes in the Wielkopolska Province. In view of the obtained range of variation in the synthetic measure most communes might be classified to the classes representing low and very low levels of financial self-sufficiency. It may be assumed that the reduction of the range of variation in synthetic measures values provided by the classical TOPSIS method, as well as the transfer of many communes to classes with a lower financial self-sufficiency level, resulted from the adoption of greater ranges of values for the positive and negative ideals of development.

The positional TOPSIS method is not affected by the outlying values of indicators or the adopted values of the positive and negative ideals of development, established in the set of all rural communes in the Wielkopolska Province. Based on the values of the synthetic measure in the positional approach, a much wider range of variation was obtained for values of the synthetic measure, thus it was possible to identify five typological classes, i.e. identification of rural communes with very high, high, medium, low and very low financial self-sufficiency (Table 4). The first type (I) was composed of the Przykona and Tarnowo Podgórne communes,

[7]The Poznań agglomeration comprising 17 communes (including the urban commune with county rights—the city of Poznań with over 500 thousand inhabitants) is economically the strongest region in the Wielkopolska Province.

Table 5 Values of homogeneity, heterogeneity and correctness indexes for the constructed types

	TOPSIS method—approach	
Specification	Classical	Positional
Homogeneity (H_0)	644.1	544.0
Heterogeneity (H_e)	5.1	8.5
Accuracy/correctness of classes (*PS*)	126.3	64.0

Source: Own calculations based on data from the Central Statistical Office (*Local Data Bank*) and the Ministry of Finance (*Indicators for assessing the financial position of local government units*)

while the second (II)—the communes of Suchy Las, Komorniki and Powidz. They are communes characterised by abundance of natural resources in the case of the Przykona commune, and mining industry, a military air base in the Powidz commune and the related recreation and tourist functions, while in the case of the other communes a higher socio-economic development, resulting from the advantageous location of these communes (communes neighbouring the city of Poznań representing the residential and service functions).

In turn, types IV and V comprised rural communes of typical agricultural character with low and very low levels of financial self-sufficiency. From both the substantive and statistical point of view, typological classes identified on this basis much better reflected inter-class differences in the level of financial self-sufficiency of rural communes in the Wielkopolska Province. This is confirmed by the analysis of values in indicators concerning the quality of classification (formulas (9), (10) and (11)) (Table 5). It may be stated that the classification based on values of the synthetic measure obtained using the positional approach is characterised by a greater reliability of classification in relation to the classification based on the classical method. This is evidenced by a greater homogeneity of the identified classes (the lower the value of H_0, the more homogeneous the classes) and a greater heterogeneity (the higher the value of H_e, the greater the inter-class differences).

4 Concluding Remarks

In the case when in a set of feature values characterising outliers, the application of classical methods may lead to excessive reduction of the variation range of values of the constructed synthetic measure of development, and as a result—to problems connected with the complete and accurate identification of development types of the investigated complex phenomenon. These problems result from the fact that at the application of an arbitrary method of identification of development types empty classes may appear. Conducted empirical studies showed that the positional approach in comparison to the classical approach provides a more accurate representation of inter-class differences in the level of development in the analysed phenomena, which is connected with a more comprehensive and more accurate identification of development types.

The application of the classical TOPSIS method to assess the level of financial self-sufficiency in the rural communes of the Wielkopolska Province, in view of the lesser variation in values of the synthetic measure, led to the identification of only typological classes (excluding the classes representing very high and high levels of development). In turn, in the case of the positional approach based on the Weber median, the obtained values of the synthetic measure covered a much greater part of the potential range from 0 to 1, and as a result they provided a more accurate identification of development types of financial self-sufficiency of the rural communes in the Wielkopolska Province. This is also evidenced by the identification of classes of communes with very high and high levels of financial self-sufficiency.

References

1. Bank Danych Lokalnych (Local Data Bank) http://www.stat.gov.pl/bdl,Cited20Mar2015
2. Hellwig, Z.: Zastosowania metody taksonomicznej do typologicznego podziału krajów ze wzgl edu na poziom ich rozwoju i strukture wykwalifikowanych kadr (Procedure to evaluating high level manpower data and typology of countries by means of the taxonomic method). Przeglad Statystyczny **4**, 307–327 (1968)
3. Hellwig, Z.: Procedure of evaluating high-level manpower data and typology of countries by means of taxonomic method. Towards a system of human resources indicators for less developed countries: papers prepared for a Unesco research project. Wrocław, Zakład Narodowy im. Ossolinskich, pp. 115–134 (1972)
4. Hwang, C.L., Yoon, K.: Multiple Attribute Decision-Making – Methods and Applications. Springer, Berlin (1981)
5. Lira, J., Wagner, W., Wysocki, F.: Mediana w zagadnieniach porzadkowania obiektów wielocechowych (The median in issues of ordering of multivariate objects). In: Paradysz, J., (ed.) Statystyka regionalna w służbie samorzadu terytorialnego i biznesu (Regional statistics in the service of local government and business), pp. 87–99. Akademia Ekonomiczna w Poznaniu, Poznań (2002)
6. Młodak, A.: Analiza taksonomiczna w statystyce regionalnej (Taxonomic Analysis in Regional Statistics). Difin, Warszawa (2006)
7. Rousseeuw, P.J., Leroy, A.M.: Robust Regression and Outlier Detection. Wiley, New York (1987)
8. Stewart, T.J.: A critical survey on the status of multiple criteria decision making theory and practise. OMEGA Int. J. Manag. Sci. **20**, 569–586 (1992)
9. Trussel, J.M., Patrick, P.A.: A predictive model of fiscal distress in local governments. J. Public Budg. Account. Financ. Manag. **21**(4), 578–616 (2009)
10. Vardi, Y., Zhang, C.H.: The multivariate L1-median and associated data depth. Proc. Natl. Acad. Sci. **97**(4), 1423–1426 (2000)
11. Wang, X., Dennis, L., Tu, Y.S.: Measuring financial condition: a study of U.S. states. Public Budg. Financ. **27**(2), 1–21 (2007)
12. Wskaźniki do oceny sytuacji finansowej jednostek samorzadu terytorialnego w latach 2011–2013 (Indicators to assessment the financial situation of local government), Ministerstwo Finansów, Warszawa http://www.finanse.mf.gov.pl/documents. Cited 17 Jun 2015
13. Wysocki, F.: Metody taksonomiczne w rozpoznawaniu typów ekonomicznych rolnictwa i obszarów wiejskich (The Methods of Taxonomy for Recognition of Economic Types in Agriculture of Rural Areas). Wyd. Uniwersytetu Przyrodniczego w Poznaniu, Poznań (2010)

A Method for Transforming Ordinal Variables

Odysseas Moschidis and Theodore Chadjipadelis

Abstract The similarity of individuals with respect to a number of ordinal variables is the main topic of this work. We consider the application of Multiple Correspondence Analysis (MCA) on k ordinal variables for N subjects. In the context of ordinary MCA, each variable is transformed into a suitable number of binary variables and the derived matrix is analyzed using the X^2 metric as the similarity measure. As a consequence, there is a loss of information from the original data, since ordinal variables are treated as nominal. In this paper, we propose a method for transforming the original variables, taking into account their ordinal nature. By applying the proposed method, a variable measured on m categories is transformed into a variable with n categories by assigning a probability to each category, instead of recoding each category into a new binary variable. We argue that the proposed transformation scheme leads to more accurate results than the one used in ordinary MCA.

1 Introduction

In several studies the subjects of the analysis (statistical units) are described by a mixture of quantitative and/or qualitative variables (nominal and/or ordinal). For the analysis of such datasets, several different strategies have been proposed. For instance, Becue-Bertaut and Pages [2] defined and combined suitable distances calculated between different groups of variables. Gower [4], Grabmeier and Rudolph [5] and Podani [12] proposed a series of distance or similarity coefficients in order to standardize the variables a posteriori.

In the case of ordinal variables, a common practice is first to transform an ordinal variable with m categories into a group of m binary variables and then apply CA to the Burt table or the indicator (0-1) matrix [3]. Lombardo and Beh [8] proposed

O. Moschidis
University of Macedonia, Egnatias str. 157, Thessaloniki, Greece
e-mail: fmos@uom.edu.gr

T. Chadjipadelis (✉)
Aristotle University of Thessaloniki, Aristotle University Campus, Thessaloniki, Greece
e-mail: chadji@polsci.auth.gr

F. Palumbo et al. (eds.), *Data Science*, Studies in Classification, Data Analysis, and Knowledge Organization, DOI 10.1007/978-3-319-55723-6_22

the use of Bivariate Moment Decomposition (BMD) instead of the Singular Value Decomposition (SVD) that is used in ordinary CA. McCullagh [9] defined a regression-based technique, under the assumption of an underlying continuous variable. In addition, Agresti and Kateri [1] proposed a logistic regression that is based on treating a qualitative variable as a set of binary variables.

The main idea in this work is to use the X^2 metric for the analysis of ordinal variables (or for common analysis of nominal and ordinal variables in the context of CA/MCA) by transforming ordinal variables that have been measured on a different scale, without the assumption of an underlying continuous variable. The transformation scheme is similar to the well-known z-score transformation of continuous variables, $(x_i - \bar{x})/s$, that leads to variables with a mean equal to 0 and a standard deviation equal to 1. In this paper we propose a standardization technique for ordinal variables. Issues arising in MCA due to the different number of categories can be solved by utilizing this approach.

Let us assume that an individual selects value 2 out of a 5-point ordinal scale. By selecting value 2, the individual accepts that value 5 is located further away than value 4, which is located further away than value 3. Accordingly, values 3 and 1 are the best alternatives to value 2.

This typical order of preference, stated by a value of 2, can be expressed in a mathematical way. We argue that the usual transformation of value 2 into a number of zeros and a value of 1, i.e., into the quintet $(0, 1, 0, 0, 0)$, as in the context of MCA, ignores the ordinal nature of the 5-point scale in question [11]. Furthermore, the X^2 metric cannot convey the similarity and may sometimes distort it, as shown in the following example (Table 1).

In this table we measure n individuals on a 3-point scale ordinal variable. Individuals a_2 and a_4, both selected value 2. Individual a_1 selected value 1, a_3 selected value 3, etc. In Table 1, k_1, k_2, and k_3 are the sums of the respective columns.

The X^2 metric is calculated as follows:

$$d_{x^2}^2(a_1, a_2) = \frac{n}{k_1} + \frac{n}{k_2},$$

$$d_{x^2}^2(a_1, a_3) = \frac{n}{k_1} + \frac{n}{k_3}.$$

Table 1 The 0-1 indicator matrix for a 3-point ordinal scale

	1	2	3
a_1	1	0	0
a_2	0	1	0
a_3	0	0	1
a_4	0	1	0
...	...	...	...
a_n	0	0	1
Weight	k_1/n	k_2/n	k_3/n

Assuming a case where k_1, k_2, and k_3 are almost equal, then

$$d_{\chi^2}^2(a_1, a_2) \approx d_{\chi^2}^2(a_1, a_3),$$

which means that individual a_1 resembles to a_3 as much as does to a_2, which does not depict reality.

Our aim, therefore, is to find an alternative transformation scheme for ordinal variables, which will more accurately describe the similarities among individuals in MCA. According to the proposed scheme, for instance, the value 2 in a 5-point scale will be transformed into the set of values $(0.167, 0.667, 0.111, 0.037, 0.019)$, which numerically expresses that the larger the distance from 2, the weaker the preference.

In order to perform such a transformation, we regard that each point of the scale has a total weight q equal to 1. By using the proposed algorithm we split the total weight q into a quintet of values that correspond to each of the 5 levels of the original variable. This way, the value selected by each individual can be classified accordingly.

Section 2 presents the algorithm in the general case of an m-point scale. Section 3 describes a software macro for applying the proposed algorithm. An application to a real dataset is provided in Sect. 4. Some results and remarks for the proposed transformation are given in Sect. 5.

2 Transformation of the Values of an m-Point Ordinal Scale into an n-Tuple of $\mathbb{R}^n$

In this section we present the transformation scheme of each one of the values $l = 1, 2, \ldots, m$ of an m-point scale into an n-tuple of the vector space $\mathbb{R}^n$. The corresponding n-tuple, $(y_{l1}, y_{l2}, \ldots, y_{ln})$ for each value l, has a sum of 1.

If $1, 2, \ldots, m$ are the values of the m-point scale, then dividing the interval $\left(\frac{1}{2}, m + \frac{1}{2}\right)$ into n equal intervals, these points are considered as the midpoints of the m-intervals

$$\left(\frac{1}{2}, \frac{1}{2} + \frac{m}{n}\right), \left(\frac{1}{2} + \frac{m}{n}, \frac{1}{2} + 2\frac{m}{n}\right), \ldots, \left(\frac{1}{2} + (n-1)\frac{m}{n}, \frac{1}{2} + n\frac{m}{n}\right),$$

respectively.

The $n + 1$ bounds of these intervals are $\{B_0, B_1, \ldots, B_n\} = \left\{\frac{1}{2}, \frac{1}{2} + \frac{m}{n}, \frac{1}{2} + \frac{2m}{n}, \ldots, \frac{nm}{n}\right\}$ and the midpoints of these intervals are

$$\{A_1, A_2, \ldots, A_n\} = \left\{\frac{1}{2} + \frac{m}{2n}, \left(\frac{1}{2} + \frac{m}{n}\right) + \frac{m}{2n}, \ldots, \left(\frac{1}{2} + (n-1)\frac{m}{n}\right) + \frac{m}{2n}\right\}.$$

Our aim, as stated above, is to transform every value of an m-point scale to an n-tuple of $\mathbb{R}^n$, with a sum equal to 1. We assume that every value of the m-point scale is associated with a mass q with a weight equal to 1. This mass $q = 1$ is decomposed into n parts, which are attributed to the n coordinates of the n-tuple in question. The decomposition process is presented below.

It is well known that point x of the center of gravity of masses m_1 and m_2, which are at points x_1, x_2, respectively, is given by

$$x = \frac{m_1 x_1 + m_2 x_2}{m_1 + m_2} \tag{1}$$

Consider the total mass q, $q = m_1 + m_2$. Then

$$m_1 = \frac{x_2 - x}{x_2 - x_1} q. \tag{2}$$

As we mentioned above there is a total mass q associated with every value l or the m-point scale. If $q = 1$ and $x = l$, then

$$m_1 = \frac{x_2 - l}{x_2 - x_1}. \tag{3}$$

Concerning the decomposition of mass $q = 1$ of value l for an m-point scale into an n-tuple of $\mathbb{R}^n$. Here we can distinguish two cases, A and B:

A. Value l lies in the first or the last interval.

We assume that l belongs to the first out of the n intervals. At first, mass $q = 1$ decomposes into masses m_1 and m_2 to the bounds B_0 and B_1, i.e., to points $x_1 = \frac{1}{2}$ and $x_2 = \frac{1}{2} + \frac{m}{n}$, respectively.

From Eq. (2)we get:

$$m_1 = \frac{\left(\frac{1}{2} + \frac{m}{n}\right) - l}{\frac{1}{2} + \frac{m}{n} - \frac{1}{2}} = \frac{\frac{1}{2} - l + \frac{m}{n}}{\frac{m}{n}} = 1 + \left(\frac{1}{2} - l\right)\frac{n}{m} \tag{4}$$

Consequently,

$$m_2 = 1 - m_1 = 1 - \left(1 + \left(\frac{1}{2} - l\right)\frac{n}{m}\right) = \left(l - \frac{1}{2}\right)\frac{n}{m} \tag{5}$$

The mass $m_2 = \left(l - \frac{1}{2}\right)\frac{n}{m}$, which lies in B_1, the upper bound of the first interval, decomposes in m_1' at A_1 and $m_2 - m_1'$ at B_2. Since A_1 is the midpoint of the first interval, B_1 is the upper bound of the first interval, and B_2 the upper bound of the second interval, the ratio of the distances B_2B_1 and B_2A_1 is equal to $\frac{2}{3}$. Thus,

$$m_1' = \left(\frac{B_2B_1}{B_2A_1}\right) m_2 = \frac{2}{3} m_2 = \frac{2}{3}\left(l - \frac{1}{2}\right)\frac{n}{m}$$

Then the total mass in the first interval (i.e., the first component y_{l1} of the n-tuple) is equal to the sum

$$m_1 + m_1' = 1 + \left(\frac{1}{2} - l\right)\frac{n}{m} + \frac{2}{3}\left(l - \frac{1}{2}\right)\frac{n}{m} = 1 + \frac{1}{3}\left(\frac{1}{2} - l\right)\frac{n}{m}.$$

The remaining mass at B_2 is equal to $\frac{1}{3}m_2$. According to the algorithm, it is decomposed at points B_3 and A_2. Since A_2 is the midpoint of the second interval and B_3 is the upper bound of the third interval, the ratio of the distances from B_2 to A_2 and from B_2 to B_3 is $\frac{1}{2}$. Therefore,

$$y_{l2} = \frac{2}{3}\left(\frac{1}{3}m_2\right) = \frac{2}{3^2}\left(l - \frac{1}{2}\right)\frac{n}{m}.$$

Similarly, the remaining mass $\frac{1}{3}\left(\frac{1}{3}m_2\right)$ which lies at point B_3 decomposes into two parts, at points A_3 and B_4.

By using the decomposition process we get $y_{li} = \frac{2}{3^i}\left(l - \frac{1}{2}\right)\frac{n}{m}$, for $i = 2, 3, \ldots, n-1$. Accordingly, the n-component is

$$y_{ln} = \frac{1}{3}\frac{1}{3^{n-2}}(l-1)\frac{n}{m} = \frac{1}{3^{n-1}}\left(l - \frac{1}{2}\right)\frac{n}{m}.$$

Remark If the value l belongs to the last interval, it is trivial to show that, due to symmetry, the corresponding n-tuple is equal to $\left(y_{ln}, y_{l(n-1)}, \ldots, y_{l1}\right)$ and can be computed beginning from the last interval.

B. Value l lies in a middle interval.

Let us assume that l lies in the $p+1$ interval. Then it holds:

$$B_p = \frac{1}{2} + p\frac{m}{n} < l < \frac{1}{2} + (p+1)\frac{m}{n} = B_{p+1}.$$

Mass $q = 1$, which lies at point l, decomposes into m_1 and m_2 to bounds B_p and B_{p+1}. It holds from Eq. (3), that:

$$m_1 = \frac{\left[\frac{1}{2} + (p+1)\frac{m}{n}\right] - l}{\left[\frac{1}{2} + (p+1)\frac{m}{n}\right]\left(\frac{1}{2} + p\frac{m}{n}\right)} = \left[\left[\frac{1}{2} + (p+1)\frac{m}{n}\right] - l\right]\frac{n}{m} = \left(\frac{1}{2} - l\right)\frac{n}{m} + (p+1)$$

$$m_2 = 1 - m_1 = \left(l - \frac{1}{2}\right)\frac{n}{m} - p.$$

Mass m_1 in point B_p decomposes into a mass $\frac{1}{3}m_1$ to point B_{p-1} and into a mass $\frac{2}{3}m_1$ to point A_{p+1}. In a similar way, mass m_2 in point B_{p+1} decomposes into $\frac{2}{3}m_2$ in A_{p+1} and $\frac{1}{3}m_2$ in B_{p+2}.

The total mass in point A_{p+1} (the midpoint of the $p+1$ interval) equals to $\frac{2}{3}m_1 + \frac{2}{3}m_2 = \frac{2}{3}(m_1 + m_2) = \frac{2}{3}$. This quantity is the $p+1$ component of the corresponding n-tuple, i.e., $y_{l(p+1)} = \frac{2}{3}$. The other components are computed exactly as in case A, where l belongs to the first interval. For example, component $y_{l(p+i)}$ is equal to $\frac{2}{3^{i-1}}\left(\frac{1}{3}m_2\right)$, while component y_{ln} is equal to $\frac{1}{3^{n-1}}\left(\left(l-\frac{1}{2}\right)\frac{n}{m}-p\right)$.

3 Suggested Software

In order to utilize the proposed algorithm, we created a macro, which is incorporated into Microsoft Excel. The macro and specific instructions for its use are available in the following website: http://www.polres.gr/dias.html.

Example Transformation of each value l of the variable E, measured on a 7-point scale into (Table 2)

(a) a 3-tuple: E_{l1}, E_{l2}, E_{l3}
(b) a 5-tuple: $E_{l1}, E_{l2}, E_{l3}, E_{l4}, E_{l5}$
(c) a 7-tuple: $E_{l1}, E_{l2}, E_{l3}, E_{l4}, E_{l5}, E_{l6}, E_{l7}$.

Table 2 Results of the transformation of a 7-point scale into a 3-tuple, a 5-tuple, and a 7-tuple

l	E_{l1}	E_{l2}	E_{l3}	E_{l1}	E_{l2}	E_{l3}	E_{l4}	E_{l5}	E_{l1}	E_{l2}	E_{l3}	E_{l4}	E_{l5}	E_{l6}	E_{l7}
1	0.929	0.048	0.024	0.881	0.079	0.026	0.009	0.004	0.833	0.111	0.037	0.012	0.004	0.001	7E-04
2	0.786	0.143	0.071	0.310	0.667	0.016	0.005	0.003	0.167	0.667	0.111	0.037	0.012	0.004	0.002
3	0.31	0.667	0.024	0.071	0.667	0.175	0.058	0.029	0.056	0.111	0.667	0.111	0.037	0.012	0.006
4	0.167	0.667	0.167	0.056	0.111	0.667	0.111	0.056	0.019	0.037	0.111	0.667	0.111	0.037	0.019
5	0.024	0.667	0.31	0.029	0.058	0.175	0.667	0.071	0.006	0.012	0.037	0.111	0.667	0.111	0.056
6	0.071	0.143	0.786	0.003	0.005	0.016	0.667	0.31	0.002	0.004	0.012	0.037	0.111	0.667	0.167
7	0.024	0.048	0.929	0.004	0.009	0.026	0.079	0.881	7E04	0.001	0.004	0.012	0.037	0.111	0.833

4 Application

A total of 386 students of a Business School were asked to evaluate some aspects and characteristics of their academic environment on a 5-point scale, from 1 = strongly disagree to 5 = strongly agree. Among others, they were asked to express their level of agreement with the following statements:

Θ_1: The curriculum helps me to develop creative thinking about business.
Θ_2: I get sufficient information about market functioning through the curriculum.
Z_1: The current economic environment in Greece promotes Entrepreneurship.

The data for the three aforementioned variables were transformed to a 0-1 matrix and also, using the proposed algorithm, to a 5-tuple matrix. The data were first analyzed by ordinary MCA of the 0-1 matrix and subsequently by using the 5-tuple matrix. The raw data, as well as the two matrices, are available in the following website: http://www.polres.gr/dias.html.

For each analysis we present the respective inertia decomposition table (Tables 3 and 4) and the first factorial plane (Figs. 1 and 2).

The total inertia of the analysis of the 0-1 matrix is equal to 4.00 (the total inertia is equal to the number of levels (15) divided by the number of variables (3), minus 1), which is greater than the total inertia in the case of the 5-tuple analysis (1.42). However, a total of 30.16% of the inertia is explained in the first factorial plane using the 5-tuple matrix, which is significantly higher than the 22.18% explained in the case of the 0-1 matrix. Also from the two planes shown in Figs. 1 and 2, it can be deduced that the ordinal nature of the variables is preserved. In the analysis of the 5-tuple matrix, if we project the variable points on the first axis, their relative positions correspond to those of the initial ordinal scale. On the contrary, this does not hold for the analysis of the 0-1 matrix.

Table 3 Inertia decomposition for the 0-1 matrix

Axis	Inertia	Relative inertia (%)	Cumulative inertia
1	0.465	11.62	11.62
2	0.422	10.55	22.18
3	0.394	9.86	32.03
...	...	...	...
12	0.215	5.38	100.00

Table 4 Inertia decomposition for the 5-tuple matrix

Axis	Inertia	Relative inertia (%)	Cumulative inertia
1	0.235	16.63	16.63
2	0.191	13.53	30.16
3	0.169	12.01	42.16
...	...	...	...
12	0.033	2.38	100.00

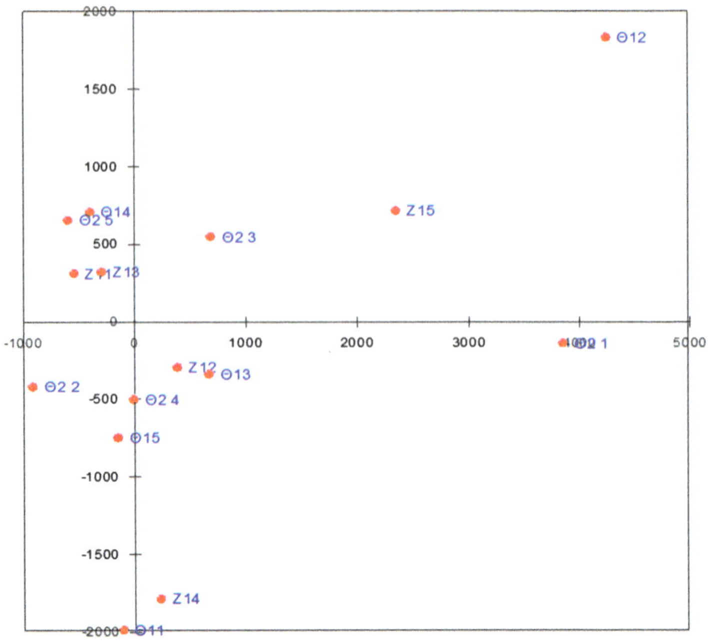

Fig. 1 MCA first factorial plane for the 0-1 matrix

5 Results and Remarks for the Proposed Transformation

Assuming that we measure values $1, 2, 3$ for individuals a_1, a_2, a_3 on a 3-point scale, respectively, by using the proposed transformation for $n = 3$ we get the 3-tuples shown in Table 5:

Computing the Euclidean metric $d_e^2(a_i, a_j)$ for every pair (a_i, a_j), $i \neq j = 1, 2, 3$ we get $d_e^2(a_1, a_2) = 0.75$, $d_e^2(a_1, a_3) = 1.19$, $d_e^2(a_2, a_3) = 0.75$. Therefore it is obvious that individual a_1 resembles more to a_2, than to a_3, which totally depicts reality.

The computed X^2 metric in the quota table, assuming that the column weights are equal, expresses the similarity of the individuals, as the Euclidean metric does. In the other case, the computed X^2 metric combines the order of values with the column weights.

It is well known that the usual 0-1 coding of the values of an m-point scale is rough and it ignores the ordinal nature of the scale. On the other hand, the suggested transformation preserves the ordinal nature of the scale and minimizes the loss of information. The suggested transformation could be seen as a generalization of the 0-1 coding in the case of ordinal data.

It is well known that the more the values of a scale, the more accurate is the estimation we get. Suppose a respondent expresses his/her view choosing the

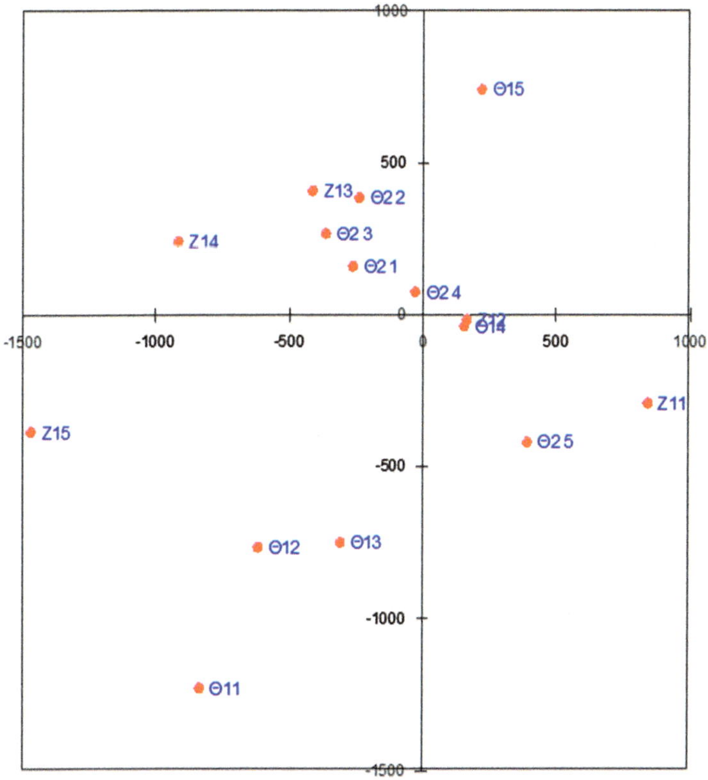

Fig. 2 MCA first factorial plane for the 5-tuple matrix

Table 5 Results for $m = 3$, $n = 3$

	1	2	3
a_1	0.83	0.11	0.06
a_2	0.17	0.66	0.17
a_3	0.06	0.11	0.83

highest value of a scale. Then applying the algorithm to the same variable measured on m_1 and m_2 scales, where $m_1 > m_2$ we get that $y_{m_1 n} > y_{m_2 n}$ which means that the proposed algorithm preserves the information of the scale. For example:

for $m = 7, n = 3$, value 7 is transformed to: $(0.02, 0.05, \mathbf{0.93})$.
for $m = 5, n = 3$, value 5 is transformed to: $(0.02, 0.08, \mathbf{0.90})$.
for $m = 3, n = 3$, value 3 is transformed to: $(0.06, 0.12, \mathbf{0.82})$.

Therefore, $y_{73} > y_{53} > y_{33}$.

Additionally, the suggested transformation allows the application of Multiple Correspondence Analysis also in the case where there are values with zero weight, since by this transformation the weight of every column is different from zero.

By using this method, we essentially propose a different solution by MCA in the case where there are variables measured on an ordinal scale but with different number of categories [6, 7, 10].

Finally, the proposed method could be applied in the case there is a subset of variables measured on an ordinal scale and the remaining variables are measured on a nominal scale. Transforming the ordinal variables, as we propose, and expressing the nominal ones by the usual 0-1 scheme, we could analyze the derived mixed matrix of transformed values and 0-1s.

Acknowledgements The authors would like to thank the anonymous reviewers for their valuable comments and suggestions.

References

1. Agresti, A., Kateri, M.: Categorical Data Analysis. Springer, Berlin, Heidelberg (2011)
2. Becue-Bertaut, M., Pages, J.: Multiple factor analysis and clustering of a mixture of quantitative, categorical and frequency data. Comput. Stat. Data Anal. **52**, 3255–3268 (2008)
3. Benzècri, J.P.: L'analyse des donnees. 2. L'analyse des correspondances. Dunod, Paris (1973)
4. Gower, J.C.: A General coefficient of similarity and some of its properties. Biometrics **27**, 857–871 (1971)
5. Grabmeier, J., Rudolph, A.: Techniques of cluster algorithms in data mining. Data Min. Knowl. Disc. **6**, 303–360 (2002)
6. Le Roux, B.: Analyse specifique d'un nuage euclidien: application a l'etude des questionnaires. Math. Sci. Hum. **146**, 65–83 (1999)
7. Le Roux, B.: Geometric Data Analysis: From Correspondence Analysis to Structured Data Analysis. Springer, New York (2004)
8. Lombardo, R., Beh, E.J.: Simple and multiple correspondence analysis for ordinal-scale variables using orthogonal polynomials. J. Appl. Stat. **37**, 2101–2116 (2010)
9. McCullagh, P.: Regression models for ordinal data. J. R. Stat. Soc. Ser. B Methodol. **42**, 109–142 (1980)
10. Moschidis, O.E.: A different approach to multiple correspondence analysis (MCA) than that of specific MCA. Math. Sci. Hum. Math. Soc. Sci. **186**, 77–88 (2009)
11. Moschidis, O.: Unified coding of qualitative and quantitative variables and their analysis with ascendant hierarchical classification. Int. J. Data Anal. Tech. Strateg. **7**, 114–128 (2015)
12. Podani, J.: Extending Gower's general coefficient of similarity to ordinal characters. Taxon **48**, 331–340 (1999)

Big Data Scaling Through Metric Mapping: Exploiting the Remarkable Simplicity of Very High Dimensional Spaces Using Correspondence Analysis

Fionn Murtagh

Abstract We present new findings in regard to data analysis in very high dimensional spaces. We use dimensionalities up to around one million. A particular benefit of Correspondence Analysis is its suitability for carrying out an orthonormal mapping, or scaling, of power law distributed data. Power law distributed data are found in many domains. Correspondence factor analysis provides a latent semantic or principal axes mapping. Our experiments use data from digital chemistry and finance, and other statistically generated data.

1 Introduction

We present new findings in regard to data analysis in very high dimensional spaces. We use dimensionalities up to around one million. A particular benefit of Correspondence Analysis is its suitability for carrying out an orthonormal mapping, or scaling, of power law distributed data. Power law distributed data are found in many domains. Correspondence factor analysis provides a latent semantic or principal axes mapping. Our experiments use data from digital chemistry and finance, and other statistically generated data.

Correspondence Analysis of an infinite (unbounded) number of rows or observations, crossed by 1000 attributes, was discussed in [2] and also in [3]. Our objective in this article is to describe useful properties of data spaces, of high dimensionality. Our particular interest is in properties that are of benefit to "big data" analytics. See [12] for further examples of application.

It was shown experimentally in [8] how points in high dimensional spaces become increasingly equidistant with increase in dimensionality. Both [5] and [6] study Gaussian clouds in very high dimensions. The former finds that "not only

F. Murtagh
University of Derby, Derby, UK

Goldsmiths University of London, London, UK
e-mail: fmurtagh@acm.org

F. Palumbo et al. (eds.), *Data Science*, Studies in Classification, Data Analysis, and Knowledge Organization, DOI 10.1007/978-3-319-55723-6_23

are the points [of a Gaussian cloud in very high dimensional space] on the convex hull, but all reasonable-sized subsets span faces of the convex hull. This is wildly different than the behavior that would be expected by traditional low-dimensional thinking."

That very simple structures come about in very high dimensions can have far-reaching implications. Firstly, even very simple structures (hence with many symmetries) can be used to support fast, and perhaps even constant time worst case, proximity search [8]. Secondly, as shown in the machine learning framework by [6], there are important implications ensuing from the simple high dimensional structures. Thirdly, [9] shows that very high dimensional clustered data contain symmetries that in fact can be exploited to "read off" the clusters in a computationally efficient way. Fourthly, following [4], what we might want to look for in contexts of considerable symmetry are the "impurities" or small irregularities that detract from the overall dominant picture.

In general, data analysis considered as the search for symmetries in data is discussed in [10]. This relates in particular to hierarchical clustering. That can be considered as a natural extension of the work described in this paper.

2 Properties of Very High Dimensional Data Spaces

2.1 *Piling and Concentration of Data, with Increase in Dimensionality*

With high dimensional, sparse data [6], there is a very strong concentration of our clouds (rows/points, columns/projection vectors) into concentrated (i.e., small variance) Gaussians. Therefore, there is a good approximation of our cloud by its mean. This in turn means that the mean random projection is a very good representative of our data cloud.

From the given, non-negative valued data, k_{IJ}, our I cloud and J cloud are converted to frequencies, denoted f_{IJ} with associated mass distributions, f_I and f_J.

We use the following notation. Our given data table values are k_{ij} where $i \in I, j \in J$, for row i and column j. The complete data table is denoted k_{IJ}. Now, the row sum $k_i = \sum_{j \in J} k_{ij}$, and analogously for the column sum, k_j. Next, we form the frequencies as follows: $f_{IJ} = k_{IJ}/k$ where the denominator, $k = \sum_{i \in I} \sum_{j \in J} k_{ij}$. In a similar manner, the row and column marginal distributions, taking the frequencies as empirical probabilities, are as follows: $f_I = k_I/k$ and $f_J = k_J/k$.

The conditional distribution of f_J knowing $i \in I$, also termed the jth profile with coordinates indexed by the elements of I, is

$$f_J^i = \{f_j^i = f_{ij}/f_i = (k_{ij}/k)/(k_i/k); f_i \neq 0; j \in J\}$$

and analogously for f_I^j. Thus our data are points in a high dimensional data cloud, defining row or column profiles.

Analysis is carried out on row profiles, i.e., vector f_J^i, with associated mass f_i. The dual space of column profiles is analogous to this.

Through high dimensional piling, i.e., concentration, we have that the profile vectors tend towards the average profile. What gives rise to this is sparsity through high dimensionality, which also implies low sample (or population) size. It implies this because we are not considering here the case of both population size and dimensionality tending to infinity at the same, or related, rate.

By the central limit theorem, and by the concentration (data piling) effect of high dimensions [6, 15], we have as dimension $m \rightarrow \infty$: pairwise distances become equidistant; orientation tends to be uniformly distributed. We find also: the norms of the target space axes are Gaussian distributed; and as typifies sparsified data, the norms of the points in our high dimensional data cloud, in the factor space, are distributed as a negative exponential or a power law.

2.2 *Relative and Absolute Contributions*

The moment of inertia of the clouds $N_J(I)$ and $N_I(J)$, relative to the α axis, is λ_α. Let ρ be the Euclidean distance from the cloud center in the factor space, and let the projection of $i \in I$ on the α factor be $F_\alpha(i)$. Decomposition of the cloud's inertia is then as follows:

$$M^2(N_J(I)) = \sum_{\alpha=1..\nu} \lambda_\alpha = \sum_{i \in I} f_i \rho^2(i) \quad (1)$$

The maximum number of non-zero factors is $\nu = \min\{|I| - 1, |J| - 1\}$ where $|I|, |J|$ are, respectively, the cardinalities of the observations or row set, and the cardinality of the attributes or column set. In greater detail, we have for this decomposition:

$$\lambda_\alpha = \sum_{i \in I} f_i F_\alpha^2(i) \text{ and } \rho^2(i) = \sum_{\alpha=1..\nu} F_\alpha^2(i) \quad (2)$$

Contributions to inertia are fundamental in order to define the mapping into the factor space. Contributions by row points, or by column points, in their respective dual spaces, define the importance of those given data elements for the constructed mapping. Supported by the experimental results to be reported on in the following sections, we will use the average contribution to the inertia as a measure of cloud concentration. The inertia is the fundamental determinant of not just relative positioning, but of essential cloud properties.

Why we use the contributions to the total inertia of the cloud, as the basis for a measure of concentration is motivated for the following reason. Consider the following hypothetical scenario. Consider where massive points in the cloud were moved towards the center or origin, leaving light points to drift away from the center. Through inertia, we would characterize such a scenario as concentration. Or consider where massive points drift apart, and their inertia contributions outpace the inertia contributions of less massive points that move closer to the origin. Again in that scenario, our inertia measure of concentration would be appropriate for quantifying the lack of concentration. In these hypothetical scenarios, we see how contribution to inertia is a key consideration for us. Inertia is more important than projection (i.e., position) per se.

We now look at absolute versus relative contributions to inertia. The former one of these is the more relevant for us. This will be seen in our experiments below. What we consider for the attributes (measurements, dimensions) holds analogously for the observations.

- $f_j \rho^2(j)$ is the absolute contribution of attribute j to the inertia of the cloud, $M^2(N_I(J))$. Therefore, from expressions (2), this absolute contribution of point j is also: $f_j \sum_{\alpha=1..\nu} F_\alpha^2(j)$.
- $f_j F_\alpha^2(j)$ is the absolute contribution of point j to the moment of inertia λ_α.
- $f_j F_\alpha^2(j)/\lambda_\alpha$ is the relative contribution of point j to the moment of inertia λ_α. Analogous to expression (1), we have that $\lambda_\alpha = \sum_{j \in J} f_j F_\alpha^2(j)$. So the relative contribution of point j to the moment of inertia λ_α is: $f_j F_\alpha^2(j)/ \sum_{j \in J} f_j F_\alpha^2(j)$. The total relative contribution of j, over all $j \in J$, is 1. The total contribution over all factors, indexed by α, then becomes ν, the number of factors. So the mean contribution (here, the mean relative contribution) of the attributes is $\frac{\nu}{|J|}$. In the evaluations below, the trivial first eigenvalue, and associated axis, is included here.

We have now the technical machinery needed to evaluate data clouds in very high dimensions. We will keep our cloud of observables, small. This is $N(I)$. It is in a $|J|$-dimensional space. That dimensionality, $|J|$, will be very large. That is to say, the cloud of what we take as attributes, $N(J)$, will be huge. While the cloud itself, $N(J)$, is huge, each point in that cloud $j \in J$ is in a space of dimension $|I|$, which is not large.

Now we will carry out our evaluations. Our choice of cloud cardinality and dimensionality are motivated by inter-study comparison. The R code used is available at the web site, www.correspondances.info.

3 Evaluation 1: Uniformly Distributed Points in Data Clouds of Dimensionality Up To One Million

Uniformly distributed values, in [0, 1], were used for five data clouds, each of 86 points in dimensionalities of: 100, 1000, 10,000, 100,000, and 1,000,000. In the usual analysis perspective, we have 86 observations, and the dimensionalities are

Table 1 Five data clouds, each of 86 points in spaces of dimensionality: 100, 1000, 10,000, 100,000, and 1,000,000

Contributions to inertia of factors by the columns				
Dim.	Contributions	Mean	Std. dev.	Median
100	Absolute	0.01322144	0.0005623589	0.01325343
	Relative	0.86	0.04588791	0.869127
1000	Absolute	0.001331763	5.440168e−05	0.001333466
	Relative	0.086	0.009729907	0.08547353
10,000	Absolute	0.0001332053	5.279421e−06	0.0001332981
	Relative	0.0086	0.0009742588	0.008577748
100,000	Absolute	1.330499e−05	5.269165e−07	1.332146e-05
	Relative	0.00086	9.783086e−05	0.0008574684
1,000,000	Absolute	1.330706e−06	5.278487e−08	1.332186e-06
	Relative	8.6e−05	9.788593e−06	8.576992e−05
Maximum factor projection				
Dim.	Projection			
100	0.3590788			
1000	0.2777193			
10,000	0.2799913			
100,000	0.3678137			
1,000,000	0.3750852			

The original coordinate values are randomly uniform in [0, 1]

associated with the attributes or features. This input data is therefore dense in value. Results obtained are shown in Table 1.

Note how increasing dimensionality implies the following. We base concentration, or compactness, on the absolute contribution to the inertia of the factors. The average absolute contribution to the factors tends towards zero. The standard deviation also approaches zero. Thus the cloud becomes more compact.

We provide median as well as mean as an indication of distributional characteristics of the absolute contribution that we are examining. We observe a relatively close match between mean and median values, implying an approximate Gaussian distribution of the absolute contributions. For all cases (including the 1,000,000-dimensional case), we checked that the distributions of absolute and relative contributions, and norms squared of the input data, are, visually, close to Gaussian.

The maximum projection values, that do not decrease, serve to show that concentration with increasing dimensionality is a phenomenon relating to the whole cloud, and therefore to the average (or median).

3.1 Computational Time Requirements

The machine used in this work was a MacBook Air, with a 2 GHz processor, and 8 GB of memory, running OS X version 10.9.4. The version of R in use was 2.15.2.

The largest, uniformly random generated, data set used was of dimensions 86 × 1000000. In order to create this data array, an elapsed time of 82.8 s was required. Carrying out the main processing, furnishing the results in Table 1, involved a basic Correspondence Analysis of this input data matrix. The projections and contributions (to inertia) of the 86 points were to be determined.

Standard processing proved satisfactory for these evaluations. For this large data set, our main processing took an elapsed time of 95.6 s.

4 Evaluation 2: Time Series of Financial Futures in Varying Embedding Dimensions

The following data were used in [9]. In that work we used the sliding window approach to embed the financial signal in spaces of varying dimensionality. The work in [9] showed, in various examples, how there may be no "curse of dimensionality," in Belman's [1] famous phrase, in very high dimensional spaces. There is no such obstacle if we seek out, and make use of, the "remarkable simplicity" [9] of very high dimensional data clouds.

We use financial futures, from circa March 2007, denominated in euros from the DAX exchange. Our data stream, at the millisecond rate, comprised 382,860 records. Each record includes: 5 bid and 5 asking prices, together with bid and asking sizes in all cases, and action.

We extracted one symbol (commodity) with 95,011 single bid values, on which we now report results. These values were continuous and avoided missing values. The data values were between 6788 and 6859.5 in value. There were either integer valued, or ending in 0.5. Very often this signal contained short sequences of successive identical values.

Similar to [9], we define embeddings of this financial signal as follows. Each embedding begins at the following time steps in the financial signal: 1, 1000, 2000, …, 85,000. The lengths of the successive embeddings were, in our three case studies: 100, 1000, 10,000. That provided matrices, in these three case studies, of sizes: $86 \times 100, 86 \times 1000, 86 \times 10000$.

Results obtained are presented in Table 2. The histograms of projections on the factors were visually observed to be Gaussian-distributed. We observe how the mean absolute contribution, as well as the median absolute contribution, decreases as the embedding dimensionality increases. The standard deviation of absolute and of relative contributions decrease too, indicating the increasing concentration. Our measure of concentration is the average (or median) contribution by the embedding dimensionality values (what we may consider as attributes or characterizing features

Table 2 Embeddings, of dimensionalities 100, 1000, and 10,000, for a financial time series

Dim.	Contribution	Mean	Std. dev.	Median
100	Absolute	0.01	9.260615e−08	0.01000002
	Relative	0.86	0.05399462	0.8672608
1000	Absolute	0.001	3.297399e−08	0.001000008
	Relative	0.086	0.0121773	0.08518253
10,000	Absolute	0.0001000001	2.168381e−08	9.999872e−05
	Relative	0.0086	0.001159708	0.008477465
Maximum factor projection				
Dim.	Projection			
100	0.0001054615			
1000	0.0002979516			
10,000	0.0008869227			

of the "sliding window" over the signal) to the inertia of the factors. We observe also how the maximum projection on the factors does not decrease. This just means that the cloud in the overall sense, and on the whole, gets increasingly compact or concentrated, as the attribute dimensionality increases.

5 Evaluation 3: Chemistry Data, Description of Its Power Law Property

5.1 Data and Determining Power Law Properties

The following data were used in our earlier work in [14]. We used a set of 1,219,553 chemical structures coded through 1052 presence/absence values, using the Digital Chemistry bci1052 dictionary of fragments [16]. That binary-valued matrix was sparse: occupancy (i.e., presence = 1 values) of the chemicals crossed by attribute values was 8.6%.

Our motivation here is to investigate the effect of greatly increasing the attribute dimension. In the next section we will develop a novel way to do this. In this section we determine the relevant statistical properties of our data.

Here, we will use 425 chemicals from this set, in 1052-dimensional space. We took 425 chemicals in order to have a limited set, $|I| = 425$, in the attribute space, J. Each chemical had therefore presence/absence (i.e., 1 or 0, respectively) values on $|J| = 1052$ attributes. The occupancy of the 425×1052 data set used was 5.9%. Since we wanted this sample of 425 of the chemicals to be representative of the larger set from which they came, we now look at the most important distributional properties.

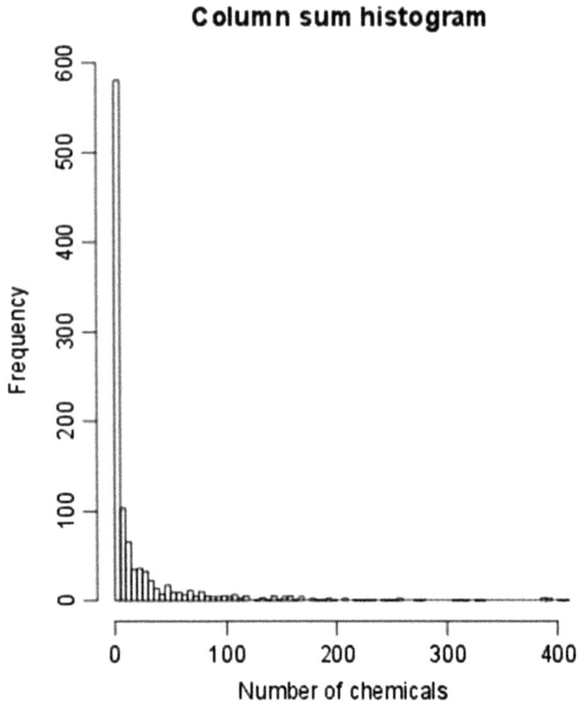

Fig. 1 Histogram of column, i.e., chemical attribute, sums

A power law (see [7]) is a frequency of occurrence distribution of the general form $x^{-\alpha}$ where constant $\alpha > 0$; whereas an exponential law is of the form e^{-x}. For a power law, the probability that a value, following the distribution, is greater than a fixed value is as follows: $P(x > x_0) \sim cx^{-\alpha}$, $c, \alpha > 0$. A power law has heavier tails than an exponential distribution. In practice, $0 \leq \alpha \leq 3$. For such values, x has infinite (i.e., arbitrarily large) variance; and if $\alpha \leq 1$, then the mean of x is infinite. The density function of a power law is $f(x) = \alpha c x^{-\alpha-1}$, and so $\ln f(x) = -\alpha \ln x + C$, where C is a constant offset. Hence a log–log plot shows a power law as linear. Power laws have been of great importance for modeling networks and other complex data sets.

The marginal distribution, shown in Fig. 1, is not unlike the marginal distribution displayed in [14]. In that previous work, we found the power law distribution of the chemical attributes to be of exponent -1.23. Let us look at the power law of the baseline distribution function used here, i.e., relating to the 425 chemicals.

Figure 2 shows a log–log plot based on the 1052 presence/absence attributes, using the 425 chemicals. In a very similar way to the power law properties of large networks (or file sizes, etc.) we find an approximately linear regime, ending (at the lower right) in a large fan-out region. The slope of the linear region characterizes

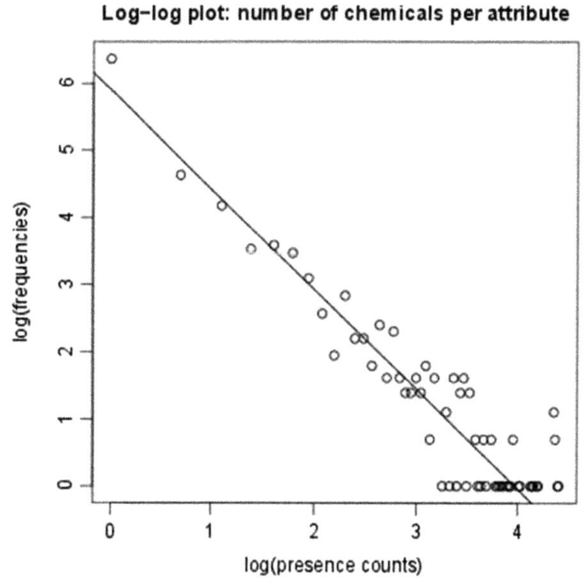

Fig. 2 Log–log plot of numbers of chemicals per attribute, based on the data set of 425 chemicals

the power law. For this data, we find that the probability of having more than n chemicals per attribute to be approximately $c/n^{1.49}$ for large n.

The histogram of attributes per chemical, on the other hand, is approximately a Gaussian. This is as observed in [14].

5.2 *Randomly Generating Power Law Distributed Data in Varying Embedding Dimensions*

In Sect. 3 we used dense uniformly distributed data. In Sect. 4, our financial futures were slow-moving, in the sense of small variation between successive values. But there too the data were dense and real-valued. Our chemistry context is sparse and boolean-valued (for presence/absence). We use this context to generate data that keep the property of the attributes (i.e., the columns or dimensions) following a power law in regard to their distribution.

To generate new random data sets that fully respect the distributional characteristics of our known data, we will use the distribution function that is displayed in Fig. 1. This is the data distribution of coding attributes that characterize the chemicals, i.e., presence of molecules.

In line with our earlier notation, the marginal distribution in Fig. 1 is f_J for attribute set, J. The chemicals set is I. The presence/absence cross-tabulation of

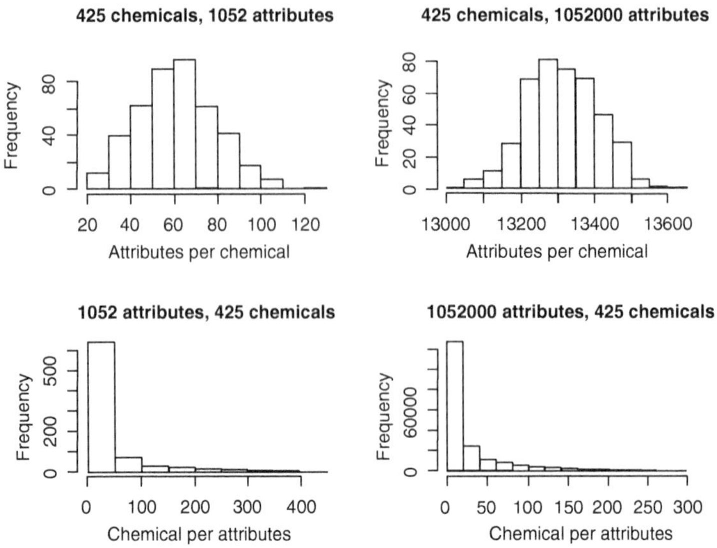

Fig. 3 Histograms of marginal distributions of the original 425 × 1052 chemicals by attributes, and the generated data with similar marginal distributions of 425 × 1052000 chemicals by attributes. Marginal distribution values greater than 0 were taken into account

chemicals by their attributes is, in frequency terms, f_{IJ}. The (i, j) elements, again in frequency terms, are $f_{i,j}$. In whole number terms, representing presence or absence, i.e., 1 or 0, the chemicals-attributes cross-tabulation is denoted k_{IJ}.

We generate a new data set that cross-tabulates a generated set of chemicals, I', crossed by a generated set of attributes, J'. Let $|.|$ denote cardinality. We randomly sample (uniformly) $|J'|$ values from k_J. Therefore we are constructing a new, generated set of attribute marginal sums. The generated values are of the same distribution function. That is, both $f_{J'} \sim f_J$ and $k_{J'} \sim k_J$. The next step is to consider the newly generated chemicals, in the set I', of cardinality $|I'|$. Given $k_{j'}$, we generate $|k_{j'}|$ values of 1 in the set of $|I'|$ elements. In this way, we generate the chemicals that contribute the $k_{j'}$ attribute presences found for attribute j'.

For the generated chemical data, we use 425 chemicals, in attribute spaces of dimensions 1052, and then, 10 times this, 100 times this, and 1000 times this dimensionality.

See the R code used at www.correspondances.info (see under "Evaluation 3"). This code shows the case of 1000 times the dimensionality. That is, for 425 chemicals with 1052 presence/absence or one/zero values, we generate a matrix of 425 chemicals × 1052000 presence/absence attributes. For the 425 × 1052 matrix, we have 26,405 presence values, and a density (i.e., presence or 1 values) of 5.9%. For the generated 425 × 1052000 presence/absence attributes, we have 5,645,075 presence values, and a density of 1.26%.

Figure 3 displays the marginal distributions. This shows visually how well our generated data approximates the original data. Let us also look at how close the

Table 3 Power law exponents for generated chemical data, with 425 chemicals, with presence/absence (respectively 1 or 0) in attribute dimensions: 1052, 10,520, 105,200, and 1,025,000

425 chemicals	
Dim.	Exponent
1052	−1.49
10,520	−1.75
105,200	−1.64
1,052,000	−1.78

Table 4 Four hundred and twenty five chemicals with presence/absence values on the following numbers of characterizing attributes: 1052, 10,520, 105,200, and 1,052,000

425 chemicals	Absolute contribution		
Dimensionality	Mean	Std. dev.	Max. projection
1052	0.01161321	0.007522956	16.27975
10,520	0.00133034	0.002798697	12.31945
105,200	0.000140571	0.0002946923	10.91465
1,052,000	1.39319e−05	2.919471e−05	11.06306

The dimensionality of the space in which the chemicals are located is given by the number of characterizing attributes

power law distributional properties are. Table 3 lists the power law exponents for our generated data sets.

Table 4 shows clearly how the absolute contribution to the inertia of the factors, which is mass times distance squared, becomes of smaller mean value, and of smaller standard deviation (hence the mean is a tighter estimate), as dimensionality increases. The degree of decrease of the mean value is approximately linear in the increase of dimensionality (i.e., tenfold for each row of Table 4). Once again, we show very conclusively how increasing dimensionality brings about a very pronounced concentration of the data cloud that is considered. As dimensionality increases, the cloud becomes much more compact, i.e., far more concentrated.

6 Conclusion

Our main objective with this work has been to demonstrate practical applicability. Beginning with methodology, we generated data sets that were modeled on various application domains.

We explored a wide range of evaluation settings. We have shown that it is easy and straightforward to analyze data that are in very high attribute dimensions (or feature dimensions, in other words, typically the number of columns of our input data matrix). Of course one needs to understand the nature of one's analysis. It is not a "black box" process. Instead it is necessary to investigate how to "let the data speak."

The benefits of this work are to be found in how the clustering of data, hierarchical clustering or partitioning, can benefit from very high dimensionality data. Furthermore the dual spaces of rows and columns can allow us to draw benefit from this same work, now though for massive sized row sets, in low to moderate dimensionality, i.e., column set. Respectively, these themes are at issue in the following work in progress: [11, 13].

References

1. Bellman, R.: Adaptive Control Processes: A Guided Tour. Princeton University Press, Princeton, NJ (1961)
2. Benzécri, J.P.: L'approximation stochastique en analyse des correspondances. Les Cahiers de l'Analyse des Données **7**(4), 387–394 (1982)
3. Benzécri, J.P.: Approximation stochastique, réseaux de neurones et analyse des données. Les Cahiers de l'Analyse des Données **22**(2), 211–220 (1997)
4. Delon, F.: Espaces ultramétriques. J. Symb. Log. **49**, 405–502 (1984)
5. Donoho, D.L., Tanner, J.: Neighborliness of randomly-projected simplices in high dimensions. Proc. Natl. Acad. Sci. **102**, 9452–9457 (2005)
6. Hall, P., Marron, J.S., Neeman, A.: Geometric representation of high dimension, low sample size data. J. R. Stat. Soc. Ser. B **67**, 427–444 (2005)
7. Mitzenmacher, M.: A brief history of generative models for power law and lognormal distributions. Internet Math. **1**, 226–251 (2004)
8. Murtagh, F.: On ultrametricity, data coding, and computation. J. Classif. **21**, 167–184 (2004)
9. Murtagh, F.: The remarkable simplicity of very high dimensional data: application of model-based clustering. J. Classif. **26**, 249–277 (2009)
10. Murtagh, F.: Symmetry in data mining and analysis: a unifying view based on hierarchy. Proc. Steklov Inst. Math. **265**, 177–198 (2009)
11. Murtagh, F.: Massive data clustering in moderate dimensions from the dual space of observation and attribute data clouds, 17 pp. http://arxiv.org/abs/1704.01871 (2017)
12. Murtagh, F., Contreras, P.: Random projection towards the Baire metric for high dimensional clustering. In: Gammerman, A., Vovk, V., Papadopoulos, H. (eds.) Statistical Learning and Data Sciences. Lecture Notes in Artificial Intelligence, vol. 9047, pp. 424–431. Springer, Heidelberg (2015)
13. Murtagh, F., Contreras, P.: Clustering through high dimensional data scaling: applications and implementations. Arch. Data Sci. Ser. A **2**(1), 1–16 (2017)
14. Murtagh, F., Downs, G., Contreras, P.: Hierarchical clustering of massive, high dimensional data sets by exploiting ultrametric embedding. SIAM J. Sci. Comput. **30**, 707–730 (2008)
15. Terada, Y.: Clustering for high-dimension, low-sample size data using distance vectors, 16 pp. http://arxiv.org/abs/1312.3386 (2013)
16. Wright, M.: Fingerprinting and dictionary generation. http://www.digitalchemistry.co.uk/prod_fingerprint.html (2006)

Comparing *Partial Least Squares* and *Partial Possibilistic Regression* Path Modeling to Likert-Type Scales: A Simulation Study

Rosaria Romano and Francesco Palumbo

Abstract Partial possibilistic regression path modeling (PPRPM) combines the principles of path modeling with those of possibilistic regression to model the net of relations among latent variables through interval-valued coefficients, in order to take into account the vagueness in the model specification. An interval valued coefficient is defined by a midpoint and a range. Through a simulation study, the paper presents a comparison between PPRPM and partial least squares path modeling (PLSPM), when these are used for analyzing questionnaire data, with responses recorded on Likert scales. The estimates of the two models have similar behaviors, with respect to the simulated scenarios. Focusing on a realistic scenario setup, the results highlight the benefit of PPRPM that allows the model to report the component-wise estimation of vagueness in the inner model.

1 Introduction

Questionnaires designed to measure latent variables, such as personality traits or consumer's behaviors, typically use Likert scales made up of several item sets, each aiming to measure the same concept [10]. The aim of such researches is generally the study of the relations among the different latent concepts that are assumed to be correlated to each other according to a predefined theory or with respect to some research hypotheses.

In this framework, structural equation modeling (SEM) [3] has become the statistical methodology of reference. Particularly, partial least squares path modeling (PLSPM) is one approach to SEM that has been developed for the analysis of the relations among blocks of items. Each block is made of a set of variables that are called manifest variables (MV) and refers to a latent variable (LV) [21, 22]. PLSPM

R. Romano (✉)
University of Calabria, Cosenza, Italy
e-mail: rosaria.romano@unical.it

F. Palumbo
University of Naples Federico II, Napoli, Italy
e-mail: fpalumbo@unina.it

F. Palumbo et al. (eds.), *Data Science*, Studies in Classification, Data Analysis, and Knowledge Organization, DOI 10.1007/978-3-319-55723-6_24

is strictly related to the traditional approach to Likert-type data analysis where each item provides an indirect measure of the LV, the use of multiple items reduces the measurement error, and the average item response becomes the better predictor of the LV [4]. In fact, PLSPM is oriented to the estimation of the LVs' scores, which are calculated as linear combinations of the corresponding MVs.

A new method named Partial Possibilistic Regression Path Modeling (PPRPM) [13, 14] has been proposed as an alternative to the classical PLSPM. As discussed in [14], PPRPM is based on the idea that different sources of uncertainty coexist in the SEM: (a) the *measurement error* related to the relations between each LV and its own block of items (measurement model); (b) the *structural error* related to the relations among the LVs (structural model). The former is generally defined as any deviation from the true value of a variable that arises in the measurement process [2]. The latter is something different: it originates from the relationships between variables that are latent and not directly measured. PPRPM assumes that the *randomness* can be referred to the *measurement error* and the *vagueness* to the *structural error*, and differently minimizes these two error components. The *randomness* is minimized in the same way as the classical PLSPM approach based on classical linear regressions, but using the least absolute values instead of the least squares. The *vagueness* is minimized by the Possibilistic Regression [16], which considers this type of uncertainty as included in the spread of model parameters, defined as interval-valued data [1], i.e., range of values denoted in terms of midpoint and spread. The estimation procedure is based on optimization techniques whose objective is to minimize the spread of the interval-valued parameters. This choice allows us to take into account the vague relations among the LVs, on the one hand, and on the other hand, the use of the least absolute values allows us to get a more robust estimate of the LV scores and ensures consistency between the minimization procedure of the two error components. In fact, PPR-PM estimation process is an L1 norm problem that independently minimizes the sum of the absolute values of the residuals in the measurement model and the sum of all the spreads of the interval-valued coefficients in the structural model.

Through a simulation study, the paper aims to investigate how the different types of uncertainty inside a SEM affect both the PLSPM and PPRPM estimates, when these are used for analyzing questionnaire data (Likert scales). Generally, data arising from such studies are described by asymmetric distribution, which is very common when working with people's perceptions. Taking this into account, focus is on the effect of various amount of measurement and structural noise, combined with different degrees of skewness of MVs and different sample sizes. The estimates are compared with respect to the bias and precision (mean square errors).

The structure of the remainder of the paper is as follows: Sect. 2 presents the PLSPM and PPRPM methodologies; the simulation study including the data-generating process is shown in Sect. 3; Sect. 4 shows the main results based on the simulations. The paper ends with the main conclusions presented in Sect. 5.

2 Methods

SEM enables the simultaneous use of LVs and MVs within one framework. The basic model can be described as

$$\mathbf{y} = \Lambda_y \eta + \epsilon, \tag{1}$$

$$\mathbf{x} = \Lambda_x \xi + \delta, \tag{2}$$

$$\eta = B\eta + \Gamma\xi + \zeta, \tag{3}$$

where $\mathbf{y}$ is a $(p \times 1)$-dimensional vector containing p endogenous observed variables, $\mathbf{x}$ is $(q \times 1)$-dimensional vector with q exogenous observed variables, η is an $(r \times 1)$-dimensional vector containing r endogenous latent variables, ξ is an $(s \times 1)$-dimensional vector containing s exogenous latent variables; ϵ and δ are error vectors, respectively, in $(p \times 1)$ dimensions and $(q \times 1)$ dimensions, and ζ is a residual vector of $(r \times 1)$ dimensions; Λ_x and Λ_y are, respectively, loading matrices in $(p \times r)$ and $(q \times s)$ dimensions, and B and Γ are, respectively, structural coefficient matrices of $(r \times r)$ and $(r \times s)$ dimensions. A LV is endogenous if it is a dependent variable in a structural equation, otherwise it is exogenous. It follows that the MVs are defined endogenous if they are connected to an endogenous LV, as well as they are defined exogenous if they are linked to an exogenous LV. Both Eqs. (1) and (2) form the measurement equation (also referred to as outer relations or measurement model), and Eq. (3) is called as the structure equation (also referred to as inner relation or structural model). Focusing on the error terms, ζ represents the error in the inner relations, i.e., disturbance in the prediction of the endogenous latent variables from their respective explanatory latent variables, whereas ϵ and δ represent imprecision in the measurement process.

2.1 *Partial Least Squares Path Modeling*

In PLSPM an iterative procedure permits the LVs' scores and the loadings to be estimated, while structural coefficients are obtained from ordinary least squares regression between the estimated LVs. Since in PLSPM notation there is no difference between endogenous and exogenous LVs or between the MVs, in the following any block of MVs is referred to as $\mathbf{X}_h$ and each LV as ξ_h, with $h = 1, \ldots, H$. The generic pth MV of the hth block is denoted by x_{ph}, where $(p = 1, \ldots, P_h)$ and P_h is the number of MVs in each block $(P = \sum P_h)$.

The algorithm computes the LV's scores alternating the *outer* and *inner* estimation till convergence. The procedure starts on centered (or standardized) MVs by choosing arbitrary weights w_{ph}. In the external estimation, the hth LV is estimated as a linear combination of its own MVs

$$\mathbf{v}_h \propto \sum_{p=1}^{P_h} w_{ph}\mathbf{x}_{ph} = \mathbf{X}_h\mathbf{w}_h, \tag{4}$$

where $\mathbf{v}_h$ is the standardized outer estimate of the LV $\boldsymbol{\xi}_h$ and the symbol $\propto$ means that the left-hand side of the equation corresponds to the standardized right-hand side. In the internal estimation, the LV is estimated by considering its links with the other h' adjacent LVs

$$\boldsymbol{\vartheta}_h \propto \sum_{h'} e_{hh'}\mathbf{v}_{h'}, \tag{5}$$

where $\boldsymbol{\vartheta}_h$ is the standardized inner estimate of the latent variable $\boldsymbol{\xi}_h$ and the inner weights $e_{hh'}$, according to the so-called *centroid scheme* [18], are equal to the sign of the correlation between $\mathbf{v}_h$ and $\mathbf{v}_{h'}$ (with $h, h' = 1, \ldots, H$). Alternative weighting schemes are provided in the original algorithm [11]. In PLSPM, these first two steps update the outer weights w_{ph}, which are the regression coefficients in the simple regressions between the pth manifest variable of the hth block $\mathbf{x}_{ph}$, on the inner estimate of the hth latent variable $\boldsymbol{\vartheta}_h$. They correspond to the covariances as $\boldsymbol{\vartheta}_h$ are standardized

$$w_{ph} = cov(\mathbf{x}_{ph}\boldsymbol{\vartheta}_h). \tag{6}$$

Even for the external weights, the original algorithm provides an alternative scheme [11], not considered in this paper as it is not related to the Likert-type data analysis.

The algorithm iterates till convergence which can be demonstrated to be convergent only for one and two-block models [12]. After convergence, structural or path coefficients are estimated through single and multiple linear regressions among the estimated LVs

$$\boldsymbol{\xi}_h = \beta_{h0} + \sum_{h':\boldsymbol{\xi}_{h'} \to \boldsymbol{\xi}_h} \beta_{hh'}\boldsymbol{\xi}_{h'}, + \zeta_h \tag{7}$$

where $\boldsymbol{\xi}_h$ is the generic dependent LV and $\beta_{hh'}$ is the generic *path coefficient* interrelating the h'th independent LVs to the hth dependent one (with $h \neq h'$).

2.2 Partial Possibilistic Regression Path Modeling

PPRPM differently treats the vagueness in the prediction of the LVs and the imprecision in the measurement of MVs. The first type of error is assumed to depend on the indefiniteness/vagueness of the parameters, which is manifested by yielding interval path coefficients [1]. The second type of error is still considered as a measurement error, but the estimation process minimizes the sum of the absolute values and not the squares, as in PLSPM.

PPRPM follows the same iterative procedure as in PLSPM by alternating the inner and outer estimation of the LV scores. However, the outer weights and the path coefficients are obtained in a different way. In PPRPM the weight w_{ph} is the regression coefficient in the Quantile Regression (QR) [9] of the pth manifest variable of the hth block $\mathbf{x}_{ph}$ on the inner estimate of the hth latent variable $\boldsymbol{\vartheta}_h$

$$\mathbf{x}_{ph} = w_{ph}\boldsymbol{\vartheta}_h + \boldsymbol{\epsilon}_{ph}. \tag{8}$$

QR is an extension of the classical estimation of the conditional mean to the estimation of a set of conditional quantiles [9]

$$Q_\tau(\mathbf{x}_{ph}|\boldsymbol{\vartheta}_h) = \boldsymbol{\vartheta}_h w_{ph}(\tau) + \boldsymbol{\epsilon}_{ph}, \tag{9}$$

where $0 < \tau < 1$ and $Q_\tau(.|.)$ denotes the conditional quantile function for the τth quantile. In particular, PPRPM considers only the case in which $\tau = 0.5$, i.e., the median is the single chosen quantile.

The path coefficients are estimated through Possibilistic Regression (PR) [16, 17] among the estimated LVs

$$\boldsymbol{\xi}_h = \tilde{\beta}_{h0} + \sum_{h':\boldsymbol{\xi}_{h'} \to \boldsymbol{\xi}_h} \tilde{\beta}_{hh'}\boldsymbol{\xi}_{h'}, \tag{10}$$

where $\boldsymbol{\xi}_h$ is the generic dependent LV and $\tilde{\beta}_{hh'}$ is the generic *interval path coefficient* in terms of midpoint and range $\tilde{\beta}_{hh'} = \{c_{hh'}; a_{hh'}\}$, or equivalently $[\underline{\beta}_{hh'}, \overline{\beta}_{hh'}] = [c_{hh'} \pm a_{hh'}]$, interrelating the hth independent variable to the h'th dependent one (with $h \neq h'$). It is worth mentioning that the range is a measure of the spread around the midpoint coefficient. The higher the midpoint coefficient the higher the contribution to the prediction of the dependent LV. At the same time, the higher the spread coefficient the higher the vagueness in the relation among the considered LVs. PPRPM differs itself from PLSPM in that path coefficients are interval-valued [1], and the residual vector ζ is no longer covered in the PR model.

The purpose of PR is to explain a dependent variable as an interval output in terms of the variation of explanatory variables. Specifically, PR defines the relation between one dependent variable $\mathbf{y}$ and a set of M predictors $\mathbf{x}_1, \mathbf{x}_m, \ldots, \mathbf{x}_M$, observed on N statistical units, through a linear function holding interval-valued coefficients

$$\mathbf{y} = \tilde{\omega}_1\mathbf{x}_1 + \ldots + \tilde{\omega}_m\mathbf{x}_m + \ldots + \tilde{\omega}_M\mathbf{x}_M, \tag{11}$$

where $\tilde{\omega}_m$ denotes the generic interval-valued coefficient. Interval-valued coefficients are defined in terms of midpoint and spread: $\tilde{\omega}_m = \{c_m; a_m\}$. There are no restrictive assumptions on the model. Unlike statistical regression, the deviations between data and linear models are assumed to be depending on the vagueness of the parameters and not on the measurement errors. This means that in PR there is no external error component but the spread of the coefficients embeds all uncertainty, such that PR minimizes the total spread of the interval coefficients

$$\min_{a_m} \sum_{m=1}^{M} \left(\sum_{n=1}^{N} a_m |x_{nm}| \right), \tag{12}$$

under the following linear constraints:

$$\begin{aligned} &\sum_{m=1}^{M} c_m x_{nm} + \sum_{m=1}^{M} a_m |x_{nm}| \geq y_n \,, \\ &\sum_{m=1}^{M} c_m x_{nm} - \sum_{m=1}^{M} a_m |x_{nm}| \leq y_n \,, \qquad \text{for all } n = 1, \ldots, N, \end{aligned} \tag{13}$$

satisfying the following conditions: (*i*) $a_m \geq 0$; (*ii*) $c_m \in R$; (*iii*) $x_{n1} = 1$.

3 The Simulation Study

Simulations are mainly based on [6, 7, 19, 20]. The focus is on the effect of various amount of measurement and structural errors, combined with different degrees of skewness in the data and different sample sizes. Sensitivity of results is investigated with respect to:

- Skewness of data (symmetric, moderately skewed, highly skewed)
- Sample sizes (50, 200, 500)
- Level of noise in structural model (ζ, 10%, 30%, 50%)
- Level of noise in measurement model (ϵ, 10%, 30%, 50%)

The PLSPM and PPRPM estimation of the structural and measurement model are compared in terms of bias and precision (mean square errors). Of special interest is also the distribution of the estimated scores of LVs.

3.1 The Data-Generating Process

Data were generated according to a structural model (see Fig. 1) that consists of two exogenous LVs (ξ_1 and ξ_2), and one endogenous LV (ξ_3). The inner model was defined as

$$\xi_3 = \beta_1 \xi_1 + \beta_2 \xi_2 + \zeta,$$

where β_1 and β_2 are the path coefficients and ζ is the random disturbance effect. The following values for parameters were assumed: $\beta_1 = 0.9; \beta_2 = 0.3$. The measurement model equations for the generic latent variables ξ_h, with $h = 1, \ldots, 3$,

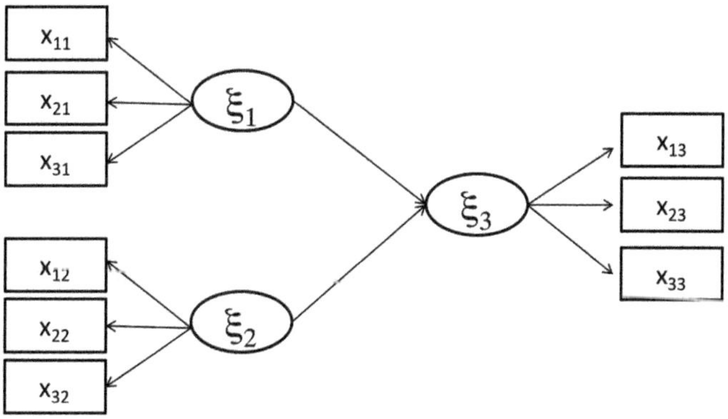

Fig. 1 Path diagram of the structural and measurement model in the simulation study

were

$$x_{1h} = \lambda_{1h}\xi_h + \epsilon_{1h},$$
$$x_{2h} = \lambda_{2h}\xi_h + \epsilon_{2h},$$
$$x_{3h} = \lambda_{3h}\xi_h + \epsilon_{3h},$$

where λ_{1h}, λ_{2h}, and λ_{3h} are the loadings and ϵ_{1h}, ϵ_{2h}, and ϵ_{3h} are the random noise effects. The following values for parameters were assumed: $\lambda_{ph} = 0.75; 0.80; 0.85$ for $p = 1, \ldots, 3$. The exogenous LVs ξ_h were generated from the beta distribution $\beta_{u,v}$: (1) B(6,6) symmetric case; (2) B(9,4) moderately right-skewed case; B(9,1) highly right-skewed case. The noises $\boldsymbol{\zeta}$ and $\boldsymbol{\epsilon}$ were realizations of the continuous uniform distribution U(a, b) with expectation zero and variance accounted to three levels of the variance of the corresponding dependent variable: (1) 10% (low noise); (2) 30% (medium noise); (3) 50% (high noise). The generated data were re-scaled in the interval [1, 9] and rounded to obtain integer (ordinal) responses, in order to reproduce a Likert-type scale. Figure 2 shows the frequency distribution of the generated data for the exogenous variables, both in the symmetric and skewed data context. For each case, samples of size n=50, 200, and 500, respectively, were generated. The sampling distributions consist of 500 replicates of the model estimations. Due to the complexity of the experimental design, the simulation study only focuses on a selective number of design cases. Table 1 shows the selected simulation settings.

Considering A, B, and C allow us to compare the two methods with increasing sample size. Considering B, D, E, F, and G allow us to compare the two methods when the level of noise in both the inner and outer model increases. Finally B, H, and I allow us to compare the two methods with increasing skewness of MVs.

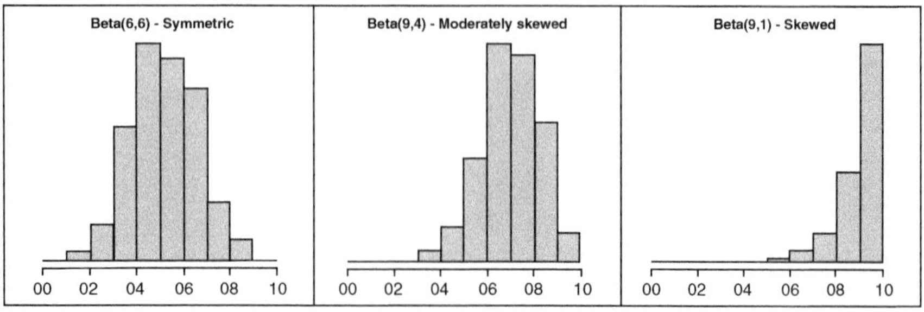

Fig. 2 Distribution of manifest variables in the simulation study

Table 1 Simulation design

		$\beta(6,6)$			$\beta(9,4)$			$\beta(9,1)$		
ϵ (%)	ζ (%)	n=50	n=200	n=500	n=50	n=200	n=500	n=50	n=200	n=500
10	10	A	B	C		H			I	
10	30		D							
10	50		E							
30	10		F							
50	10		G							

4 Results

Results in Table 2 compare the two methods with each other with increasing sample size ($n_A = 50$, $n_B = 200$, $n_C = 500$). The two methods present nearly the same results. The bias does not seem to be influenced by the increasing sample size. This finding was expected for PLSPM approach which is known to provide estimates that are only *consistent at large* [8], i.e., the bias is reduced when increasing the number of MVs in different blocks. The bias of the measurement model coefficients (λ_{ph}) is a minimum as the level of noise is low, both in the inner and outer model, in all three simulation scenarios ($\epsilon = \zeta = 10\%$). The path coefficients ($\beta_{hh'}$) show a larger bias: the estimate of the small coefficient shows a positive bias, while that of the large coefficient shows a negative bias. Note that PLSPM is known to underestimate the inner model parameters [15].

Results in Table 3 allow us to compare the bias and the precision when the level of noise in both the inner and outer model increases. The two approaches have similar results in this respect also, but with some slight differences. The higher the noise level in the inner models (D and E), the higher the average bias and MSE of the structural coefficients, while the measurement coefficients remain unchanged. The bias of the large coefficient is larger, while the estimates of the smaller coefficient are close to the true value. The increase in the level of noise in the outer model (F and G) not only increases the distortion of loadings, but also has a proportional direct effect on the estimation of the structural model. In particular, in the F and

Table 2 Comparison of simulation settings A, B, and C

	A		B		C	
	PLSPM	PPRPM	PLSPM	PPRPM	PLSPM	PPRPM
Bias						
λ_{11}	0.01	0.01	0.01	0.01	0.00	0.01
λ_{12}	0.01	0.01	0.01	0.01	0.01	0.00
λ_{13}	0.01	0.01	0.01	0.01	0.00	0.00
λ_{21}	0.01	0.01	0.01	0.00	0.00	0.00
λ_{22}	0.01	0.00	0.01	0.01	0.00	0.01
λ_{23}	0.01	0.01	0.01	0.00	0.01	0.00
λ_{31}	0.01	0.01	0.01	0.01	0.00	0.00
λ_{32}	0.01	0.01	0.01	0.01	0.00	0.00
λ_{33}	0.01	0.01	0.01	0.01	0.00	0.00
Average (abs)	0.01	0.01	0.01	0.01	0.00	0.00
β_1	−0.07	−0.07	−0.07	−0.07	−0.08	−0.08
β_2	0.08	0.07	0.07	0.07	0.07	0.06
Average (abs)	0.07	0.07	0.07	0.07	0.07	0.07
Mean square error						
λ_{11}	0.00	0.00	0.00	0.00	0.00	0.00
λ_{12}	0.00	0.00	0.00	0.00	0.00	0.00
λ_{13}	0.00	0.00	0.00	0.00	0.00	0.00
λ_{21}	0.00	0.00	0.00	0.00	0.00	0.00
λ_{22}	0.00	0.00	0.00	0.00	0.00	0.00
λ_{23}	0.00	0.00	0.00	0.00	0.00	0.00
λ_{31}	0.00	0.00	0.00	0.00	0.00	0.00
λ_{32}	0.00	0.00	0.00	0.00	0.00	0.00
λ_{33}	0.00	0.00	0.00	0.00	0.00	0.00
Average	0.00	0.00	0.00	0.00	0.00	0.00
β_1	0.01	0.02	0.01	0.01	0.01	0.01
β_2	0.01	0.02	0.01	0.01	0.00	0.01
Average	0.01	0.02	0.01	0.01	0.01	0.01
Skewness						
ξ_1	−0.01	−0.01	−0.01	−0.01	0.00	0.00
ξ_2	−0.01	−0.01	−0.01	−0.01	0.02	0.02
ξ_3	0.01	0.01	−0.01	−0.01	0.00	0.00

G the PPRPM shows greater precision (lower average bias) but a worse accuracy (higher average MSE), for both the measurement and the structural model.

Results in Table 4 show how both approaches seem to be unaffected by the skewness of the MVs. On the other hand, the robustness of PLSPM to deviations from symmetry has already been highlighted [7, 19]. The results related to the asymmetry of LVs show that the distribution of MVs is impacting on the distribution of the corresponding LVs. Since the scores are weighted averages of the related

Table 3 Comparison of simulation settings B, D, E, F, and G

	B		D		E		F		G	
	PLSPM	PPRPM	PLSPM	PPRPM	PLSPM	PPRPM	PLSPM	PPRPM	PLSPM	PPRPM
Bias										
λ_{11}	0.01	0.01	0.01	0.01	0.01	0.01	0.05	0.04	0.10	0.10
λ_{12}	0.01	0.01	0.01	0.01	0.01	0.01	0.05	0.05	0.10	0.10
λ_{13}	0.01	0.01	0.01	0.01	0.01	0.01	0.05	0.05	0.10	0.10
λ_{21}	0.01	0.00	0.01	0.01	0.01	0.00	0.04	0.04	0.09	0.09
λ_{22}	0.01	0.01	0.01	0.00	0.01	0.00	0.05	0.04	0.09	0.07
λ_{23}	0.01	0.00	0.01	0.00	0.01	0.00	0.05	0.04	0.10	0.07
λ_{31}	0.01	0.01	0.01	0.01	0.01	0.01	0.04	0.04	0.10	0.10
λ_{32}	0.01	0.01	0.01	0.01	0.01	0.01	0.04	0.04	0.10	0.10
λ_{33}	0.01	0.01	0.01	0.01	0.01	0.01	0.05	0.05	0.10	0.10
Ave. (abs)	0.01	0.01	0.01	0.01	0.01	0.01	0.05	0.04	0.10	0.09
β_1	−0.07	−0.07	−0.17	−0.17	−0.29	−0.28	−0.16	−0.14	−0.26	−0.25
β_2	0.07	0.07	0.02	0.02	−0.02	−0.03	0.03	0.04	−0.01	−0.01
Ave. (abs)	0.07	0.07	0.09	0.09	0.15	0.15	0.10	0.09	0.14	0.13
Mean square error										
λ_{11}	0.00	0.00	0.00	0.00	0.00	0.00	0.00	0.00	0.01	0.01
λ_{12}	0.00	0.00	0.00	0.00	0.00	0.00	0.00	0.00	0.01	0.01
λ_{13}	0.00	0.00	0.00	0.00	0.00	0.00	0.00	0.00	0.01	0.01
λ_{21}	0.00	0.00	0.00	0.00	0.00	0.00	0.00	0.00	0.01	0.02
λ_{22}	0.00	0.00	0.00	0.00	0.00	0.00	0.00	0.00	0.01	0.01
λ_{23}	0.00	0.00	0.00	0.00	0.00	0.00	0.00	0.00	0.01	0.02
λ_{31}	0.00	0.00	0.00	0.00	0.00	0.00	0.00	0.00	0.01	0.01
λ_{32}	0.00	0.00	0.00	0.00	0.00	0.00	0.00	0.00	0.01	0.01
λ_{33}	0.00	0.00	0.00	0.00	0.00	0.00	0.00	0.00	0.01	0.01
Average	0.00	0.00	0.00	0.00	0.00	0.00	0.00	0.00	0.01	0.01
β_1	0.01	0.01	0.03	0.04	0.08	0.09	0.03	0.04	0.07	0.09
β_2	0.01	0.01	0.00	0.01	0.00	0.01	0.00	0.02	0.00	0.03
Average	0.01	0.01	0.02	0.02	0.04	0.05	0.01	0.03	0.04	0.06
Skewness										
ξ_1	−0.01	−0.01	−0.02	−0.02	0.02	0.02	0.00	0.00	−0.01	−0.01
ξ_2	−0.01	−0.01	0.01	0.01	0.01	0.01	0.01	0.01	0.01	0.01
ξ_3	−0.01	−0.01	0.00	0.00	0.01	0.01	−0.01	−0.01	0.00	0.00

MVs, such a result was expected. However, it should be noted that the skewness of the indicators is reflected to a lesser extent on the endogenous LV. To further deepen the comparison between the two methods, let us consider a realistic scenario with 200 statistical units, data distribution moderately skewed and medium level of noise

Table 4 Comparison of simulation settings B, H, and I

	B		H		I	
	PLSPM	PPRPM	PLSPM	PPRPM	PLSPM	PPRPM
Bias						
λ_{11}	0.01	0.01	0.01	0.01	0.00	0.00
λ_{12}	0.01	0.01	0.01	0.01	0.00	0.00
λ_{13}	0.01	0.01	0.01	0.01	0.00	0.00
λ_{21}	0.01	0.00	0.01	0.01	0.01	0.00
λ_{22}	0.01	0.01	0.01	0.00	0.00	0.00
λ_{23}	0.01	0.00	0.01	0.01	0.00	0.00
λ_{31}	0.01	0.01	0.01	0.01	0.00	0.01
λ_{32}	0.01	0.01	0.01	0.01	0.00	0.00
λ_{33}	0.01	0.01	0.01	0.01	0.00	0.00
Average (abs)	0.01	0.01	0.01	0.01	0.00	0.00
β_1	−0.07	−0.07	−0.07	−0.08	−0.08	−0.08
β_2	0.07	0.07	0.07	0.07	0.07	0.07
Average (abs)	0.07	0.07	0.07	0.07	0.07	0.08
Mean square error						
λ_{11}	0.00	0.00	0.00	0.00	0.00	0.00
λ_{12}	0.00	0.00	0.00	0.00	0.00	0.00
λ_{13}	0.00	0.00	0.00	0.00	0.00	0.00
λ_{21}	0.00	0.00	0.00	0.00	0.00	0.00
λ_{22}	0.00	0.00	0.00	0.00	0.00	0.00
λ_{23}	0.00	0.00	0.00	0.00	0.00	0.00
λ_{31}	0.00	0.00	0.00	0.00	0.00	0.00
λ_{32}	0.00	0.00	0.00	0.00	0.00	0.00
λ_{33}	0.00	0.00	0.00	0.00	0.00	0.00
Average	0.00	0.00	0.00	0.00	0.00	0.00
β_1	0.01	0.01	0.01	0.01	0.01	0.02
β_2	0.01	0.01	0.01	0.01	0.01	0.02
Average	0.01	0.01	0.01	0.01	0.01	0.02
Skewness						
ξ_1	−0.01	−0.01	−0.38	−0.38	−1.31	−1.31
ξ_2	−0.01	−0.01	−0.38	−0.38	−1.31	−1.30
ξ_3	−0.01	−0.01	−0.27	−0.27	−0.92	−0.92

in both the structural and measurement model ($\epsilon = \zeta = 10\%$). The generated data are graphically represented in Fig. 3. Plots on the diagonal show the distribution of the MVs, which appear to be moderately skewed, consistently with the simulation plan. Plots outside of the diagonal correspond to the scatter plots between all possible pairs of MVs. It can be seen how the correlations between MVs of the same block (same color) are greater than those with variables of other blocks. Plot in Fig. 4 allows us to highlight a substantial difference between the path coefficients of

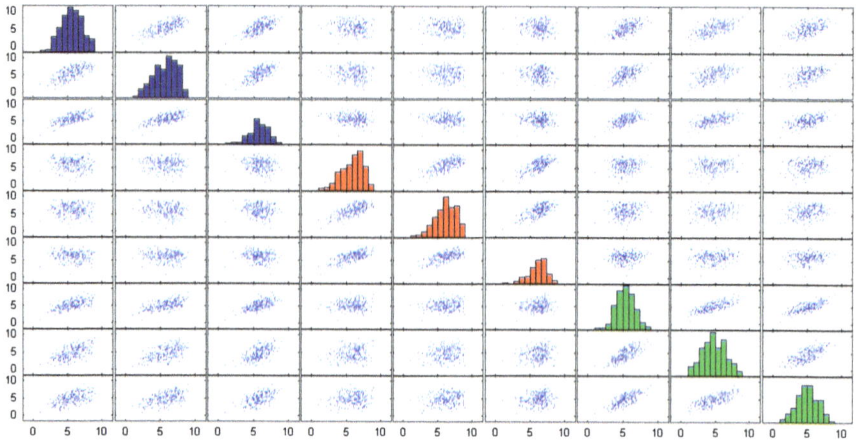

Fig. 3 Scatter matrix among the manifest variables from the realistic scenario

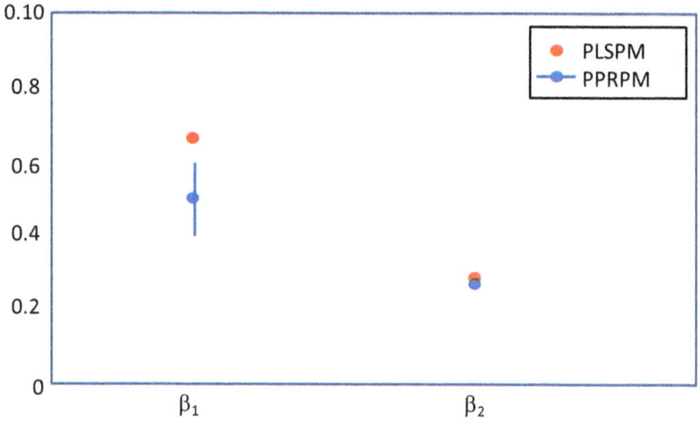

Fig. 4 Path coefficients from the realistic scenario

PPRPM and those of PLSPM. Consistent to the plane of simulation, the value of β_1 is higher than that of β_2 and this information is reported by both methods. However, PPRPM also shows how the relationship between the latent variables ξ_1 and ξ_3 is characterized by a higher level of vagueness, which is expressed by a larger spread of the β_1 coefficient. It is important to note that this information is provided only by PPPRM. The uncertainty on the model parameters can be measured in PLSPM using bootstrap confidence intervals. But that is a different kind of uncertainty related to sampling variability. The uncertainty taken into account in PR and PPRPM is related to the vagueness in the relations among the variables.

5 Conclusion

Many studies on the nature of Likert scales (as opposed to single Likert items) have shown that the Likert response format produces empirically interval data, i.e., the Latent Variables, and that SEMs are consistent approaches for analyzing such data. This work presented a comparison between the PLSPM and PPRPM for the analysis of Likert scales questionnaires and it showed that the estimates of the two models have similar behaviors with respect to the simulated scenarios consistently with [5]. However, PPRPM structural model parameters are defined by interval-valued data: each parameter is defined by a midpoint and a range. The realistic scenario results highlighted the advantages of having interval-valued inner model parameters that embed the inner model error terms. In fact, each range refers to the *imprecision* of each independent LV in each structural equation and provides a component-wise estimation of the inner model error.

References

1. Alefeld, G., Mayer, G.: Interval analysis: theory and applications. J. Comput. Appl. Math. **121**, 421–464 (2000)
2. Asher, H.B.: Some consequences of measurement error in survey data. Am. J. Polit. Sci. **18**, 469–485 (1974)
3. Bollen, K.: Structural Equations with Latent Variables. Wiley, New York (1989)
4. Camparo, J., Camparo, L.B.: The analysis of Likert scales using state multi-poles: an application of quantum methods to behavioral sciences data. J. Educ. Behav. Stat. **38**(1), 81–101 (2013)
5. Carifio, J., Perla, R.: Resolving the 50-year debate around using and misusing Likert scales. Med. Educ. **42**(12), 1150–1152 (2008)
6. Cassel, C., Hackl, P., Westlund, A.: Robustness of partial least-squares method for estimating latent variable quality structures. J. Appl. Stat. **26**(4), 435–446 (1999)
7. Cassel, C., Hackl, P., Westlund, A.: On measurement of intangible assets: a study of robustness of partial least squares. Total Qual. Manag. **11**, 897–907 (2000)
8. Chin, W.: The partial least squares approach for structural equation modeling. In: Marcoulides, G.A. (ed.) Modern Methods for Business Research, pp. 295–236. Lawrence Erlbaum Associates, London (1998)
9. Koenker, R., Basset, G.: Regression quantiles. Econometrica **46**, 33–50 (1978)
10. Likert, R.: A Technique for the Measurement of Attitudes. Archives of Psychology. Columbia University Press, New York (1931)
11. Löhmoller, J.: Latent Variable Path Modeling with Partial Least Squares. Physica-Verlag, Heidelberg (1989)
12. Lyttkens, E., Areskoug, B., Wold, H.: The convergence of NIPALS estimation procedures for six path models with one or two latent variables. Technical Report, University of Goteborg (1975)
13. Romano, R., Palumbo, F.: Partial possibilistic regression path modeling for subjective measurement. J. Methodol. Appl. Stat. **15**, 177–190 (2013)

14. Romano, R., Palumbo, F.: Partial possibilistic regression path modeling. In: Abdi, H., Vinzi, V.E., Russolillo, G., Saporta, G., Trinchera, L. (eds.) The Multiple Facets of Partial Least Squares Methods. Springer Proceedings in Mathematics & Statistics. Springer, New York (2016)
15. Schneeweiss, H.: Consistency at large in models with latent variables. In: Statistical Modelling and Latent Variables, pp. 299–320. Elsevier, Amsterdam (1993)
16. Tanaka, H., Guo, P.: Possibilistic Data Analysis for Operations Research. Physica-Verlag, Wurzburg (1999)
17. Tanaka, H., Watada, J.: Possibilistic linear systems and their application to the linear regression model. Fuzzy Sets Syst. **27**, 275–289 (1987)
18. Tenenhaus, M., Vinzi, V.E., Chatelin, Y.M., Lauro, C.: PLS path modeling. Comput. Stat. Data Anal. **48**(1), 159–205 (2005)
19. Vilares, M., Almeida, M., Coelho, P.: Comparison of likelihood and PLS estimators for structural equation modeling: a simulation with customer satisfaction data. In: Esposito Vinzi, V., et al. (eds.) Handbook of Partial Least Squares, pp. 289–305. Springer, Berlin (2010)
20. Westlund, A., Cassel, C., Hackl, P.: Structural analysis and measurement of customer perceptions, assuming measurement and specifications errors. Total Qual. Manag. **12**(7–8), 873–881 (2001)
21. Wold, H.: Estimation of principal component and related models by iterative least squares. In: Krishnaiah, P. (ed.) Multivariate Analysis, pp. 391–420. Academic, New York (1966)
22. Wold, H.: Modelling in complex situations with soft information. In: Third World Congress of Econometric Society, Toronto (1975)

Cause-Related Marketing: A Qualitative and Quantitative Analysis on Pinkwashing

Gabriella Schoier and Patrizia de Luca

Abstract In recent years a growing number of companies have decided to demonstrate their sensitivity to the subject of health. Therefore cause-related marketing campaigns that involve a partnership between companies and associations engaged in scientific research have increased significantly. Considering a particular initiative of cause-related marketing connected to those pink marketing projects aimed to fund breast cancer research, this paper intend to test the effects of the phenomenon of pinkwashing (pink = rose + washing = wash), a term proposed by the American Breast Cancer Action Campaign "Think before you pink," launched in 2002, on trust, on the perceived risk, and consumer confusion. Based on an analysis of the literature on a similar subject, i.e., greenwashing it has developed a qualitative and quantitative research to explore the effects of pinkwashing. The starting point has been a structured questionnaire. Through this a descriptive analysis also using a two step cluster analysis and a hypothesis development has been performed.

1 Introduction

Marketing literature has paid attention above all on environmental issue. Until now it has paid little attention to those initiatives of cause-related marketing tied to women, and especially to those projects financing research on breast cancer, usually highlighted with a pink packaging or a pink ribbon, and for this reason known as pink marketing [2, 12].

This paper has the aim to verify the effects of the pink marketing and the related problem of pinkwashing on trust, perceived risk, and consumer confusion.

The term pinkwashing is used to describe the practice of companies over claiming their support to the research on breast cancer through the sale of their products, but this support cannot be substantiated. It was proposed by the American Breast Cancer Action during the campaign "Think before you pink" launched in 2002 to draw attention to the phenomenon that occurs when an organization says

G. Schoier (✉) • P. de Luca
Dipartimento di Scienze Economiche Aziendali Matematiche e Statistiche, Università di Trieste, via dell'Università 1, 34100 Trieste, Italy
e-mail: gabriella.schoier@deams.units.it; patrizia.delca@deams.units.it

F. Palumbo et al. (eds.), *Data Science*, Studies in Classification, Data Analysis, and Knowledge Organization, DOI 10.1007/978-3-319-55723-6_25

it will fight breast cancer, offering products tagged with a pink ribbon, but actually produces and sells products in various ways at the same time connected to the same disease.

In 2012 a survey conducted on a sample of American consumers demonstrated that the 86% of respondents represent that they have a positive impression of companies and brands that support the cause of breast cancer and almost half (45%) declared they had purchased a product in support of the cause in the month of October, the National Breast Cancer Awareness month [7]. However, some of the messages may be ambiguous and limit socially responsible behaviors of the consumers. The same research has revealed some skepticism among respondents: the 77% believes that some of the companies engaged in support of the cause are involved only for corporate interests, 68% said that only a few of the campaigns, undertaken by companies, stand out from the others due to the large number of similar programs available on the market, 30% say they don't know if buying a pink product actually benefits the cause.

Based on an analysis of the literature on a similar subject, i.e., greenwashing it has developed a qualitative and quantitative research to explore the effects of pinkwashing.

On the bases of a structured questionnaire a descriptive analysis in order to individuate clusters of units has been considered.

A subsequent statistical analysis to test relations among pinkwashing, perceived risk, trust, and consumer confusion has been performed.

It is important to point out that all these analyses can be reconducted in a Data Mining approach which uses utilities for the management [25, 26, 34].

2 The Theoretical Framework

Cause-related marketing (CRM) is "the process of formulating and implementing marketing activities that are characterized by an offer from the firm to contribute a specified amount to a designated cause when customers engage in revenue-providing exchanges that satisfy organizational and individual objectives" [30].

Over the last years, CRM has become a relevant marketing tool for companies and their differentiation strategy in a competitive context. Despite this growing trend, CRM programs have been criticized and may run the risk of consumers' backlash. Indeed, regarding specific CRM programs, the attributions that consumers make of the company's reasons for conducting the program could influence how they respond [22]. Consumers are becoming increasingly more critical and skeptical concerning the authenticity of the various proposals in the environmental and social field [24], above all because some companies try to alter their image through actions of whitewashing, that is an attempt to hide unpleasant facts, in different contexts [16]. For example, companies could try to make themselves and their products look like they are really helping the environment, but if they do not actually help it, it is a case of greenwashing.

Marketing literature has paid attention above all on environmental issue, with studies on greenmarketing and greenwashing and on their effects on consumer behavior [3, 5, 13, 23], while until now it has paid little attention to those initiatives of cause-related marketing tied to women, and especially to those projects financing research on breast cancer, usually highlighted with a pink packaging or a pink ribbon, and for this reason known as pink marketing [2, 12]. Also in this case some companies could try to pinkwash their products, with effects on trust, perceived risk, and confusion of the consumers.

In marketing literature trust has been defined as the assumption that the other party will behave as expected [11] or as an intention to accept the vulnerability that comes from good intentions or expectations to the behavior of the other party. Ganesan [10] also stressed how trust is based on the expectation of skill, reliability, and kindness of the others.

The consumer confusion, conscious or unconscious, could also affect the purchasing process decisions [19]. Turnbull et al. [29] and Leek [15] define confusion as the failure of the consumer to develop a correct understanding of the various aspects of a product/service, during the information process. Three main groups of stimuli can influence the consumer confusion: excessive information, similar information, and ambiguous information [8]. Indeed it can be the result of an information overload [15], a series of stimuli such as advertising and products so similar to each other to be confused [19], or the use of misleading information. In the case under study, using exaggerated pink marketing campaigns, with ambiguous information about the amount of the donation or the composition of products, may confuse the consumer. Since the state of confusion is often associated with feelings of uncertainty, anxiety, disorientation, and indecisiveness, the buying process could be inconclusive and frustrating [20, 31].

In literature the idea that consumer confusion negatively influence his confidence is widely spread. For example, Mitchell and Papavassiliou [19] argue that confused consumers may quit or postpone their purchasing decisions and stop believing in the company offer. Singh and Sirdeshmukh [27] affirm that if consumers are confused suspicion tends to grow and confidence to decline, because of vague and unclear advertisements. For some authors, this impact is significantly negative [17, 31]. To reduce consumer confusion and, conversely, increase cognitive clarity, according to [32], it could be useful to increase consumer confidence. Morgan and Hunt [21] believe that confused consumers have less confidence in a market that provides them with product information ambiguous and sometimes conflicting. Kalafatis and Pollard [13] also argue consumers are reluctant to believe in the products of those companies promoting their offer through confusing and ambiguous statements.

According to [4], there is also a strong correlation between negative emotions and perceived risk; for this reason pinkwashing, as a phenomenon tied to negative emotions of consumption, could have effects on the perceived risk by the consumer, where the perceived risk is a subjective expectation of a loss.

According to the opinion of different authors such psychological state can influence consumer behavior [4, 9, 19] and act negatively on the purchasing process. This condition makes it very difficult for people to assess the real quality of the product before purchase and can facilitate unethical behavior in business. Consumers will be therefore more reluctant to buy a certain product because of their lack of confidence in the seller due to the asymmetry of the knowledge in transaction. As Mitchell and Papavassiliou [19] says, if consumers perceive a high level of risk with respect to a certain product or brand, they lose confidence in it. Koehn [14] also highlights a relationship between these two aspects, because the reduction of perceived risk can increase consumer confidence.

3 The Methodology

In the present section we present the questionnaire, the descriptive analysis, the hypothesis development, and the results of the analysis.

3.1 The Questionnaire

From the methodological point of view a quantitative research has been developed using a structured questionnaire based on a seven point Likert scale [28]. The questionnaire consists of two parts and is created using Google Forms. It has been tested on a small group of Italian consumers, and then revised for the final form, made via Internet on a randomly selected group. In January–February 2015, 173 questionnaires were collected.

The questionnaire has the following structure:

1. Part 1 Opinion and knowledge of pink marketing
    a. Evaluation of general opinions on ethics of companies to support social causes (three items).
    b. Evaluation on the effects on consumers of the cause-related marketing (six items).
    c. Evaluation and knowledge on pink marketing (three items).
2. Part 2 Misuration of the constructs
    a. Pinkwashing perception (five items).
    b. Pink-perceived risk (four items).
    c. Trust in pink products (five items).
    d. Consumer confusion (four items).

3.2 Descriptive Analysis

In this section we consider a preliminary descriptive analysis followed by a two step cluster analysis.

3.2.1 A Preliminary Descriptive Analysis

Part 1 Opinion and knowledge of pink marketing

(a) + (b) Opinions on cause-related marketing

Opinions on cause-related marketing have been analyzed through nine items. The former (three questions) have as an objective to include the opinions of consumers concerning cooperation between a company and a non-profit organization. The latter, (six questions), wanted to investigate consumer attitudes to firms engaged in operations of CRM. Specifically, it was considered interesting to evaluate the possible change of judgments in case the cause supported by the undertaking enterprise is a generic or a theme particularly important for respondents as they are directly involved.

The majority of respondents (82.1%) quite agree, agree, or strongly agree in believing ethically correct that companies support social causes.

The point of view changes slightly when talking about social causes, advertising and increase of profits. In fact, respondents show perplexity with regard to the use of the social causes to advertise of companies. The percentage of people who agrees are 41.1% in the case of advertising for the support of social causes and 31.2% when speaking of increased profits through social causes but 63% declares themselves in favor of the support to social causes.

The other six items of the variable "opinions on cause-related marketing" concern the attitude of consumers in the event of a transaction of CRM. The percentage of respondents who agree in saying that there is an improvement in the opinion of the company in the case of support to a generic social cause is 65%, and increases when the cause is individually significant (69.4%). The most obvious thing is the increase in the propensity to purchase in the case of support for a personally relevant cause. As regards the willingness to pay a higher price for the products involved in operations of generic CRM, more than 50% of respondents not agree to bear a higher cost, symptom that the donation according to consumers must be fully borne by the company. However in the case of a social cause personally relevant the 49% is in favor to pay a higher price (against 34.7% in the case of a generic cause).

These data are extremely useful as they stress the importance for companies to choose significant causes for most of the target since they are able to increase both the corporate reputation and the sales, with the possibility of applying a slightly higher price.

(c) Knowledge and opinions on pink marketing

As regards pink marketing, consumer behavior is quite favorable. The majority of respondents declares an increase in the propensity to purchase these products (56.7%), symptom that the cause of breast cancer is important for the majority of respondents.

More interesting are the statements regarding the level of knowledge of the pink logo as international symbol of the support to the cause. The 84% of respondents pretend to be aware of this fact.

Part 2 Misuration of the constructs

(a) Perception of pinkwashing

In order to analyze the perception of pinkwashing, five items have been used.

There is a slight tendency to believe that the words used are ambiguous (39.9% is favorable, 20.8% does not know the answer, and the 39.2% does not believe that there is ambiguity), and there is an equally favorable trend to believe that images are not ambiguous (41% do not think that there is ambiguity, 18.5% do not know, 40.5%, thinks that there is ambiguity). More effective were the other three items, 73.4% declare that the statements of the campaigns of pink marketing are hardly verifiable, 63.6% declare that the involvement of the company to the cause is not great, and 70.5% fail or conceal important information.

(b) Risk perception (pink-perceived risk)

In the case of the pink-perceived risk, respondents have a certain distrust of the campaigns of pink marketing as 47.3% believes that some companies do not really support the cause of breast cancer, with respect to the 28.3% that believes companies are sincere (the remaining 24.4% does not know the answer). In addition, the 61.3% believed that some products are not compatible with the image that the company wants to transmit of itself (20.2% instead does not know the answer).

As regards the possibility that pink products can be risky for the health, the majority do not think that this is possible. The phenomenon of pinkwashing in Italy is, in fact, as previously discussed, only to the origins and it is therefore normal that there is a general tendency of consumers to believe that the product is not dangerous for the health. The debate on the use of chemical substances which are potentially carcinogenic in the cosmetic industry, for example, arose only in recent times and there are still few websites or consumer associations which analyze the INCI products and explain the danger of chemical substances used.

(c) Consumer confidence

As regards the assessment of consumer confidence with respect to the businesses which implement campaigns of pink marketing, five different items were used. The research has in this regard highlighted as consumers are skeptical of such marketing operations. Specifically, 64.7% of respondents do not know the answer or believe that the reputation of these companies is not reliable. The 65.9% do not know judge or believe that the products used are not reliable while almost all (84.9%) believe that the advertising are not trusted or are in doubt. Most of the sample addition does not believe that companies support the fight against breast cancer with adequate mode or believed not to be able to judge on the basis of the information provided (76.3%). The 75.2% still did not believe that businesses maintain their promises and pledges to support the cause of breast cancer or is in doubt.

(d) Consumer confusion

In order to analyze the confusion of the consumer four items have been considered. Considering the statement "I believe that the great variety of products present on the market with pink packages or bearing a pink ribbon might create confusion," 58.4% of respondents declared that they agree. More than 60% believe that it is difficult to identify how you can support the cause more effective due to the diffusion of products with same characteristics (60.7%).

3.2.2 Two Step Cluster Analysis

Various clustering algorithms have been developed to group data into clusters, however, they work effectively either on numerical or categorical data but most of them perform poorly on mixed categorical and numerical data. Two step cluster analysis allows to avoid this problem [1, 18].

SPSS two step clustering developed by Chiu et al. [6], Xu and Wunsch [33] for the analysis of large data sets consists of two steps:

- Step 1: Pre-clustering of cases.
  A sequential approach is used to pre-cluster the cases. The aim is to compute a new data matrix with fewer cases for the next step; in order to reach this aim, the computed pre-clusters and their characteristics (cluster features) are used as new cases. The pre-clusters are defined as dense regions in the analyzed attribute space. The results may depend on the input order of cases therefore it is recommended to use random order.

- Step 2: Clustering of cases.
  A model based hierarchical technique is applied. Similar to agglomerative hierarchical techniques, the pre-clusters are merged stepwise until all clusters are in one cluster. In contrast to agglomerative hierarchical techniques, an underlying statistical model is used. The model assumes that the continuous variables are within clusters independent normal distributed and the categorical variables are within clusters independent multinomial distributed. Two distance measures are available: euclidean distance and a log-likelihood distance in case of mixed types of attributes.

In the next table, cf. Table 1, some results related to the collected data are presented. As one can see three clusters can be individuated.

Cluster 2 is composed mainly of young employees with a high school diploma. This group is almost always neutral in its assessments. Only in certain cases, it has demonstrated a propensity to perceive the phenomenon of pinkwashing, units in this cluster agree in saying that statements of pink marketing are hardly verifiable, that advertising seems to exaggerate the involvement of enterprises to the cause and finally that seems there is a voluntary omission of important information.

As regards the confusion of the consumer, units belonging to Cluster 1 declare to be confused because of the great variety of products pink or bearing a pink ribbon and believe that this could make difficult the identification of companies with support more effective to the cause.

Units in this cluster do not agree to declare that the choice of purchase is more difficult, while they are not able to express themselves on the difficulty to distinguish between products involved in the fight against breast cancer by others.

Cluster 1 and Cluster 3 are formed by units who take a more precise position with respect to the analyzed variables.

Specifically, the group composed mainly of adults between 45–54 years and over (Cluster 3) are more skeptical about the enterprise actions. These units appear to be the most receptive in perceiving the phenomenon of pinkwashing, the risk regarding the purchase of pink products, and to be confused by the operations of pink marketing.

On the contrary, Cluster 1 is composed mainly of young graduates, it is the one with more confidence in the commercial operations of the companies, it has the highest number of consensus as regards the credit in respect of the operations of pink marketing and the lowest perception of pinkwashing, the perceived risk, and confusion for the consumer.

The results of the cluster analysis just exposed before show a certain tendency to the division of the opinions between the two generations identifiable as parents and children. The cluster "parents" is the most skeptical and cynical about the commercial operations of the companies, probably due to a greater experience in life.

Young people instead are those with a more positive vision of marketing and advertising, probably due to a greater propensity to believe in the goodness of commercial operations.

3.3 Hypothesis Development

Based on the analysis of the literature (see [5]), we developed five research hypotheses:

1. pinkwashing is negatively correlated with consumer confidence (H1);
2. pinkwashing is positively correlated with the perceived risk (H2);

Table 1 Clusters identification

Cluster	Age	Study title	Profession	Pinkwashing	Perceived risk	Trust	Consumer confusion
1	18–24 25–34	High school Bachelor's deg.	Student	−, −, −, −, −	−, −, −, −	+, +, +, +, +	−, −, −, −
2	25–34	High school Magisters deg.	Employee	N, N, +, +, +	N, N, N, N	N, N, N, N, N	+, −,+ = N,N
3	45–54 65–	High school Magisters deg.	Employee retired	+, +, +, +, +	+, +, +, +	−, −, −, −, −	+, +, +, +

N = neutral, + = agree, − = do not agree

Table 2 Correlation coefficients

Hypothesis	Effect	Spearman coefficient	Kendall coefficient	Results
H1	−	−0.174	−0.208	Hp confirmed
H2	+	+0.564	+0.455	Hp confirmed
H3	+	+0.427	+0.337	Hp confirmed
H4	−	−0.474	−0.557	Hp confirmed
H5	−	−0.208	−0.162	Hp confirmed

3. pinkwashing is positively correlated to consumer confusion (H3);
4. consumer confusion is negatively associated with consumer confidence (H4);
5. perceived risk is negatively associated with confidence against products pink (H5)

From the statistical analysis of the data (correlation analysis between the four analyzed variables: pinkwashing, perceived risk, trust, and consumer confusion—by coefficients of Spearman and Kendall) all the hypotheses are verified, cf. Table 2

1. pinkwashing is negatively correlated with consumer confidence (H1);
2. pinkwashing is positively correlated with the perceived risk (H2);
3. pinkwashing is positively correlated to consumer confusion (H3);
4. consumer confusion is negatively associated with consumer confidence (H4);
5. perceived risk is negatively associated with confidence against products pink (H5)

The phenomenon of pinkwashing is adversely affect with regard to three relevant aspects—confidence, perceived risk, and confusion—able to influence consumer purchasing choices. The value of the coefficients shows, however, that, even in the case of a perceived risk already quite high, this influence is still slight. The reason is probably to be sought in a perception of the phenomenon still in its early stages and therefore rather low.

It emerges clearly that campaigns unclear or inconsistent with the business have negative effects on consumers that companies cannot overlook. The decline in consumer confidence, increasing the perception of risk and consumer confusion, can produce dangerous consequences, as unpleasant emotions of purchase, distrust in the enterprise and in its cause-related marketing campaigns, dissatisfaction [4, 29, 31, 32].

4 Conclusion

Pinkwashing is a very spread phenomenon due to the increasing of pink marketing.

This research enhanced some of the effects of pinkwashing on consumers that is: less trust, more perceived risk, and more confusion.

This can be a problem for companies in particular as regards the reputation and a less efficacy of cause-related marketing. In order to avoid this it is necessary to control advertisement and to control the use of the pink ribbon.

Because of the increasing role of pink marketing in the communication strategy of numerous companies, also pinkwashing is unfortunately becoming a very spread phenomenon. This research enhanced some of the effects of pinkwashing on consumers, that is to say less trust, more perceived risk, and more confusion in the buying process. As highlighted in previous studies on whitewashing, and greenwashing in particular, these emotional states of consumers can be a problem, in particular as regards the reputation and a less efficacy of cause-related marketing, both for companies implementing the process of pink marketing and for institutions benefiting from cause-related marketing campaigns. Indeed an unclear and incoherent advertising campaign could have negative effects on consumers, such as detrimental word of mouth, doubt on the firm reliability, dissatisfaction, cognitive dissonance, and at the end postponement or reject of buying decision. For these reasons, companies should pay more attention to their cause-related marketing decisions and to their communication process. At the same way, also institutions benefiting from these campaigns should evaluate and control more deeply the source of funds, in order to avoid negative consequences in long time.

References

1. Bacher, J., Wenzing, K., Vogler, M.: SPSS Two Cluster A First Evaluation, Universitat Erlangen-Nurnberg, pp. 1–20. www.statisticalinnovations.com/products/twostep.pdf (2004)
2. Broderick A., Amandeep J., Garry T.: Tickled pink: the personal meaning of cause related marketing for customers. J. Mark. Manag. **19**(5–6), 583–610 (2003)
3. Chamorro, A., Rubio, S., Miranda, F.J.: Characteristics of research on green marketing. Bus. Strateg. Environ. **18**, 223–239 (2009)
4. Chaudhuri, A.: Consumption emotion and perceived risk: a macro-analytic approach. J. Bus. Res. **39**(2), 81–92 (1997)
5. Chen, Y., Chang, C.: Greenwash and green trust: the mediation effects of green consumer confusion and green perceived risk. J. Bus. Ethics **114**(3), 489–500 (2012)
6. Chiu, T., Fang, D., Chen, J., Wang, Y., Jeris, C.: A robust and scalable clustering algorithm for mixed type attributes in large database environment. In: Proceedings of the 7th ACM SIGKDD International Conference on Knowledge Discovery and Data Mining 2001, pp. 263–268 (2001)
7. Dailey, W.: Consumers Question Intent and Impact of Breast Cancer Cause Marketing. New Research Shows Americans Increasingly Skeptical. Cone Communication, Boston (2012)
8. de Luca, P., Aramendia Muneta, E.: Analysis of consumer confusion in the mobile phone market. Exploratory research in Italy and Spain. In: 8th International Conference Marketing Trends, Paris-Venice (2009)
9. Engel, J.F., Blackwell, R.D., Miniard, P.W.: Consumer Behavior. The Dryden Press, Chicago (1986)
10. Ganesan, S.: Determinants of long-term orientation in buyer-seller relationships. J. Mark. **58**, 1–19 (1994)
11. Hart, P., Saunders, C.: Power and trust: critical factors in the adoption and use of electronic data interchange. Organ. Sci. **8**(1), 23–42 (1997)

12. Harvey, J.A., Strahilevitz, M.A.: The power of pink: cause-related marketing and the impact on breast cancer. J. Am. Coll. Radiol. **6**(1), 26–32 (2009)
13. Kalafatis, S.P., Pollard, M.: Green marketing and Ajzen's theory of planned behaviour: a cross-market examination. J. Consum. Mark. **16**(5), 441–460 (1999)
14. Koehn, D.: The nature of and conditions for online trust. J. Bus. Ethics **43**, 3–19 (2003)
15. Leek, S., Chansawatkit, S.: Consumer confusion in the Thai mobile phone market. J. Consum. Behav. **5**(6), 518–532 (2006)
16. Lyon, T.P., Maxwell, J.W.: Greenwash: corporate environmental disclosure under threat of audit. J. Econ. Manag. Strateg. **20**(1), 3–41 (2011)
17. Matzler, K., Stieger, D., Fuller, J.: Consumer confusion in internet-based mass customization: testing a network of antecedents and consequences. J. Consum. Policy **34**, 231–247 (2011)
18. Ming-Yi, S., Jar-Wen, J., Lien-Fu, L.: A two-step method for clustering mixed categorical and numeric data. Tamkang J. Sci. Eng. **13**(1), 11–19 (2010)
19. Mitchell, V.W., Papavassiliou, V.: Marketing causes and implications of consumer confusion. J. Product Band Manage. **8**(4), 319–339 (1999)
20. Mitchell, V.W., Walsh, G., Yamin, M.: Towards a conceptual model of consumer confusion. In: Menon, G., Rao, A.R. (eds.) Advances in Consumer Research, pp. 143–150. Association for Consumer Research, Minneapolis (2005)
21. Morgan, R.M., Hunt, S.D.: The commitment-trust theory of relationship marketing. J. Mark. **58**(3), 20–38 (1994)
22. Naseri, H.: Consumers' response to cause-related marketing (a case study among customers of Iranian chain stores). Interdisciplinary J. Contemp. Res. Bus. **5**(6), 498–508 (2013)
23. Pegan, G., de Luca, P.: Can implicit and explicit attitudes predict green product choice under time pressure? an experimental research. J. Manag. World Bus. Res. **9**(1), 21–36 (2012)
24. Rozensher, S.: The growth of cause marketing: past, current, and future trends. J. Bus. Econ. Res. **11**(4), 181–186 (2013)
25. Schoier, G., Borruso, G.: Spatial data mining for highlighting hotspots in personal navigation routes. Int. J. Data Warehouse. Min. **8**, 45–61 (2012)
26. Schoier, G., Melfi, G.: A different approach for the analysis of web access logs. In: Vichi, M., Monari, P., Mignami, S., Montanari, A. (eds.) New Developments in Classification and Data Analysis, pp. 211–216. Springer, Berlin (2004)
27. Singh, J., Sirdeshmukh, D.: Agency and trust mechanisms in consumer satisfaction and loyalty judgements. J. Acad. Mark. Sci. **28**(1), 150–167 (2000)
28. Tullis, T., Albert, B.: Measuring the user experience. Elsevier, Amsterdam (2013)
29. Turnbull, P.W., Leek, S., Ying, G.: Customer confusion: the mobile phone market. J. Mark. Manag. **16**(1–3), 143–163 (2000)
30. Varadarajan, R.P., Menon, A.: Cause related Marketing: a coalignment of marketing strategy and corporate philantrophy. J. Mark. **52**(3), 58–74 (1998)
31. Walsh, G., Mitchell, V.W.: The effect of consumer confusion proneness on word of mouth, trust, and customer satisfaction. Eur. J. Mark. **44**(6), 838–859 (2010)
32. Walsh, G., Hennig-Thurau, T., Mitchell, V.W.: Consumer confusion proneness: scale development, validation, and application. J. Mark. Manag. **23**(7), 697–721 (2007)
33. Xu, R., Wunsch, D.: Survey of clustering algorithms. IEEE Trans. Neural Netw. **16**(3), 645–678 (2005)
34. Zani, S., Cerioli, A.: Analisi dei dati e Data Mining per le Decisioni Aziendali. Giuffré Editore, Milano (2007)

Predicting the Evolution of a Constrained Network: A Beta Regression Model

Luisa Stracqualursi and Patrizia Agati

Abstract Social network analysis allows to map and measure relationships and flows (links) between people, groups, computers, URLs, or other connected knowledge entities (nodes). In this context, a relevant issue is the treatment of constrained scale-free networks such as the network of student transfers between degree courses offered by an University, that are strongly influenced by a number of institutional decisions. In the analysis of such a system, special attention has to be paid to identify current or future "critical points", that is nodes characterized by a high number of outcoming or incoming links, on which to act in order to optimize the network. To predict the evolution of a constrained system over time in dependence of constraint modifications, a beta regression model is proposed, that fits links represented by quantities varying between 0 and 1. The algorithm was successfully applied to the network of student transfers within the University of Bologna: the link was defined by the out-transfer rate of the degree course (computed as the ratio of the number of out-transfers to the number of students enrolled) and the critical points of the system were defined by the courses characterized by a high out-transfer rate.

1 Introduction

1.1 Some Issues About Complex Networks

In complex systems the principal characteristics are not always demonstrated by points but by the relationships between them. For example, in biology the cell is best described as a complex network of elements connected by chemical reactions [5]; the Internet is a complex network of routers and computers linked by various physical or wireless links; fads and ideas spread on the social network, whose nodes are human beings and edges represent various social relationships; the WorldWideWeb is an enormous virtual network of Web pages connected by hyperlinks [1].

L. Stracqualursi (✉) • P. Agati
Department of Statistical Sciences, University of Bologna, Bologna, Italy
e-mail: luisa.stracqualursi@unibo.it; patrizia.agati@unibo.it

F. Palumbo et al. (eds.), *Data Science*, Studies in Classification, Data Analysis, and Knowledge Organization, DOI 10.1007/978-3-319-55723-6_26

Traditionally, the studies of complex networks have been the territory of graph theory that has its origins in the eighteenth century in an essay by Leonhard Euler, an early work on graphs with a high degree of regularity [6]. In mathematical terms, a graph is a pair of sets $\mathbf{S}\{\mathbf{A},\mathbf{E}\}$, where $\mathbf{A}$ is a set of N nodes (or vertices) $A_1, A_2, \ldots, A_N$ and $\mathbf{E}$ is a set of edges (or links) that connect two elements of $\mathbf{A}$. Graphs are usually represented as a set of dots, each corresponding to one node, two of these dots being joined by a line if the corresponding nodes are connected.

In the twentieth century, graph theory has become more statistical and computational. Over time, a rich source of ideas has arisen from the studies of random graphs in which the edges are distributed randomly. A random network consists of N nodes where each node pair is connected with probability p. Two mathematicians, Erdős and Alfréd Rényi, have played an important role in understanding the properties of these networks [9].

In a given realization of a random network some nodes gain numerous links, while others acquire only a few or no links. These differences are captured by the *degree distribution*, which describes the probability $P(k)$ that a randomly chosen node has exactly k links (the degree of the node). The exact form of the degree distribution of a random network is the binomial distribution. For $N \gg k$ the binomial is well approximated by a Poisson distribution (Fig. 2). Consequently most nodes have the same number of links, and nodes with a very large number of links don't exist. Real networks are not static and the number of links rarely remains unvaried. Indeed for many real networks the probability that a node has exactly k links is

$$P(k) \sim k^{-\gamma} \tag{1}$$

where the value of γ is between 2 and 3 for most networks: that is, the number of nodes with a given degree k follows a power law [4]. This implies that nodes with only few links are numerous, but few nodes (called "hubs") have a very large number of links. Networks with a power-law degree distribution are called scale-free [3].

This model is based on two principal assumptions:

1. *incremental growth*: networks grow with the addition of new nodes;
2. *preferential attachment*: the probability of linking to a node with a large number of connections is high. Therefore a new node prefer to attach to the more connected nodes (this principle is also known as "the rich get richer")[2].

In networks having this topology the distribution remains the same when the number of network nodes increases, thus it is independent from the magnitude and scale of the network, hence the name *scale-free*.

1.2 Constrained Networks

A relevant issue is the treatment of *constrained* networks, whose links are strongly influenced by rules, institutional decisions, or system boundaries. Even if the number of nodes with a given degree follows a power-law distribution, hence the distribution remains the same also increasing the number of network nodes (typical of scale-free networks), several constraints inhibit free growth and preferential attachment. Even if "hubs" are found, their presence is strongly influenced by different decisions and rules. Therefore, the evolution of the network is not in terms of growth, but in terms of linkage structure. For this reason, predicting the evolution of such a constrained system in dependence on modifications of constraints requires to assess a micro model of the network which works on the links between nodes rather than on the nodes themselves: in other words, a function could be defined for each node, that captures the network linkage structure and allows to predict changes in the weight of the link, in dependence on perturbations of the initial state of suitable covariates. In this context, we analyze University network of degree courses.

Particularly, the transfers by students between degree courses within the University of Bologna can be viewed as a network, where the nodes are the courses, the links identify directed connections between two nodes and denote the inter-course transfers of students. In this context, the weight of each link is measured by the number of students that transfer.

Figure 1 shows the network, where different colors represent different geographic locations: courses outside Bologna are represented by a white small circle, while courses within Bologna by a grey small circle.

The degree distribution of the Bologna University network appears to be well described by a power law with exponent $\gamma \simeq 1.7$ (see Fig. 2). Indeed the analysis of the different degree courses indicates that most degree courses have few "enter" or "leave" of students and only a few courses have most students that "enter" or "leave."

Unlike to scale-free topology, in such a network the assumption of incremental grow is not fulfilled, due to the influence exercised by law restrictions and institutional decisions: therefore, the network does not evolve in terms of "growth," but rather in terms of "linkage structure."

For this reason, predicting the evolution of such a constrained system requires to define a function for each degree course, that captures the network linkage structure and allows to predict changes in the weight of the link, in dependence on perturbations of the initial state of suitable covariates, such as the geographical location of the course, the existence of degree courses having similar exam subjects, and the admission typology (free/restricted access).

A special attention has to be paid to identify current or future "critical nodes," on which to intervene in order to optimize the network. A very intuitive measure of the criticality level of a node is the transfer-out rate of the degree course, computed as the ratio of the number of students who are known to have transferred out of the

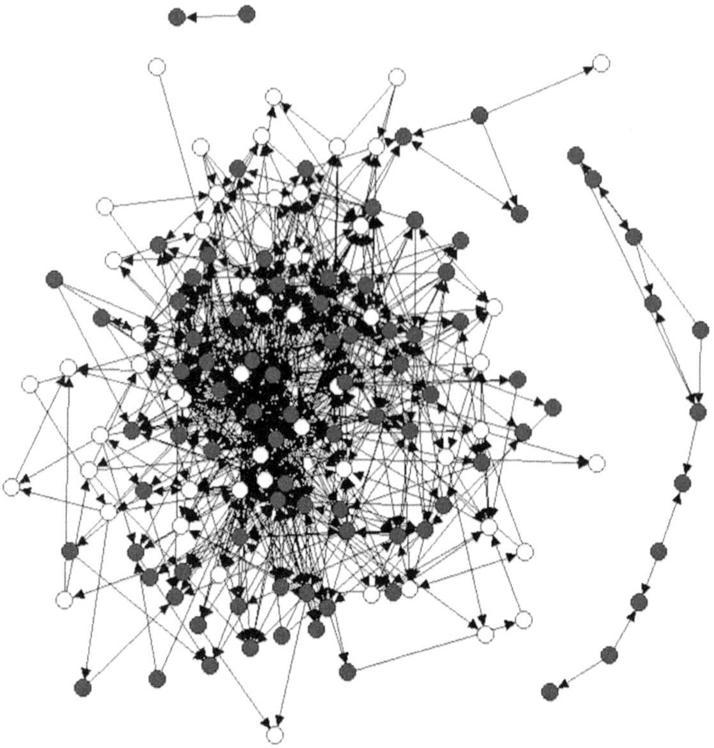

Fig. 1 Course network in year 2007–2008

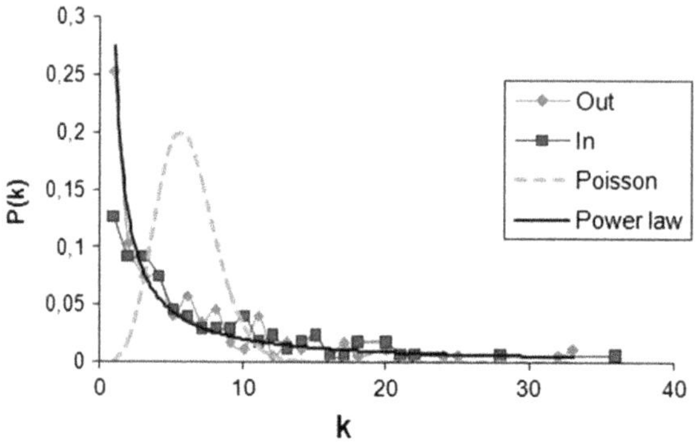

Fig. 2 Degree distribution of University degree courses network

course to the number of students that were enrolled [12]. Since this measure ranges from 0 (no transfer out of the course) to 1 (the totality of enrolled students have transferred out)—the larger the transfer-out rate, the larger the criticality level of the degree course—a Beta regression model is proposed to fit the linkage structure of the network.

The paper is organized as follows: Sect. 2 outlines a synthetic review of beta regression model; the case study is described and the beta model is fitted in Sect. 3; finally, Sect. 4 contains a simulation study and some concluding remarks.

2 A Brief Review of Beta Regression Model

Beta regression model is tailored for situations where the response variable y takes on values within the real open interval (0, 1). For such variables that may stem from rates, proportions, concentrations, etc., the normal assumption underlying the linear regression model is not supported: bounded range continuous variables usually display heteroscedasticity (the variance is smaller near the extremes) and asymmetry; linear fitted values could exceed the lower and upper bounds of y, resulting in invalid and misleading outcomes.

Instead, the beta distribution is a very flexible model for variables within the standard unit interval: its density, given by

$$f(y\,;p,q) = \frac{\Gamma(p+q)}{\Gamma(p)\,\Gamma(q)} y^{p-1} (1-y)^{q-1}, 0 < y < 1 \tag{2}$$

can have quite different shapes depending on the values of the two parameters $p > 0$, $q > 0$, and can accommodate skew and asymmetry. The expected value and the variance of y are $E(y) = p/(p+q)$ and $V(y) = pq/[(p+q)^2(p+q+1)]$. For modelling purposes, a different parameterization of the beta density was proposed by Ferrari and Cribari-Neto [10] by setting $\mu = p/(p+q)$ and $\phi = p+q$, i.e., $p = \mu\phi$ and $q = (1-\mu)\phi$:

$$f(y\,;\mu,\varphi) = \frac{\Gamma(\varphi)}{\Gamma(\mu\varphi)\,\Gamma((1-\mu)\,\varphi)} y^{\mu\varphi-1} (1-y)^{(1-\mu)\varphi-1}, 0 < y < 1 \tag{3}$$

where $0 < \mu < 1$ and $\varphi > 0$. The expected value and the variance of y, in the new parameterization, are $E(y) = \mu$ and $V(y) = \mu(1-\mu)/(1+\varphi)$, so that μ is the mean of y and φ can be regarded as a precision parameter: for fixed μ, the larger the value of φ, the smaller the variance of y.

Let $y_1, \ldots, y_j, \ldots, y_n$ be independent random variables, where each y_j, $j = 1, \ldots, n$ is Beta distributed with mean μ_j and unknown precision φ, and $(x_{j1}, \ldots, x_{ji}, \ldots, x_{jk})$ be observations on k covariates, which are assumed as fixed and known. The beta regression model can be written as

$$g(\mu_i) = \sum_{i=1}^{k} x_{ji}\beta_i \tag{4}$$

where $\beta_1, \ldots, \beta_i, \ldots, \beta_k \in \Re^k$ is a vector of unknown regression parameters and $g(\cdot)$ is a strictly monotonic and twice differentiable link function that maps $(0, 1)$ to $\Re$. Several choices are possible for the link function $g(\mu)$, such as the logit function $log[\mu/(1-\mu)]$ (the inverse cumulative distribution function of the logistic) or the probit function $\Phi^{-1}(\mu)$ (the inverse cumulative distribution function of the standard normal variable).

Extensions of the beta regression model outlined above were proposed by Smithson and Verkuilen [14], Simas et al. [13], Cook et al. [7] among others.

3 The Case Study: Dataset Description and Beta Model Fit

Dataset consists of the $n = 130$ undergraduate and postgraduate degree courses of the University of Bologna that were operating in the academic year 2007–2008.

For each course, in addition to the number of transfers in and transfers out by students within the University of Bologna network, the value of several variables was observed or computed:

- *liking* represents the student satisfaction (revealed by the questionnaires that the University submits to its students every year) on the degree course which students are currently enrolled in: it is a continuous variable whose values were that takes values within the standard unit interval [0, 1];
- *similar* refers to the existence of degree courses having similar exam subjects: this could result in an advantage to the student in case of transfer from a course to the other one. The value of this variable was determined as follows: for each course, exam subjects that share the same SSD (Settore Scientifico Disciplinare) were grouped and the total number of credits was computed; *similar* takes on the value 1 if the number of credits they validate in a potential student transfer is not less than 25, value 0 otherwise;
- *location* is a dummy variable that refers to the geographical location of the course: it assumes value 0 if the course is located in Bologna city and 1 otherwise (course located in campus of the University of Bologna in other cities of the region);
- *open* refers to the typology of access: it is a dummy variable that takes on value 0 if the course is "free access" and 1 if it is "restricted access";
- *enroll* is a discrete variable indicating the number of students enrolled in the course in the academic year;
- *indn* is the transfer-in rate of the course, a variable computed as the ratio of the number of students who have transferred-in to the total number of students who are enrolled in the degree course;
- *range* is a discrete variable computed as the modulus of the difference between the number of transfers in and the number of transfers out of the degree course.

The first empirical measurements which allow us to discover "critical" nodes are *out*-degrees: the greater the number of nodes out links (*out*-degree), the smaller the tendency of the courses to retain students.

Since these quantities depend on the size of each course, all degrees are normalized by the number of the students in the respective courses.

In a previous paper [12] some "critical" courses such as "Biology" and "Pharmacy" were pointed out. Students often enroll in these courses if they have failed the admittance test for "Medicine" for which only a fixed number of students can enroll. Students who do not pass the admittance test tend to enroll temporarily in similar courses and successively request transfer. Table 1 shows transfer-out rates for some critical nodes of the network.

The R package `betareg` [8] was used to fit a Beta regression model where transfer-out rate depends on all variables listed above, employing the probit link. Bias corrected ML estimates [11] of the i parameters are numerically obtained by using the BFGS method and are shown in Table 2, as well as their statistical significance.

Except the variable liking, all of the considered covariates are significant and enter the model, so that the estimated transfer-out rate $y_j, j = 1, , n$, can be written as

$$\hat{y}_j = g\left(\hat{\mu}_i\right) = \sum_{i=1}^{k} x_{ji}\hat{\beta}_i \tag{5}$$

The explained variation is 32% of the total variation (pseudo R-squared = 0.32). A tenfold cross-validation yielded a reasonable RMSE, equal to 0.07.

Table 1 Critical degree courses within the University of Bologna (a.y. 2007–2008)

Degree courses	No. of transfers out	No. of transfers in	No. of enrolled students	Transfer-out -rate (%)
Pharmacy	81	29	286	28.3
Biology	75	25	275	27.3
Foreign Languages and Literature	67	38	610	11.0
Management Process Engineering	14	9	39	35.9
Telecommunications Engineering	8	2	28	28.6

Table 2 Beta regression model: bias corrected ML estimates

	$\hat{\beta}_i$ estimates	Significance
(Intercept)	−1.3214	0.001
Similar	+0.2552	0.001
Location	−0.3758	0.001
Open	+0.2454	0.01
Indn	−1.0499	0.01
Enroll	−0.0007	0.05
Range	+0.0075	0.05

4 Simulation Study and Conclusions

Once the beta model was estimated, a simulation study can be useful to investigate the effect of perturbations of the initial conditions on the transfer-out rates $\hat{y}_i$.

Table 3 shows the values of the covariates that enter the model 5 and the estimates $\hat{y}_i$ for the critical nodes displayed in Table 1, while Table 4 shows the new estimates $\hat{y}_i$ after generating the following (quite hypothetical) perturbations of the covariates:

- "Pharmacy" and "Biology": elimination of *similar* exams at the first course year, transfer of the degree course outside Bologna, decrease of *range* by 20
- "Foreign Languages and Literature": introduction of restricted access;
- "Management Process Engineering": transfer of the course outside Bologna
- "Telecommunications Engineering": transfer of the course outside Bologna and increase of *indn* from 0.071 to 0.250 (for example, by implementing strategies of facilitation of earned credits validation).

The comparison of the estimates in the two tables shows a sensible reduction of transfer-out rates for all degree courses.

Figures 3 and 4 display, respectively, the ego networks of Medicine and Foreign Languages and Literature. As well known, an ego network consists of a focal node "ego" and the nodes to whom ego is directly connected to (these are called "alters") plus the ties, if any, among the alters. The displayed networks present cut-off 3: it means that links smaller than 3 units have been deleted.

The estimated out degree is obtained as

$$\text{out degree}_{\text{est}} = \frac{\hat{y}_j \cdot \text{out degree}}{y_i}$$

Table 3 Critical nodes: values of the covariates and model estimates $\hat{y}_i$

Critical nodes	Similar	Location	Open	Indn	Enroll	Range	$\hat{y}_i$
Pharmacy	1	0	1	0.101	286	52	0.230
Biology	1	0	1	0.091	275	50	0.231
Foreign Languages and Literature	1	0	1	0.062	610	29	0.137
Management Process Engineering	1	0	1	0.231	39	5	0.146
Telecommunications Engineering	1	0	1	0.071	28	6	0.192

Table 4 Simulation studies: perturbed values in bold and model estimates $\hat{y}_i$

Critical nodes	Similar	Location	Open	Indn	Enroll	Range	$\hat{y}_i$
Pharmacy	**0**	**1**	1	0.101	286	**42**	0.074
Biology	**0**	**1**	1	0.091	275	**40**	0.075
Foreign Languages and Literature	1	0	**0**	0.062	610	29	0.090
Management Process Engineering	1	**1**	1	0.231	39	5	0.076
Telecommunications Engineering	1	**1**	1	**0.250**	28	6	0.076

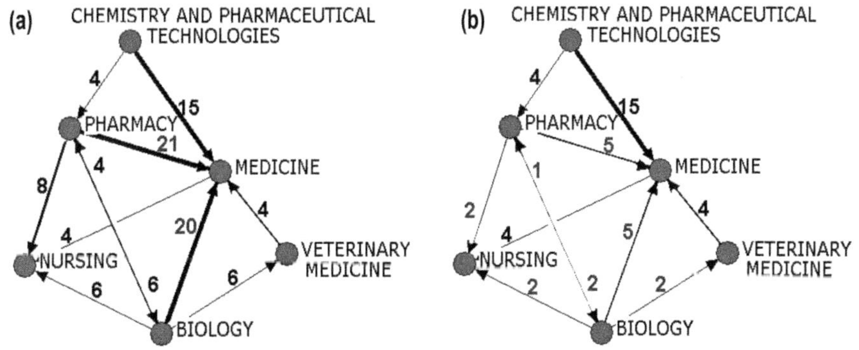

Fig. 3 Simulation studies egonet of Medicine with cut-off = 3: (**a**) out degree of the network, (**b**) estimated out degree after perturbations

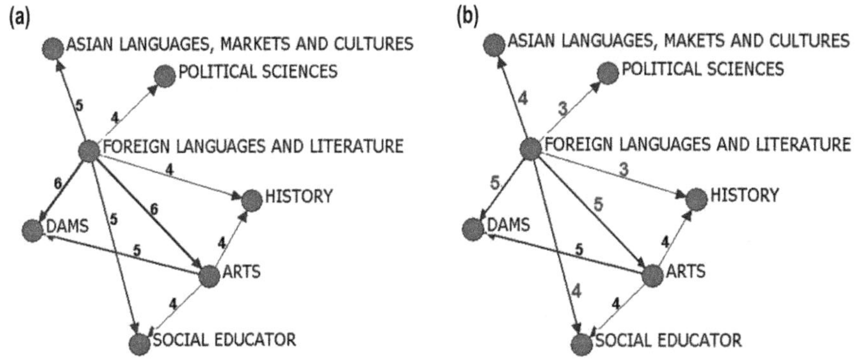

Fig. 4 Simulation studies egonet of Foreign Languages and Literature with cut-off = 3: (**a**) out degree of the network, (**b**) estimated out degree after perturbations

Figure 3 revealed that "Biology" and "Pharmacy" are transit nodes. Particularly, students often enroll in these courses if they have failed the admittance test for "Medicine" for which only a fixed number of students can enroll: students who do not pass the admittance test tend to enroll temporarily in similar courses and successively they request transfer. So the main problem of the critically of Pharmacy and Biology is the similar exam to Medicine. Furthermore Medicine, Biology, and Pharmacy have the same geographical location, Bologna city.

The simulation shows that decreasing the validate credits in a transfer to Medicine discourages the attitude of students to use Pharmacy and Biology as transit courses: so the estimated out degrees decrease.

Similarly a reduction of estimated out degrees could be obtained for Foreign Languages by introducing the restrict access (Fig. 4).

Although additional studies are needed to explore the potentiality of the algorithm and to improve its performance, the beta regression model seems to be a valid tool to investigate the evolution of a constrained network whose critical points can be characterized in terms of a measure ranging from 0 to 1: the model allows to identify the critical points in a complex system and optimize the network by studying the impact of new possible institutional choices on the behavior of its linkage structure.

References

1. Barabási, A.L.: The physics of the web. Phys. World **14**, 33–38 (2001)
2. Barabási, A.L.: Linked: The New Science of Networks. Perseus, London (2002)
3. Barabási, A.L., Albert, R.: Emergence of scaling in random network. Science **286**, 509–512 (1999)
4. Barabási, A.L., Albert, R.: Statistical mechanics of complex networks. Rev. Mod. Phys. **74**, 47–97 (2002)
5. Barabási, A.L., Oltvai, Z.N.: Network biology: understanding the cell's functional organization. Nat. Rev. Genet. **5**, 101–113 (2004).
6. Bollobás, B.: Random Graphs. Academic, London (1985)
7. Cook, D.O., Kieschnick, R., McCullough, B.D.: Regression analysis of proportions in finance with self selection. J. Empir. Financ. **15**, 860–867 (2008)
8. Cribari-Neto, F., Zeileis, A.: Beta regression in R. J. Stat. Softw. **34**(1), 1–24 (2010)
9. Erdős, P., Rényi, A.: On random graphs. Publ. Math. **6**, 290–297 (1959)
10. Ferrari, S.L.P., Cribari-Neto, F.: Beta regression for modeling rates and proportions. J. Appl. Stat. **31**, 799–815 (2004)
11. Grun, B., Kosmidis I., Zileis, A.: Extended beta regression in R: shaken, stirred, mixed, and partitioned. J. Stat. Softw. **48**, 11, 1–25 (2012)
12. Monari, P., Stracqualursi, L.: On the statistical analysis of intra course dynamics in university system networks. Quad. Stat. Università di Napoli **11**, 1–20 (2009)
13. Simas, A.B., Barreto-Souza, W., Rocha, A.V.: Improved estimators for a general class of beta regression models. Comput. Stat. Data Anal. **54**(2), 348–366 (2010)
14. Smithson, M., Verkuilen, J.: A better lemon squeezer? maximum likelihood regression with beta-distributed dependent variables. Psychol. Methods **11**, 54–71 (2006)

CPI Antony Rowe
Chippenham, UK
2018-06-18 22:45